ECONOMICS, SUSTAINABILITY, AND NATURAL RESOURCES

# Economics, Sustainability, and Natural Resources

## Economics of Sustainable Forest Management

*Edited by*

SHASHI KANT
*Faculty of Forestry, University of Toronto, Canada*

and

R. ALBERT BERRY
*Munk Centre for International Studies, University of Toronto, Canada*

A C.I.P. Catalogue record for this book is available from the Library of Congress.

ISBN 1-4020-3465-2 (HB)
ISBN 978-1-4020-3465-7 (HB)
ISBN 1-4020-3518-7 (e-book)
ISBN 978-1-4020-3518-0 (e-book)

---

Published by Springer,
P.O. Box 17, 3300 AA Dordrecht, The Netherlands.

*www.springeronline.com*

*Printed on acid-free paper*



Printed in the Netherlands.

Dedicated to Hoshwati Yadav, Rachel Carson, and Maurizio Merlo

**Companion volume:**

*Institutions, Sustainability, and Natural Resources: Institutions for Sustainable Forest Management*

# CONTENTS

## PART FOUR: NON-LINEARITIES, MULTIPLE EQUILIBRIA AND THE ECONOMICS OF SUSTAINABLE FOREST MANAGEMENT



## PART FIVE: EPILOGUE

# FIGURES AND TABLES

## FIGURES

**TABLES**

# ABOUT THE CONTRIBUTORS

**Geir B. Asheim** is Professor of Economics at the University of Oslo. In his research, he has addressed questions of intergenerational justice, in particular whether one can provide normative justifications for sustainable development. He has also worked on green national accounting, asking whether sustainable development can be indicated by means of comprehensive national accounting aggregates. Game theory is another main research interest.

**Albert Berry** is Professor Emeritus of Economics and Research Director of the program on Latin America and the Caribbean at the University of Toronto's Center for International Studies. He holds a Ph. D. from Princeton University. His main research interests, with focus on Latin America, are labor markets and income distribution, agrarian structure, the economics of small and medium enterprise, and the economics of forests and of sustainability. Apart from his academic positions at Yale University, the University of Western Ontario and the University of Toronto, he has worked with the Ford Foundation, the Colombian Planning Commission and the World Bank. He is currently directing a research program on an employment strategy for Paraguay.

**Wolfgang Buchholz** is Professor at the University of Regensburg. His main topics of research lie in the field of environmental economics, where he in particular addresses problems of intertemporal allocation and issues of international cooperation. Other main research interests are public finance and the economics of the welfare state.

**Milindo Chakrabarti** is a member of the Economics faculty of St. Joseph's College, Darjeeling and the founding Director of the Centre for Studies in Rural Economy, Appropriate Technology and Environment (CREATE), a collaborative research centre with the International Forestry Resources and Institutions (IFRI) Research Program, Indiana University, Bloomington. He has also initiated collaborative research works with colleagues at the Natural Resources Institute, Manitoba University, Winnipeg. His research interests are centered around sustainable development with special emphasis on intra-generational justice. He recently completed a study on the social cost- benefit analysis of public-private partnership in re-greening degraded forest, revenue and private lands in India. His earlier works include studies on the implications of WTO agreements for Indian agriculture, and sustainable management of Indian fisheries.

**David Colander** is the Christian A. Johnson Distinguished Professor of Economics at Middlebury College, Middlebury, Vermont. His latest work focuses on economic education, complexity, and the methodology appropriate to applied policy economics. He has authored, co-authored, or edited 30 books and over 100 articles on a wide range of topics. These include Principles of Economics, History of Economic Thought, Macroeconomics, and Why Aren't Economists as Important as

Garbagemen? He has taught at Columbia University, Vassar College, and the University of Miami. In 2001-2002 he was the Kelley Professor for Distinguished Teaching at Princeton University. He has been Vice President and President of the Eastern Economic Association and the History of Economics Society. He is currently on the editorial boards of the Journal of the History of Economic Thought, Eastern Economic Journal, and Journal of Economic Education. He is also series editor, with Mark Blaug, of Twentieth Century Economists for Edward Elgar Publishers.

**Samar K. Datta**, a product of Presidency College, Calcutta (Honors in Economics: 1965-68), Calcutta University (M.A. in Economics: 1968-70) and University of Rochester, USA (Ph.D. in Economics: 1976-79), has more than 30 years of teaching, research and administrative experience at Calcutta University (1971-76), University of Southern California, Los Angeles (1979-83), Visva Bharati University at Santiniketan (1983-90) and Indian Institute of Management, Ahmedabad (since 1990), where he is currently a senior Professor at the Centre for Management in Agriculture. His research areas cover rural institutional economics, transaction costs and contracts, stakeholder cooperation, agro-business trade and competitiveness analysis under the WTO regime, natural resources management and rural credit. So far he has published seven research monographs and two edited volumes, besides writing more than 30 cases and publishing more than 40 articles in peer-refereed books and journals.

**Lance Howe** is Assistant Research Professor of Economics at the Institute of Social and Economic Research, University of Alaska, Anchorage; he completed his PhD in Economics at the University of Southern California in 2002. His current research interests center on topics in rural economic development that include risk coping and management mechanisms, migration and rural settlement patterns, and community resource management. He is also interested in applying experimental methods to questions in economic development where such methods may be appropriate, for example, in exploring the effects of heterogeneity, under varying institutional arrangements, on common property resource management.

**Shashi Kant** is Associate Professor of Forest Resource Economics and Management at the Faculty of Forestry, University of Toronto, Canada; he completed his Ph. D. in Forest Resource Economics at the University of Toronto in 1996. He is a recipient of Premier's Research Excellence Award for his research on economics of sustainable forest management. His research interests include market as well as non-market signals of resource scarcity, institutional and evolutionary aspects of economics of sustainable forest management, quantum theory and behavioral economics, game theoretic and agent-based models, social choice theory and forest management. He is an Associate Editor of Journal of Forest Economics, and Canadian Journal of Forest Research. He has published more than fifty refereed papers and he has worked as a consultant to the Food and Agriculture Organization (FAO), United Nations Environment Program (UNEP), Swedish International Development Agency

(SIDA), International Network for Bamboo and Rattan (INBAR), and Ontario Ministry of Natural Resources (OMNR).

**M. Ali Khan** joined the Johns Hopkins University in 1973 and has held the position of the Abram Hutzler Professor of Political Economy in its economics department since 1989. He graduated from the London School of Economics (LSE) in 1969 with First Class Honors, and obtained his M. Phil and Ph. D. degrees (Ph. D dissertation: Large Exchange Economies) from Yale University in 1972 and 1973 respectively. In addition to Johns Hopkins, he has taught at the LSE, Northwestern, Cornell and the University of Illinois. His research interests lie in mathematical economics and in the interface between economics and philosophy.

**Jack Knetsch** is Professor Emeritus, School of Resource and Environmental Management and the Department of Economics, Simon Fraser University, Canada. He received his M. Sc. (Agriculture Economics) from Michigan State University and his M.P.A. and Ph. D. (Economics) from Harvard University. Jack is a distinguished environmental and behavioral economist. He has published various books and papers on economics of outdoor recreation, pollution control strategies, economics of water resources, the endowment effect, non-reversible indifference curves, environmental damage schedules, and fairness as a constraint on profit seeking. Knetsch collaborated with Daniel Kahneman on several papers about behavioral economics

**Marc Lavoie** is Professor of Economics at the Department of Economics, University of Ottawa, Canada. His research interests include non-orthodox economic theories (more specifically post-Keynesian theory), economic growth, macroeconomic theory, monetary theory and policy, and economics of sports. He has edited two books and is the author of five others, including Foundations of Post-Keynesian Economic Analysis (1992). Besides having been an Associate editor of the Encyclopedia of Political Economy (1999), he has published over 125 articles or chapters in various scholarly journals and books.

**Tapan Mitra** is Professor of Economics at Cornell University. He received his Ph.D. from the University of Rochester in 1975, and taught at the University of Illinois and the State University of New York at Stony Brook before joining Cornell University in 1981. He served as the Chairman of the Department of Economics at Cornell University from 1993 to 1998 and from 1999 to 2002. His research work has focused on the theory of efficient and optimal inter-temporal allocation, with applications to economic growth and development and the economics of natural resources. He was awarded the Alfred P. Sloan Fellowship in 1981, and was elected a Fellow of the Econometric Society in 1997.

**Jeffrey B. Nugent** is Professor of Economics at USC in Los Angeles. His research has focused on trade, FDI, economic integration, household decision-making and income distribution issues making use of new institutional economics and political economy perspectives. He has written five books and edited five others and authored or co-authored well over one hundred articles in professional journals. He serves on the editorial boards of nine scholarly journals and has served on the Board of Directors of the Western Economic Association International, the Economic Research Forum (ERF) and the Middle East Economic Association, the latter as Executive Secretary and President, and worked for or consulted with numerous international agencies.

**Matthew D. Potts** is a National Science Foundation Post-Doctoral Fellow in Bioinformatics in the Institute on Global Conflict and Cooperation at the University of California, San Diego. Before joining UCSD he was a research fellow in the Center for International Development at Harvard University and a Visiting Fellow in the Institute of Biodiversity and Environmental Conservation at the University of Sarawak, Malaysia. He holds Master's and Ph.D. degrees in Applied Mathematics from Harvard University. His publications have appeared in the Proceedings of the National Academy of Sciences, Ecology, Ecology Letters, Journal of Theoretical Biology, and Journal of Tropical Ecology.

**Colin Price** after periods of teaching urban, regional and agricultural economics at Oxford University, moved to the University of Wales, Bangor, where he is currently Professor of Environmental and Forestry Economics. He is author of nearly 200 assorted publications, including books on landscape valuation, forestry economics, and discounting. Most of the other publications are concerned with the economics of silvicultural and harvesting decision-making, the evaluation of environmental benefits and costs, cost--benefit analysis and investment appraisal.

**J. Barkley Rosser, Jr.** is Professor of Economics and Kirby L. Kramer, Jr. Professor of Business Administration at James Madison University. He serves as editor of the Journal of Economic Behavior and Organization and also on the editorial boards of several other journals. Besides his more than 100 published articles his books include From Catastrophe to Chaos: A General Theory of Economic Discontinuities; Comparative Economics in a Transforming World Economy; Complexity in Economics; and The Changing Face of Economics.

**Jeffrey R. Vincent** is Professor of Environmental and Resource Economics in the Graduate School of International Relations & Pacific Studies at the University of California, San Diego. Before joining UCSD he was a fellow at Harvard Institute for International Development. His research focuses on natural resource and environmental policy in developing countries, in particular countries in Asia. He is co-editor of the North-Holland Handbook of Environmental Economics and author of numerous articles in economics, development, and forestry journals. In addition to his research, Vincent has extensive experience with international policy advising and capacity building projects.

# Preface of the Series

The increasing support for and dedication to the concept of Sustainable Development (SD), expressed through various international conventions, reflect an evolution in the human value system, which in turn reflects the social, cultural, economic, and environmental conditions of the late twentieth and early twenty-first century. The concept of 'Sustainable Development' reflects the challenge humanity now faces in managing our global natural resources in such a way as to sustain and enhance human welfare well into the future. An understanding of the finite nature of natural resources and their distribution among economic agents is at the heart of economics, but in the past economists rarely gave much thought to the question of sustainability. Unfortunately, main-stream economists, not all but most of them, have remained one of the most reluctant groups within the scientific community to accept sustainability as a serious economic challenge; some even feel that it is not an appropriate topic for economics. What an irony! How can the discipline of economics, with its basis in the analysis of scarce resources afford not to discuss sustainability issues, which are basically related to scarcity and allocation of natural resources, human welfare, and inter-generational equity?

The main stream economists' reluctance to take sustainability issues seriously probably results in part from the intellectual limitations imposed by neo-classical theory which sticks rigidly to the assumption of the economically rational or selfish agent, and thus gets caught up in a way of thinking which has been called a "rational fool's trap", and all efforts to take it out of the trap have faced almost impenetrable resistance. The experimental observations on human behavior, markets, and institutions, reported by so-called heterodox economists, have been termed anomalies, and behavior that violates the stringent canons of economic rationality or selfishness have been treated as idiosyncratic or irrational. For that reason, and others, an economic theory capable of effectively encompassing and integrating the concept of sustainability must be broader and different than the neo-classical theory which is at the root of most sustainability issues and currently used to address such issues (along with everything else). A new economic theory, rather than a new public policy based on the old theory, will be needed to guide humanity toward sustainability.

Simply put, sustainability involves ensuring opportunities for a desirable "quality of life" for all future generations as well as for the present one. The quality of human life includes not only the economic dimension but also at least two others—the ecological and the social. Hence, the economic analysis of sustainability will be much more complex than the traditional concentration on efficiency and equity, and it will have to be based on a different set of principles, in which economic, ecological, and social dimensions are inseparable elements of the same whole. Thus the challenge to economists, together with other social scientists, is to build a new dominant economic paradigm—based on an approach which is more organic, holistic, and integrative than the current reductionist approach of the neo-classical paradigm. We will refer to this paradigm as Post-Newtonian Economics.

In the last two decades, new streams of economics--such as agent-based modelling, behavioral economics, complexity theory, ecological economics, the economics of increasing returns, experimental economics, evolutionary economics and evolutionary game theory have challenged the basic foundations of the neo-classical paradigm, but these streams have not focused on sustainability issues. The book series *Sustainability, Economics, and Natural Resources* aims to integrate the concept of sustainability fully into economics and to provide a foundation for the new economic paradigm. The series is designed to reflect the multi- and interdisciplinary nature of the paradigm and will cover and integrate concepts from the new streams of economics mentioned above, other streams of economics such as old and new institutional economics, post-Keynesian consumer theory, and social choice theory, and concepts from relevant streams of physical and biological sciences such as S-matrix theory, quantum mechanics, the theory of relativity, and the theory of evolution. The series will be a forum for new ideas, concepts, theories, and analytical tools associated with the economic analysis of sustainability and the applications of these ideas and tools for sustainable management of natural resources.

Forest ecosystems are important components of almost all the international agreements related to sustainability, and interactions between human systems and forest ecosystems can provide an experimental setting for the study of interactions between the ecological, social, and economic dimensions of human welfare. Hence, forest ecosystems are an excellent and unique starting point in the effort to integrate the concept of sustainability within economic theory and to build a new economic paradigm. Accordingly the first two volumes of the series focus on forest ecosystems—Volume 1: Economics, Sustainability, and Natural Resources: Economics of Sustainable Forest Management; and Volume 2: Institutions, Sustainability, and Natural Resources: Institutions for Sustainable Forest Management.

Shashi Kant

# PREFACE AND ACKNOWLEDGEMENTS

In the short-term human beliefs and values are heavily influenced by existing social, cultural, economic, and environmental conditions, while in the long-term these conditions are in turn influenced by human behavior. These continuous interactions underlie the dynamic nature of human beliefs and values, as well as the surrounding social, cultural, economic, and environmental conditions. The increasing support for and dedication to Sustainable Forest Management (SFM) reflects an evolution in the human value system, which in turn reflects the social, cultural, economic, and environmental conditions of the late twentieth and early twenty-first century, conditions which are quite different from those of the nineteenth and early twentieth century. The economic principles, theory, and models of SFM need to reflect the realities of the twenty-first century.

The concept of SFM incorporates human preferences for timber and non-timber products, preferences for marketed as well as non-marketed products and services, the preferences of industrial as well non-industrial agents, including Aboriginal and other local people, and the preferences of future generations as well as the present one. It takes account of diversity of preferences across agents, communities, time, and generations, and incorporates preferences that are revealed through the market as well as through non-market mechanisms. Forests, in the context of SFM, are valuable for their contributions to ecosystem functioning as well as their physical outputs. However, the existing paradigm of forest economics, which is focused on sustained yield timber management and has its roots in the conventional neoclassical paradigm of economics, is based on the combination of utility maximizing rational agents and the 'invisible hand' leading to an efficient general equilibrium. In this framework, peoples' preferences are internally consistent, static and revealed through the market only; public inputs are selected on the basis of market signals; all systems, including ecosystems, can be commoditized, which converts them into functionally-disjointed and discrete units; and there are no commitments and moral judgments attached to the domains of forest values. It is evident that the basic premises of the existing paradigm of forest economics are in serious contradiction of the realities and expectations of SFM, and the economics of SFM will thus require an extension of the boundaries of forest economics.

Keeping the unique features of SFM and the need to extend the boundaries of forest economics in perspective, Shashi Kant published, "Extending the boundaries of forest economics" in Volume 5 (2003) of Forest Policy and Economics. Response to the publication of this article revealed that there were many other forest and resource economists who shared our vision of extending the boundaries of forest economics. We then planned an International Conference on the Economics of Sustainable Management, at the University of Toronto, on May 22-24, 2003, but due to the outbreak of Severe Acute Respiratory Syndrome (SARS) in Toronto, the conference had to be rescheduled to May 20-22, 2004. In fact, the SARS outbreak was a good example and a reminder to economists of natural uncertainties.

We are pleased to announce that this volume is the first of the new series "Sustainability, Economics and Natural Resources". The papers in this volume and its companion "Institutions, Sustainability, and Natural Resources: Institutions for Sustainable Forest Management" were originally presented at the conference. (In addition, a special edition of Forest Policy and Economics, Volume 6, Issues 3-4, also includes papers from the conference.) The volume is not a mere re-printing of conference papers, however. The original selection of papers and the rewriting, and reworking of them after the conference have been designed to cover the issues related to SFM in an integrated and reasonably comprehensive way. We are thankful to the authors for responding positively to our suggestions.

In this volume, leading economists from behavioral economics, complexity theory, resource economics, post-Keynesian economics, and social choice theory discuss selected key aspects of the economics of SFM, including complexity, ethical issues, consumer choice theory, intergenerational equity, non-convexities, and multiple equilibria. The companion volume mentioned above focuses on institutions for sustainable forest management, markets for environmental services, deforestation and specialization, and some country experiences related to institutions for carbon emissions and sequestration (Kyoto Protocol), international trade, biodiversity conservation, and sustainable forest management in general.

The conference was organised by the Faculty of Forestry, University of Toronto in collaboration with the Groups 4.04.02 and 4.13.00 of the International Union of Forestry Research Organizations (IUFRO). We are thankful to the late Prof. Maurizio Merlo and to Prof. Hans A. Joebstl, Group Leaders of IUFRO Groups, for their support.

The conference was made possible through the financial and overall support of the Canadian Forest Service, Ford Foundation, Forest Products Association of Canada, International Paper, Living Legacy Trust, Ontario Ministry of Natural Resources, Sustainable Forest Management Network, and Weyerhaeuser Canada. Along with our thanks to these organisations we would like to specifically recognize the contributions of – Gordon Miller, Jeffrey Campbell, Michael L. Willick, Paul K. Perkins, Sharon G. Haines, Karan Aquino, and Mark Hubert – who supported us throughout the period of about two years. We also express our thanks to Prof. Robert J. Birgeneau, Ex-President, University of Toronto, Prof. Rorke Bryan, Dean, Faculty of Forestry, University of Toronto, and Brian Emmett, Assistant Deputy Minister, Canadian Forest Service for their support and participation in the conference.

Special thanks are due to Amalia Veneziano and Sushil Kumar who were instrumental in the organization of the conference, with the assistance of other students and staff members of the faculty.

Finally, we would like to thank Springer Publishers and their staff members, specifically Paul Ross, Henny Hoogervorst, and Esther Verdries for taking up this project and Sushil Kumar who did a great job turning the manuscript into camera ready copy.

Shashi Kant
R. Albert Berry

# CHAPTER 1

# ECONOMICS, SUSTAINABILITY, AND FOREST MANAGEMENT

SHASHI KANT

*Faculty of Forestry, University of Toronto*
*33 Willcocks Street, Toronto, Canada M5S 3B3*
*Email: shashi.kant@utoronto.ca*

R. ALBERT BERRY

*Munk Centre for International Studies, University of Toronto*
*1 DevonshirePlace, Toronto, Canada M5S 3K7*
*Email: berry2@chass.utoronto.ca*

**Abstract.** This chapter provides an overview of the contents of the volume. To put those contents in perspective, it first reviews developments related to the concepts of sustainability and sustainable development, the reactions of some main stream economists, the main problematic features of traditional economics, and the resulting need for a new paradigm within economics if sustainability issues are to be adequately handled. Next, an overview of the economics literature on sustainability and sustainable forest management is provided. Finally, each chapter included in the five parts of this volume is briefly reviewed.

## 1. INTRODUCTION

The word "sustainable" is not as new to the forestry profession, including forest economists, as it may be to some mainstream economists. The Faustmann Formula, one of the main pillars of conventional forest economics, is based on the idea of a sustained supply of timber for an infinite number of rotations. In the eighteenth and nineteenth centuries several other social scientists expressed their concerns about sustainability of certain products in Britain –for example, Malthus (1798) about food output and Jevons (1865) about coal supplies. However, the recent concerns about sustainability, which were signaled by the publication of 'The Limits to Growth' by Meadows et al. (1972) and 'Our Common Future' by WECD (1987), are not limited to a specific product but include all natural systems and human life. Sustainability concerns have been reinforced by the Rio Earth Summit in 1992 and the



*Kant and Berry (Eds.), Economics, Sustainability, and Natural Resources: Economics of Sustainable Forest Management,* 1-22.

Johannesburg Summit in 2002. The importance of the concept of sustainability has already been acknowledged by the world community through numerous international conventions such as the Convention for Climate Change, the Biodiversity Convention, and Agenda 21 (Pearce, 1994). Unfortunately, main-stream economists, not all but most of them, have remained one of the most reluctant groups within the scientific community to accept the concept of sustainability as an (economic) issue (Ikerd, 1997); some, specifically Rust Belt economists, feel that sustainability is not an appropriate topic to be discussed by economists (Colander, 2004). In this regard, Dasgupta and Mäler (1994) write:

> ".. most writings on sustainable development start from scratch and some proceed to things hopelessly wrong. It would be difficult to find another field of research endeavor in the social sciences that has displayed such intellectual regress." (Dasgupta & Mäler, 1994, quoted in Beckerman, 1994, p. 192)

Beckerman (1994) follows Dasgupta and Maler:

> " 'sustainable development' has been defined in such a way as to be either morally repugnant or logically redundant. 'Strong' sustainability, overriding all other considerations, is morally unacceptable as well as totally impractical; and 'weak' sustainability, in which compensation is made for resources consumed, offers nothing beyond traditional economic welfare maximization." (Beckerman, 1994, p. 191)

One factor contributing to the prevalence of such observations about sustainability is that traditional, simple economic theory is built on the assumption of a representative "rational economic agent" who is close to being a "social moron" or a "rational fool" in the words of none other than Nobel Laureate Amartya Sen (1977), or a "mindless individual" in the opinion of Hegel (1964, 1967); sustainability cannot be achieved through the choices of "rational fools" or "mindless individuals". Another factor is the discomfort some economists feel with the variety of definitions of sustainability, though this variety does not seem out of context keeping with the fact that economics itself means different things to different people and that there is a broad spectrum of "heterodox" economists, who approach economic issues differently from the so-called main stream "neo-classical" economists.

The intellectual scope of main stream economics has been tragically limited by its working assumption that the world is a simple, homogeneous, and static unit, rather than being full of complexity, diversity, and dynamism. Natural science, specifically physics, has continuously demonstrated, for about the last 100 years, the existence of natural processes and phenomena which do not mesh readily with this world vision of the main stream, while these economists continue to live in the economic equivalent of a Newtonian world. Quantum theory demonstrated that even sub-atomic particles were nothing like the solid objects of classical physics, but instead are abstract entities with a dual aspect. Depending on how we look at them, they appear sometimes as particles and sometimes as waves; in fact, both pictures are needed to give a full account of the atomic reality, and both have to be applied within the limitations set by Heisenberg's Uncertainty Principle. Thus modern physics is governed by the principle of complementarity rather than the principle of substitution. In addition, the isolated material particles are abstractions, their properties being definable and observable only through their interactions with other systems (Bohr, cited in Capra, 1982, p. 137). Similarly, according to S-matrix theory, also known as the bootstrap approach, nature cannot be reduced to fundamental entities, like fundamental building blocks of matter, but has to be understood entirely through self-consistency – consistent with one another and with themselves (Capra, 1982). This transition from Newtonian physics to modern physics was not easy; even the fathers of modern physics found it difficult to accept.

> " I remember discussions with Bohr which went through many hours till very late at night and ended almost in despair; and when at the end of the discussion I went alone for a walk in the neighboring park, I repeated to myself again and again the question: Can nature possibly be so absurd as it seemed to us in these atomic experiments." ...."The violent reaction to the recent development of modern physics can only be understood when one realizes that here the foundations of physics have started moving, and that this notion has caused the feeling that the ground would be cut from science." (Heisenberg, 1963, p. 43)

> "All my attempts to adapt the theoretical foundations of physics to this knowledge failed completely. It was as if the ground has been pulled out from under one, with no firm foundation to be seen anywhere, upon which one could have built." (Einstein, 1949, p.45)

Viewed from this perspective the derogatory remarks about sustainability coming from some well known economists are not surprising. More remarkable, however, is the continuation of their strong belief (implemented in practice) that all preferences of all human beings for all time to come can be adequately captured in a single-modulus discounted utility function even as many experiments, conducted by experimental and behavioral economists, and a common world view provide contrary evidence. These economists continue to base their analysis on the conceptualization of a "rational economic agent", who has only one, and that a static, preference ordering which reflects, as per need, his interests, welfare, actual choices, and behavior (Sen, 1977), and who uses the same preference ordering for all goods, whether public or private, and sources of different types of satisfaction – ethical, spiritual, commercial, and sexual.

The words of "commitment" and "moral" are missing from the vocabulary of the "economically rational agent" but not from the vocabulary of a "human being" or a "socially rational agent". Commitment and morality would involve, in a very real sense, counter-preferential choice, that would destroy the crucial assumption that a chosen alternative must be better than (or at least as good) the other options in terms of the narrowly defined self-interest of the person choosing it; destruction of that assumption renders consumer theory different and much more complex. The traditional narrow approach of mainstream economics on this point does not mean that economists, as a group, are unaware of more realistic preference systems: Harsanyi (1955) proposed a dual structure of preferences—'ethical preferences' and 'subjective preferences', Sen (1973) suggested three categories—Prisoner's Dilemma (PD), Assurance Game (AG), and Other Regarding (AR) – of preferences, and there are many other categories of preferences available in the social choice literature. Similarly, many streams of economics (often termed "heterodox" streams), such as post-Keynesian economics, evolutionary economics, and ecological economics, along with recent developments in behavioral economics, social choice theory, experimental economics, agent-based modeling, evolutionary game theory, and complexity theory have recognized complexity, multiplicity, dynamism, and inter-connectedness as characteristics of the real world. In fact, an economic agent conceptualized by Kahneman and Tversky (Kahneman and Tversky, 2000), known as K-T man (McFadden, 1999), is close to a "socially-rational agent".

The challenge, therefore, to the current and future generations of economists is to build a new dominant economic paradigm – based on a more organic, holistic, and integrative approach than the reductionist neo-classical paradigm. The now high-profile concept of sustainability offers a challenge to economists to bring the profession closer to the real world. It is now up to the current and next generation of economists to meet this challenge. As Einstein once observed, problems cannot be solved at the same level of thinking that lead to their creation (Ikerd, 1997). Hence, the economic theory of sustainability cannot be based on neo-classical economic theory that is at thc root of most sustainability issues, and a new economic theory, rather than a new public policy based on old theory, will be needed to guide humanity toward sustainability or sustainable development.

In simple words, sustainability involves ensuring opportunities for a desirable "quality of life" for all future generations as well as for the present one. It is thus a concept related to the very long-run and, accordingly, one involving considerable uncertainty—"a direction without a precise destination" (Ikerd, 1997). However, the concept of sustainability is quite consistent with the root-word for economics, "oikonomia"—management of the household. Human's quality of life includes not only the economic dimension but at least two others—the ecological and the social. Over the very long-run, human and natural systems cannot be economically viable unless they are also ecologically sound and socially responsible; nor can they be ecologically sound unless they are economically viable and socially responsible; nor can they be socially responsible unless they are also ecologically sound and economically viable (Ikerd, 1997). However, a main pillar of the neo-classical economic theory is the condition of "ceteris paribus" which means that the theory deals with the outcomes of economic activities when "social" and "ecological" conditions are kept constant. The economics of sustainability will have to be based on a different set of principles, in which economic, ecological, and social dimensions are inseparable dimensions of the same organism.

In the efforts of developing economic theory of sustainability, forest ecosystems can be of enormous use due to numerous reasons. First, forest ecosystems are important components of almost all the international agreements related to sustainability – convention for climate change, biodiversity convention, and Agenda 21. Second, interactions between human systems and forest ecosystems can provide an experimental setting to study interactions between ecological, social, and economic dimensions of human welfare. Third, the concept of sustainability, even though in a limited sense (related to timber), has existed for about 150 years in the thinking about forestry, including forest economics. Finally, there have been serious efforts, all around the world, to transform forest management from sustained yield timber management to sustainable forest management. The contributors to this volume have taken up a challenge to contribute to the development of a new paradigm of the economics of sustainable forest management..

In this volume, leading economists from different streams—behavioral economics, complexity theory, forest resource economics, Post-Keynesian economics, and social choice theory—provide basic foundations for an economics of sustainable forest management. In future there will, we assume, be many other volumes dedicated to these issues, some focused on specific aspects of the economics of SFM. While the main purpose of this chapter is to provide an overview of the contents of the volume, to put those contents in perspective, overviews of the economics literature related to sustainability and sustainable forest management are also included.

## 2. SUSTAINABILITY AND ECONOMICS

A number of economists, such as Ciriacy-Wantrup (1952), Krutilla (1967), and Ayres and Knesse (1969), had aired their concerns about issues related to sustainability even before the publication of *The Limits to Growth* (Meadows et al., 1972). T*he Limits to Growth* inspired an interest among economists to incorporate natural resources into growth models, and this interest, among some main stream economists but mainly among heterodox economists, has been sustained by the publication of the WECD Report, by the Rio and Johannesburg Summits, and by other similar events. As a result, an enormous volume of literature, from different streams of economics, has emerged on sustainability issues. Some useful sources for the review of this literature are Pezzey (1989, 1992), Pezzey and Toman (2002, 2003), and special volumes of Ecological Economics, September 1997, volume 22, issue 3 and Land Economics, November 1997, volume 73, issue 4. While we are not in a position to summarize this whole literature, we do review some key papers which contributed theoretical foundations for some economic aspects of sustainability.

In 1974, just after the publication of The Limits to Growth, the Review of Economic Studies published a special issue (Volume 41, Issue 128) on The Economics of Exhaustible Resources. Three papers in this volume—Dasgupta and Heal (1974), Stiglitz (1974), and Solow (1974)—provided basic foundations for future debate on the subject. In all three papers, natural resources are assumed to be finite, nonrenewable, essential to production, and (human-made) capital can substitute indefinitely for natural resources. Dasgupta and Heal (1974, 1979) and Stiglitz (1974) use a familiar formulation of an economic problem - the maximization of the present value (PV optimality) of the representative agent's instantaneous utility, using a constant discount rate. Dasgupta and Heal's main finding was that the implications of this PV-maximization approach have grim implications for future generations, as a direct consequence of a positive discount rate and the inherent scarcity of the nonrenewable resources. Stiglitz (1974) assumes the rate of exogenous technical progress to be large enough to offset the effects of resource depletion, and demonstrates the PV-optimal path can have sustained increases in per capita consumption even with a growing population. Solow (1974) includes Rawl's max-min principle of intergenerational equity in his analysis, and draws two conclusions. First, the max-min criteria seems to be reasonable criterion for intertemporal planning decisions except that it requires a big initial capital stock to support a decent standard of living, and it seems to give foolishly conservative injunctions when there is stationary population and unlimited technical progress. Second, the finite pool of exhaustible resources should be used up optimally according to the general rules that govern the optimal use of reproducible capital; this conclusion depends on the presumption that the elasticity of substitution between natural resources and labor-and-capital goods is no less than unity.

The next contribution in this sequence is commonly known as Hartwick's rule or the Weak Sustainability approach (Hartwick, 1977, 1978a, 1978b). According to Hartwick's rule, in an economy with depletable resources, the rent derived from resource depletion is exactly the level of capital investment that is required to achieve constant consumption over time. Solow (1986) shows that Hartwick's rule is equivalent to maintaining aggregate wealth or appropriately defined stock of capital, including natural resources, at a constant level over time. However, Solow's result assumes a constant interest rate, as pointed out by Svensson (1986) in the same journal, and thus does not actually apply to the economies modeled by Dasgupta and Heal (1974) and Solow (1974)[1]. Later, Asheim (1986) demonstrated that Hartwick's rule cannot be applied to closed economies[2], and in the open economy case, the rule requires resource-rich economies to invest less than their own resource rents, and resource-poor economies to invest more than their own resource rents.[3] Krautkraemer (1985) extended the PV optimality formulation by including resource amenity (resource stock) and consumption in the utility function, and demonstrated that depending on society's discount rate, the initial capital stock, and the nature of the resource amenity, the economy may converge over time to either a low-resource-use equilibrium or a high-resource-use one.

After the publication of Our Common Future (WCED 1987), active discussion of sustainability issues began in the economics literature as well. WECD (1987), however, uses concepts of needs, or lack of compromise or trade off, that cannot be readily included in the framework of conventional economics. Barbier (1987), Pearce (1988), Daly and Cobb (1989), Pearce, Markandya, and Barbier (1989), and Costanza (1991) propelled this debate forward. Pezzey (1989, 1992) and Ahmad, El Serafy, and Lutz (1989) tried to incorporate these within the framework of conventional economics. Daly (1990) highlights three principles of sustainable development: (i) harvest rates should equal regeneration rates (sustained yield); (ii) waste emission rates should equal the natural assimilative capacities of the ecosystems into which the wastes are emitted; and (iii) renewable energy sources should be exploited in a quasi-sustainable manner by limiting their rate of depletion to the rate of creation of substitutes for those renewable resources. This approach is commonly known as Strong Sustainability.

An important contribution in the debate on conventional PV optimization and the sustainability constraint is Pezzy (1997) in which the author defends the possible use of different variants of sustainability as a priory constraint on PV optimality. He argues that such a constraint is not self-contradictory, redundant, or inferior as claimed by Beckerman (1994) and Dasgupta (1995). Pezzey questions Koopman's

(1960) axiomatic foundation, specifically the validity of the stationary axiom of PV maximization that was questioned also by Page (1997). Pezzy proposes an empirical approach that relies on psychological experiments on time preferences to extend the intertemporal welfare function to include a finite value of sustainability in some way. This extension might involve replacing the instantaneous utility function with a more complex function that includes the individual's value of improvements in consumption. An important feature of this approach is that it may result in Pareto-inefficient consumption paths being preferred.

Another common issue in debates on sustainability is the intergenerational distribution of resources. Howarth and Norgaard (1990) was seminal in showing that different endowments of resource rights—a nonrenewable resource stock and labor—across two overlapping generations (OLGs) result in different distributions of wealth, all of them efficient but obviously different in their equity implications, and with no a priori way of judging which is "optimal." Howarth and Norgaard (1992) extend their 1990 model to include many generations, and demonstrate that, even in theory, there is no fixed notion of "correctly" valuing an environmental cost: the value varies with society's view of the future, whether expressed as a discount rate or some sustainability criterion. Several other papers by Howarth (1991a, 1991b) and Howarth and Norgaard (1993) show the full analytical power of the OLG approach to sustainability. Howarth (1995) develops the theme that moral obligations to future generations are distinct from altruistic individualistic preferences for the well-being of future generations, and explores, among other topics, the "precautionary principle." The sustainability literature on intergeneration distribution of resources clearly demonstrates that an adequate treatment of intergenerational equity calls for a framework going well beyond the scope of conventional welfare economics.

Green national accounting is another stream which has attracted many scholars including Repetto (1989), and Pearce and Atkinson (1993). However, as Asheim (1994) and Pezzey (1994) point out, this approach has a common flaw. Shifting an economy from non-sustainability to sustainability changes all its prices. Sustainability prices and sustainability itself are thus related in a circular fashion. Without sustainability prices, we cannot know whether the economy is currently sustainable; but without knowing whether the economy is currently sustainable, currently observed prices tell us nothing definite about sustainability. This theoretical caveat does not imply that green accounting is not useful, but rather that it cannot at this time be carried out in the technically ideal way, and hence requires judgment in the way it is applied.

Unfortunately, the contributions from many other streams of economics such as behavioral economics, complexity theory, and social choice theory, which appear to imply the most serious challenges to the conclusions of neo-classical economics on sustainability issues, have not attracted much attention in the sustainability literature. The over-taking criterion (Atsumi, 1965; von Weizsäcker, 1965), the Suppes-Sen grading principle (Suppes, 1968; Sen, 1970), and the general theory of inter-temporal resource allocation (Radner, 1961; Gale, 1967; Brock, 1970; and McKenzie, 1983, 1986) are highly relevant to sustainability issues, but the social choice literature based on these criteria and principles has not intersected much with the dominant economic literature on sustainability. Mitra and Wan (1986), using the general theory of inter-temporal resource allocation, addressed the problem of forest management when future utilities are undiscounted, and found that if the utility function is increasing and strictly concave, an optimal solution converges to the maximum sustained yield solution. Chichilnisky (1997) introduces two axioms for sustainable development or sustainable preferences: the first requires that the present should not dictate the outcome in disregard for the future or it requires sensitivity to the welfare of generations in the distant future; and the second requires the welfare criterion should not be dictated by the long-run future or it requires sensitivity to the present. Chichilnisky proves the existence of sustainable preferences[4] and demonstrates that sustainable optima can be quite different from discounted optima, no matter how small is the discount factor[5]. Asheim, Buchholz, and Tungodden (2001) observe that there is a technical literature on inter-generational social preferences including Koopmans (1960), Diamond (1965), Svensson (1980), Epstein (1986), and Lauwers (1997) which essentially presents the finding that complete social preferences that treat an infinite number of generations equally need not admit optimal solutions, and resolves this apparent conflict. Asheim et al. prove that in the framework of ethical social choice theory, sustainability is justified by efficiency and equity as ethical axioms which correspond to the Suppes-Sen grading principle. In technologies that are productive in a certain sense, the set of Suppes-Sen maximal utility paths is shown to equal the set of non-decreasing and efficient paths. Since any such path is sustainable, efficiency and equity can thus be used to deem any unsustainable path as ethically unacceptable. Asheim and Tungodden (2004) propose a new approach, by imposing some conditions on the social preferences, to the problem of resolving distributional conflicts between an infinite and countable number of generations. Pezzey and Toman's (2002) observations about Asheim's work "though the uncompromising rigor of the papers limits their readership to the technical, well-motivated few." are interesting and provide important clues for economist's approach towards sustainability. It seems that economists are looking for simple solutions—maximization of all encompassing discounted utility—for complex problems, unfortunately there are no such solutions for the sustainability dimension of human welfare. A similar unrealism on the part of economists may account for the neglect of complexity theory, behavioral economics, theories of multiple equilibria, evolutionary game theory, and multi-disciplinary approaches in general, and specifically with respect to sustainability questions.

## 3. SUSTAINABLE FOREST MANAGEMENT AND ECONOMICS

In the last two decades, sustainable forest management (SFM) has emerged as a new paradigm of forest management. This paradigm is in the process of transforming forest management from sustained yield timber management (SYTM) to forest ecosystem management and from forest management by exclusion of user groups to management by inclusion of user groups. The SFM paradigm recognizes three dimensions of human welfare—economic, social, and ecological. In economic terms, the main distinguishing features of SFM are the recognition of diverse and dynamic preferences of local people (heterogeneous agents), the incorporation of multiple sources of value and utility from the forests (including non-market values), the incorporation of multiple products and services in the production process, intergenerational equity, and a systems approach to forest management. In short, SFM involves a complex matrix of interactions between social, economic and natural systems, and it implies the need for a significant shift in the dominant paradigm of forest economics.

The recognition of distinguishing economic features of SFM brings to the fore the potential conflict between the concept of SFM and the neo-classical economic framework of forest management which has been used for sustained yield timber management (Toman, Mark, & Ashton, 1996). The main response from forest economists to SFM has been the use of direct or indirect valuation techniques for non-marketed "goods" and "services", so that these values can be made comparable with the values of traditional wood products; this, however, is a controversial application of market concepts. The economic literature has already identified numerous problems with the application of these methods for valuation of environmental and forestry attributes. Anther noticeable development has been in the area of multiple criteria decision making, and some examples of this in forestry are Bare and Mendoza, 1992; Gong, 1992; Kangas, 1993; and Liu and Davis, 1995.

Kant (2003a), the first overall review of the forest economics literature from the perspective of economics of SFM, proposed a set of basic principles for the economics of SFM. He argues that the basic idea behind SFM, to manage forests in such a way that the needs of the present are met without compromising the ability of future generations to meet their own needs, demands elements of altruistic and cooperative behavior among social agents in contrast to the self-interest-maximizing rational agent of neo-classical economics. Hence, economic models of SFM should be able to capture both orientations—individualistic as well as altruistic and/or commitment—of an individual's behavior; neo-classical economics, which is guided by the "either-or" principle, is unable to incorporate such dualistic behavior of social agents[6]. Incorporation of such behavior may be possible in economic models that are based on the "both-and" principle that has been accepted by post-Newtonian physicists of the twentieth century. Under the umbrella of the "both-and" principle, Kant (2003a) proposes four sub-principles of SFM economics: existence, relativity, uncertainty, and complementarity.

The 'principle of existence' suggests that we cannot ignore the relevance of situations which have survived for a long time. Hence, we should focus first on achieving an economic understanding of the existing human-forest interactive systems, in order to be able to predict whether the effects of proposed changes would be, on balance, positive or negative. The 'principle of relativity' suggests that optimal solutions are not universal but rather situation specific; in many cases they will involve important non-market forces. The 'principle of uncertainty' suggests that due to uncertainties in natural and social systems, a social agent may typically not be in a position to maximize his outcomes, but will rather search for positive outcomes and learn by experience, such that resource allocation will be improved by adaptive efficiency, whose cumulated effects over time are likely to be more important than the achievement of allocative efficiency at each point of time. The 'principle of complementarity' suggests that human behavior combines both selfish and altruistic elements, that people have both economic and moral values, and that people need forests to satisfy both lower level and higher level needs. Kant (2003a) concludes that the two main additional elements for the economic analysis of SFM are the economics of multiple equilibria and a consumer choice theory that incorporates context-specific and dynamic preferences, heterogeneous agents, and a distinction between needs and wants.

A Special Issue of Forest Policy and Economics (Volume 6, Issues 3-4, 2004) focuses on the economics of sustainable forest management. In this issue, Wang (2004) contrasts SFM with conventional forest management (CFM), and argues that the conventional economic tools cannot be applied satisfactorily to SFM analysis. He proposes an integrative and contextualized knowledge-based two-tier approach for the economics of SFM, in which economic incentives and trade-offs dictate resource allocation and management decisions when sustainable products are involved, but precautionary principles prevail when the integrity of ecosystems is at stake. Kant and Lee (2004) argue that multiple forest values are closer to the concept of 'social states' than market price or monetary value, and the decisions related to SFM are decisions of social choice and not decisions to be guided by conventional benefit-cost analysis, based on the monetization of all costs and benefits. Cardenas (2004), based on the outcomes of economic experiments in rural communities of Colombia, argues that individuals do not seem to follow entirely the conventional economic prediction about externally imposed rules, and people in rural communities can develop norms based on non-enforceable rules of cooperation which may prove as effective as externally imposed rules in SFM. Subhrendu et al. (2004) included heterogeneity of preferences of forest landowners in a regional timber supply model and examined the impact on timber supply in the southern USA. Misra and Kant (2004) suggest an analytical framework for the production analysis of collaborative forest management, and use this framework for the analysis of Joint Forest Management, in Gujarat state of India. Other papers in the issue address various issues related to sustainable forest management such as carbon sequestration, foreign direct investment, and forest valuation.

The economics of sustainable forest management appears to be attracting the interest of a few economists, specifically resource economists, but it has not received the same level of attention from the discipline as has the economics of sustainability in general. Sustainable forest management, the topic of much discussion over the past two decades among a wide range of people involved in the forestry sector—researchers, managers, policy-makers, international agencies, donor agencies, and non-government organizations—has received much less attention from economists. In addition, many economists have not been able to accept the basic differences in economic features between SYTM and SFM, and hence continue to use the traditional but inappropriate economic tools.

The volume starts with chapters on complexity theory, ethics, and sustainable forest management and closes with the basic principles of economics of SFM and new paradigm of economics. In between, three other major themes—consumer choice theory and SFM, social choice theory and SFM, and non-linearities, multiple equilibria and SFM—are highlighted.

## 4. COMPLEXITY, ETHICS, AND THE ECONOMICS OF SFM

The previous section was designed to give a broad introduction to the economics of SFM. This section draws on the other chapters of the volume to delve deeper into SFM economics. David Colander identifies the economics of SFM as part of a broader trend within economics, that he defines as a switching from the efficiency and control story to the complexity and muddling through story. The efficiency story is about the *state of competition*, is static, and fits well into a calculus framework, while the complexity story is about the *process of competition,* and is a dynamic and evolutionary story. In the complexity story, the invisible hand of the market takes apparent chaos and turns it into an elegantly complex structure that fits together, not perfectly or efficiently, but sustainably. Colander argues that the traditional work in forest economics falls within the efficiency story line. Textbook presentations, like traditional work in forest economics, avoid discussing the fact that efficiency is not an end in itself but rather a means to an end. Sustainability fits much better into the complexity story in which one does not talk about equilibrium; one talks about basins of attractions. Nonlinearities are accepted, and one can expect phase transition jumps as the system evolves. Sustainability means remaining either in the existing basin of attraction or going to a more desirable basin but avoiding less desirable basins.

Colander sees a clear parallel between the shift towards SFM, within thinking about forestry, and the current changes occurring within the economics profession: a change in the allowable assumptions, from the holy trinity of rationality, greed and equilibrium to a broader set which might be called a new holy trinity of purposeful behavior, enlightened self-interest, and sustainability. Acceptance of these changes is apparent in behavioral economics, evolutionary game theory, agent-based modeling, experimental economics, and the new institutional economics. Colander continues his discussion with the outcomes and causes of the changes, and the policy implications of the two stories, concluding with some predictions of how the complexity story will affect future research in SFM.

In the second chapter in this section, M. Ali Khan looks at the economics of SFM through an inter-disciplinary approach involving the ethics of theorizing and modern capital theory. On the basis of his reading of the texts of Kant, Laslett, Bourdieu, Cowen-Parfitt, and Mitra-Wan-Ray-Roy, he locates the general theory of inter-temporal allocation within political scientists' and sociologists' conversations about intergenerational justice. Khan relates Kant's (2003a) four sub-principles of the economics of SFM—existence, relativity, uncertainty, and complementarity to the work of Burke, Hegel, Laslett, Keynes, Marshall, Rawls, and Wittgenstein, noting how they reflect the broad interdisciplinary approach that the subject demands, and put the focus on the principles that go into its theorizing—the "ethics of theorizing", rather than on a particular theory. Next, using the work of Laslett as a guide, Khan situates the vocabulary of inter-temporal ethics and sustainability within that of another conversation being conducted in the space of political theory, a conversation including Laslett's notions of *inter-temporal tricontract* and *intercohort trust,* which he feels go to the heart of the economics of forestry, but which must be used without *hubris,* as a basis for a theoretical opening of a conversation rather than a closing of it and for a minimizing rather than a maximizing of the distance between the theorist and the theorized.

Khan argues that these larger issues of inter-temporal obligation and submission, when conceived within the relatively narrower frame of economics, specifically forest economics, inevitably revolve around the notions of capital and the rate of discount. Relevant, holistic, conceptions of the former variable include Kant's (2003b) *ecosystem capital and* Bourdieu's (1983) *symbolic capital.* Khan observes that if the words *sustainability* and *inter-temporal equity* are to have any analytical thrust, sustainable policies cannot be rejected, or decided upon, on criteria that have already incorporated in them some form of inter-generational *myopia* or *impatience.* However, even though this idea is simple and well-understood, mainstream economic research has bypassed and ignored it on two grounds: analytical tractability and a commitment to methodological individualism as typified by the analytical construct of the *representative agent.* The current conventional wisdom is to see research incorporating the assumption of a zero time-preference as "dispensable and misdirected", and the effects of this conventional wisdom are pervasive.

From the literature on capital theory and the general theory of inter-temporal resource allocation, Khan draws on the Mitra-Wan (1986) tree farm and the Mitra-Ray-Roy (1991) orchard for a "folk theorem". According to this theorem: "for any dynamic problem falling within the rubric of the theory, there is a threshold discount factor such that the stability properties of the optimal paths are qualitatively the same as those obtained for the undiscounted case for all discount factors above that threshold, and that complicated and rich dynamics, possibly including chaos, obtain for all discount factors below that threshold". Khan identifies the next order of business for both the economics of forestry and that of orchards as the integration of the discounted and undiscounted cases. He sees much merit in an inter-disciplinary approach in which various facets and factors are examined not only in isolation, but also in such a way as to enhance the potential for mutual reinforcement and global insight.

## 5. CONSUMER CHOICE THEORY AND THE ECONOMICS OF SFM

Second part of the volume addresses the relevance of some recent developments in consumer choice theory to the economics of SFM. Some of the many such developments have already been noted. Here we limit the discussion to the main elements of Post-Keynesian consumer choice theory, some developments from behavioral economics, and theory of discounting.

In his chapter on Post-Keynesian consumer choice theory and the economics of SFM, Marc Lavoie notes that this body of theory reflects a variety of influences (e.g. socio-economists, psychologists, marketing specialists, and individuals such as Herbert Simon and Georgescu-Roegen) whose common point was recognition of the complexity of our world. He identifies four key presuppositions of Post Keynesian economics: epistemology based on realism, ontology based on organicism, rationality being procedural, and a focus on production and growth issues; these pillars contrast with the symmetric presuppositions of neoclassical theory: instrumentalism, atomism, hyper rationality, and a focus on exchange and optimal resource allocation. The multiplicity of equilibria—the belief that models must be open-ended, is a characteristic feature of post-Keynesian economics, and true uncertainty, historical time, and the importance of aggregate demand help to distinguish it from other heterodox schools.

Lavoie highlights seven principles of post-Keynesian consumer choice theory: the principles of satiation, separability, subordination, growth of needs, and non-independence, procedural rationality, and the heredity principle. A key consequence of these principles, in particular that of subordination, is that the individual's utility index cannot (as in neoclassical theory) be represented by a scalar, but only by a vector, and that the notions of gross substitution and trade-offs—so important in neoclassical economics—are reduced to a secondary role, and operate only within narrow boundaries. The Archimedes principle that "everything has a price" is not part of this theory.

Lavoie notes that ecological economists have used all seven of these principles in their efforts to improve on standard neo-classical consumer choice theory. Such common themes of post-Keynesian economists as the precautionary principle associated with fundamental uncertainty, the heredity principle, weak comparability, incommensurability, and multidimensional choice (similar to the principle of the separability of needs) are emphasized by ecological and forest economists. Both groups entertain the idea of lexicographic choices (tied to the principle of the subordination of needs) in which substitution effects can play no role. The axiom of continuity also ceases to hold under lexicographic preferences, which cancels the validity of the Archimedes axiom that every thing has a price. In reality forest-related preferences are often lexicographic—a substantial proportion of individuals refuse to make trade-offs with material goods when biodiversity, wildlife, or forests are concerned. This has implications for contingency value analyses, based on willingness to pay or willingness to accept compensation, that attempt to take into account the non-market value of ecology or forestry preservation. In sum, Lavoie concludes that post-Keynesian consumer choice theory is highly relevant to forest economics, and could be used as a basis for consumer choice models in the economics of sustainable forest management.

In the second chapter of this part, Chapter 5, Jack Knetsch highlights the relevance of behavioral economics to SFM. Since SFM involves a wider array of uses and benefits from forest land management decisions, this multiplies the need to worry about tradeoffs among them and the associated problems of identification and quantification, and of weighing or valuation. With respect to the valuation of some forest benefits, the findings of behavioral economics provide a more realistic view of people's preferences than does the standard economic theory that forms the basis for most current economic practice and analyses. The often observed differences between behavioral findings and standard theory are, in Knetsch's view, far more than random deviations from an expected outcome; they are, instead, systematic and often large. Some are the result of bounded rationality but many—and those of most interest here—reflect real preferences that are not well modeled by the axioms of standard theory. For example, people often make choices in terms of separate mental accounts or budgets. And the empirical evidence sharply contradicts the standard equivalence assumption—that the maximum sum people would be willing to pay (WTP) to gain an entitlement is, except for a normally trivial difference due to an income effect, equal to the minimum sum they would be willing to accept (WTA) to give it up. Knetsch suggests that the choice of appropriate measure in such cases depends, among other things, on what people regard as the reference state, and suggests that the appropriate choice of measure may more usefully be determined by "psychological ownership" rather than legal entitlements. Along the same lines, Knetsch observes that people systematically discount the value of future losses at a lower rate than they use to discount the value of future gains. Knetsch concludes that though most economic analyses of resource issues, including those that guide forest management and policy decisions, could be markedly improved by including the insights from behavioral economics, this potential for improvement remains largely unrealized.

In the last chapter of this part, Chapter 6, Colin Price discusses discounting issues, with focus on the plausibility of the tempting (to some people) concept of a declining discount rate (i.e. a discounting procedure whereby the discount rate applied falls the farther into the future is the point of time under consideration). He notes that use of a declining discounting regime does indeed raise the relative attractiveness of slow-growing timber and is also likely to promote environmental interests; but simply lowering the discount rate at all points of time would be even more favorable to such distant future products and benefits, as well as being more defensible from a theoretical point of view. Hence, the declining discount regime requires critical examination.

A basic challenge in deciding on a discounting regime lies in the facts that (a) different people at a given time discount future benefits at different rates and (b) it is not obvious how the benefits accruing to future generations should be discounted. Though observed real interest rates provide some evidence on the discount rates applied by current members of a society (especially the wealthier ones), they do not give us a simple answer on how to discount. A basic complication is that the appropriate discount rate for a given person or group may not be the right one for society as a whole. Against this challenging backdrop, Price first discusses various discounting protocols—for example different intra and inter-generational discounting rates and different discounting rates for different circumstances. Next, he analyses the different aspects of diminishing marginal utility (DMU)—DMU and the basket of goods, DMU and inelastic supply, and DMU and related aggregation scenarios. On this basis of this analysis, he observes: (i) averaging of initial discount rates (across incomes, goods, scenarios) is a crude and inaccurate mode of aggregation; (ii) it is feasible to aggregate the separate discount factors which result from applying different discount rates to different income groups, goods and scenarios, but the resulting composite discount factors correspond to a whole period discount rate which changes through time; (iii) while in most (but not all) cases the whole period discount rate declines through time, the profile may differ according to the underlying reasons for discounting; and (iv) the circumstances which generate the lowest rate of diminishing marginal utility eventually dominate any discount rate derived from aggregate discount factors.

Finally, Price critically examines whether any specific form of discounting can be defended logically; in this respect, he evaluates the compensation argument, the time preference argument, and the diminishing marginal utility argument. In short, discounting for the circumstances of each product, scenario or income group is a time-intensive and controversial task. Using a schedule of discount rates which varies only by time period represents a relatively manageable alternative for project evaluation.

Price concludes that despite the fundamental weaknesses of the declining-rate protocol, governments will be eager to embrace it, because of its superficially plausible intellectual justifications, because it represents a nod in the direction of sustainability, and because in practice it does not change things much. By contrast, the protocol implicitly approved here—giving equivalent present values according to predicted circumstances, not according to the passage of time as such—is demanding procedurally. Perhaps purists should not let the perfect (not discounting at all for the passage of time) be the enemy of the marginal improvement implicit in declining-rate discounting (where benefits accruing in the distant future are less heavily penalized than in present practice). But neither should they let governments —or citizens—settle into a complacent belief that some lowering of the discount rate applied to distant-future benefits constitutes a full and satisfactory solution.

## 6. SOCIAL CHOICE THEORY AND THE ECONOMICS OF SFM

Our review of the economic literature on sustainability (section 2 above) provided a look at social choice theoretic literature related to the concept of sustainability in general. In Chapter 3, Ali Khan discusses the general theory of optimal resource allocation and reviews the forest management work by Mitra-Wan. This part continues that discussion with two papers with similar motivation.

As discussed in Section 3 above, Mitra and Wan (1986) formulated the problem of forest management as one of optimizing the sum of (undiscounted) utilities from harvests of timber according to the well-known overtaking criterion. In the first chapter of this section, Chapter 7, Tapan Mitra re-examines the foundations of intertemporal preferences which involve intergenerational equity and proposes and provides an axiomatic basis for a social welfare relation (SWR) which is weaker (less restrictive) than that required by the overtaking criterion. The axioms drawn on are Weak Pareto, Anonymity, Completeness and Continuity for finite horizon comparisons, and Independence, but no continuity property on the preference relation in the infinite dimensional space containing the set of consumption streams, a property which characterizes the more restrictive SWR induced by the overtaking criterion.

Mitra applies the new SWR to rank consumption streams generated by the model of forestry used in Mitra and Wan (1986). He calls a consumption stream *maximal* if it is a maximal point in the feasible set in terms of the SWR, and studies properties of maximal paths. Mitra finds that maximal paths converge over time to the forest with the maximum sustained yield, demonstrating that this notion of maximality is enough to provide a theoretical basis for the forest management tradition of emphasizing maximum sustained yield. In fact, Mitra demonstrates the somewhat surprising result that *all* the qualitative properties of optimally managed forests that one can obtain by applying the more restrictive overtaking criterion can be obtained by applying the weaker and more acceptable SWR he proposes. Mitra, using duality theory, shows that maximal paths have generalized intertemporal profit maximizing (bounded) shadow prices associated with them, just like optimal paths do. Mitra combines the above two findings to establish the result that the set of maximal paths coincides exactly with the set of optimal paths. This leads to a conclusion that in the context of the forestry model, one can completely dispense with the more restrictive overtaking criterion. Mitra notes that, in principle, his analysis can be extended to other forest products than timber.

In the second chapter of this section, Chapter 8, Geir Asheim and Wolfgang Buchholz address a widely debated aspect of sustainability the idea that stocks of natural resources be kept intact, also termed "strong sustainability". Neo-classical economists, with their strong belief in discounted utilitarianism, have rejected this notion of sustainability. Asheim and Buchholz argue that the stock specific sustainability criterion may be defensible not only from instrumental and moral perspectives, but also from a purely economic perspective, when the natural resource cannot be substituted by man-made capital and when further reduction would push the level below the size corresponding to MSY. Heal's work has shown that utility from the resource stock itself and equal treatment of all generation (the Weak Anonymity condition) favor the proposition that optimal paths will involve non-decreasing resource stocks. Asheim and Buchholz extend that work, demonstrating that how stock-specific sustainability constraints can be obtained from rather weak ethical axioms. In particular, the Suppes-Sen grading principle, obtained by combining the Weak Anonymity and Strong Pareto conditions, leads to stock-specific sustainability constraints as long as the resource is renewable or utility is derived directly from the resource stock. Though one must keep in mind that, all models, including the models proposed in this paper, abstract from some important real-world factors, they provide a new thought-provoking economic justification for stock-specific sustainability constraints.

## 7. NONLINEARITIES, MULTIPLE EQUILIBRIA, AND THE ECONOMICS OF SFM

The possibility of multiple equilibria is a theme running through the volume. The chapters making up the forth part focus specifically on multiple equilibria due to nonlinearities in production processes or management systems.

In the first of these chapters, Barkley Rosser notes that, as forest management comes to incorporate multiple values such as biodiversity, carbon sequestration, and timber it thus comes to involve multiple issues and stakeholders, and many difficult-to-resolve conflicts between groups and goals. Any serious effort to resolve such conflicts ultimately involve the dynamic ecology of forests, including elements such as the role of fire, pest management, and the methods and techniques of cutting trees, especially the patch size of the cuts. Complex dynamics, resulting for example from different time patterns of the various forest products, services, and management practices can imply multiple equilibria and the possibility of sudden and discontinuous changes in the nature of a forest.

Deep tradeoffs can exist between the local stability of forest ecosystems and their global resilience, tradeoffs that manifest themselves in such contradictions as efforts to prevent forest fires that make forest fires worse, and efforts to eradicate pests that make their attacks worse and more destructive. This idea that in ecosystems there might exist such tradeoffs has become very widespread and influential. Rosser concludes with the observation that the existence of these nonlinearities and the related thresholds and discontinuities complicates policy making in ecologic-economic systems. Policymakers must be especially aware of the interaction among policies and of the threat of system collapse when key thresholds are crossed.

Whereas Rosser focuses on nonlinearities related to dynamics, the paper by Jeffrey Vincent and Matthew Potts looks at the implications of nonlinearities for spatial aspects of forest management, especially in the context of tropical rainforests. They argue that the conservation of biological diversity is one of the most important dimensions of SFM, and that a variety of factors--economic, institutional, and ecological, may cause a nonlinear relationship between the amount of biodiversity conserved in a forest and the amount of timber harvested. Some nonlinearities favor segregated approaches (e.g. some stands dedicated to timber production and others to biodiversity maintenance) while others favor integrated approaches to forest management. Nonlinearities can lead to the counterintuitive result that segregated management may be superior to integrated management even in forest estates comprised of identical stands; the justification for segregated management thus does not hinge only on some forest stands being richer in biodiversity than others. But their analysis also shows that a nonconvex production set does not necessarily imply that segregated management is superior to integrated management, since the relative values of biodiversity and timber also matter. Nonlinearities resulting from species' populations being clumped instead of randomly distributed across the forest favor more integrated management, in the sense of having a large number of small reserves spread across the forest (in the extreme, a refugium within each annual cutting block). Nonlinearties involving species with minimum viable populations favors more segregated management, in the sense of having a small number of large reserves (in the extreme, just a single reserve in one location in the forest).

The diversities and complexities inherently involved in SFM policy-making are reflected in a different way by the general equilibrium model of Joint Forest Management (JFM) presented by Milindo Chakrabarti, Samar Datta, Lance Howe, and Jeffrey Nugent They note that the most optimistic observers see JFM as a creative and potentially optimal property regime combining the separate strengths inherent in property regimes of private ownership, direct state control, and communal property, whose common characteristic across settings is for local communities to receive greater property rights and influence over local natural resources than under the preceding regimes. The experience seems to have varied from place to place, depending on institutional and other characteristics, so the jury is still out on the overall success rate. One approach to a better understanding of what matters to potential success of JFM is to undertake a modeling exercise, eventually tested through simulation or other empirical verification.

The authors' simple general equilibrium model, which includes five sectors – two community groups, the Forest Department, the government, and a residual sector, incorporates key stylised facts derived from the existing literature. It highlights four important environmental and institutional features of JFM, namely, (i) the heterogeneous character of, and inequality within, forest user groups, (ii) the influence of such heterogeneity on the degree of user group dependence on forest resources, the sustainability of forest production and the degree of inequality between the groups, (iii) the effect of JFM on each of these relationships and considerations, and (iv) the importance of the quality of the forest and the externalities thereof, and the possible effect of JFM on the effectiveness of regulatory control and property rights over forest land. The authors discuss some preliminary possible outcomes on the basis of first order conditions and suggest that it would be useful to conduct simulations for (i) the impact of JFM on forest biomass and forest community's welfare; (ii) a comparison of Pareto optimal conditions to the benchmark case; and (iii) the impact of inequalities on model outcomes. They also suggest several possible extensions of the model.

## 8. EPILOGUE

In the last chapter, Shashi Kant puts together a brief synthesis of all the ten chapters in the framework of four sub-principles of economics of SFM, and provides an overview of new paradigm of economics, which he terms as Post-Newtonian Economics. He attributes the current status of Newtonian or neoclassical economics to the increasing returns due to information contagion, and establishes direct and indirect correspondences between the different concepts discussed in the ten chapters of this volume and Kant's basic principles of the economics of sustainable forest management. He also identifies the basic differences between the Newtonian and Post-Newtonian economics.

## NOTES

[1] In these economies, capital accumulation through never-ending natural resource depletion cause a falling interest rate, and therefore aggregate wealth should rise over time so that the product of the interest rate and aggregate wealth can maintain constant output and consumption (Pezzey & Toman, 2002) .

[2] Asheim divides a closed economy into three classes of people – capital owners, workers, and nonrenewable resource owners, and demonstrates that natural resource owners use a rising resource price to offset their diminishing stocks and achieve constant consumption without any investment. In contrast, interest rate (price faced by capital owners) is falling, and, therefore, this class has to augment its capital stock to maintain constant consumption. Hence, in general, resource-rents in different parts of the economy need to be invested in proportion to ownership of man-made capital, and not in proportion to ownership of natural resource stock.

[3] The more significant of these qualifications is the latter, since if a resource rich economy invests equal to or more than its resource rents its consumption level will rise over time, not the source of concern comparable to that of falling consumption in the resource poor economy which invests equal to or less than its resource rents.

---

[4] The two axioms imply a more symmetric treatment of generations in the sense that neither the "present" nor the "future" should be favored over the other. The axioms provide internal consistency and ethical clarity, and lead to a complete characterization of sustainable preferences, which are sensitive to the welfare of all generations. The welfare criterion, which these axioms imply, is complete, analytically tractable, and represented by a real valued function. Chichilnisky also proves that many welfare criteria, including the sum of discounted utilities for any discount factor, Ramsey's criterion, the overtaking criterion, Rawlsian rules, and basic needs are not sustainable preferences.
[5] In the case of renewable resources, discounted utilitarian optimum involves a lower long-run stock and a higher long-run level of consumption than the sustainable optimum, and hence it is less conservative.
[6] Some economists may argue that altruistic behavior, moral values, or commitment have been (or can be) incorporated in the conventional economic models by including an appropriate variable in utility function. In such models, however, economic agent remains utility (self) maximizer, and true features of altruism are not captured. A human being, or a socially rational agent, depending on how we look at them, may appear sometimes as selfish and sometimes as altruistic; in fact, both pictures are needed to give a full account of reality, and both have to be applied within the limitations set by the uncertainty of human behavior. Such dualistic nature of human behavior and true uncertainty related to human behavior cannot be captured in conventional economic models, and it will require approach similar to quantum mechanics or S-matrix theory as discussed in the previous section. Similar to quantum theory, such economics will be based on the principle of complementarity rather than the principle of substitution.

## REFERENCES


Ahmad, Y., Salah El S., & Ernst L. (1989). *Environmental accounting for sustainable development: A UNDP–World Bank Symposium* (Eds.). Washington, DC: World Bank.

Asheim, G.B. (1986). Hartwick's rule in open economies. *Canadian Journal of Economics,* 19 (3), 395–402.

Asheim, G.B. (1994). Net national product as an indicator of sustainability. *Scandinavian Journal of Economics,* 96 (2), 257–65.

Asheim, G.B., & Tungodden, B. (2004). Resolving distributional conflicts between generations. *Economic Theory*, 24, 221-230.

Asheim, G.B., Buchholz, W., & Tungodden, B. (2001). Justifying sustainability. *Journal of Environmental Economics and Management,* 41, 252–268.

Atsumi, H. (1965). Neoclassical growth and the efficient program of capital accumulation. *Review of Economic Studies*, 32, 127-136.

Ayres, R.U., & Kneese, A.V. (1969). Production, consumption, and externalities. *American Economic Review*, 69, 282-297.

Barbier, E.B. (1987). The concept of sustainable economic development. *Environmental Conservation,* 14 (2),101–10.

Bare, B.B., & Mendoza, G.A. (1992). Timber harvest scheduling in a fuzzy decision environment. *The Canadian Journal of Forest Research,* 22, 424-428.

Beckerman, W. (1994). "Sustainable development": Is it a useful concept? *Environmental Values,* 3 (3), 191–209.

Bourdieu, P. (1983). Ökonomisches kapital, kulturelles kapital, soziales kapital. In K. Reinhard (Ed.), Soziale ungleichheiten (pp. 183-198). (Soziale Welt, Sonderheft 2) (Goettingen: Otto Schartz & Co.). English translation available as, Forms of capital, in J. G. Richardson (Ed.), *Handbook of theory and research for the sociology of education*. New York: Greenwood.

Brock, W.A. (1970). On existence of weakly maximal programmes in a multi-sector economy. *Review of Economic Studies*, 37, 275-280.

Cardenas, J.C. (2004). Norms from outside and from inside: an experimental analysis on the governance of local ecosystems. *Forest Policy and Economics*, 6, 229-241.

Capra, F. (1982). *The turning point: Science, society, and rising culture*. New York: Simon and Schuster.

Chichilnisky, G. (1997). What is sustainable development. *Land Economics*, 73(4), 467-491.

Ciriacy-Wantrup, S.V. (1952). *Resource conservation*. Berkeley, CA: University of California Press.

Colander, D. (2005). *Complexity, muddling through, and sustainable forest management*. Chapter 2, in this volume.

Costanza, R. (1991). *Ecological economics: The science and management of sustainability*, (ed.) New York: Columbia University Press.

Daly, H.E. (1990). Toward some operational principles of sustainable development. *Ecological Economics,* 2 (1),1–6.

Daly, H.E., & Cobb, J.E. (1989). *For the common good: Redirecting the economy toward community, the environment, and a sustainable future*. Boston, MA: Beacon.

Dasgupta, P.S. (1995). Optimal development and the idea of net national product. In I. Goldin, & L.A. Winters (Eds.), *The economics of sustainable development.* Cambridge, U.K.: Cambridge University Press.

Dasgupta, P. S., & Geoffrey M.H. (1974). The optimal depletion of exhaustible resources. *Review of Economic Studies, Symposium on the Economics of Exhaustible*

Dasgupta, P. S., & Geoffrey M.H. (1979). *Economic theory and exhaustible resources.* Cambridge, U. K.: Cambridge University Press.

Dasgupta, P., & Mäer, K. (1994). Poverty, institutions, and the environmental resource-base. In J Behrman and T. N. Srinivasan (Eds.), *Handbook of Development Economics*, Volume 3, Amsetrdam: North Holland.

Diamond, P. (1965). The evaluation of infinite utility streams. *Econometrica*, 33, 170-177.

Einstein, A. (1949). *Autobiographical Notes.* In P. A. Schilpp (ed.) Albert Einstein: Philosopher – Scientist, (p.45). Evanston, Illinois: The Library of Living Philosophers Inc.

Epstein, L. G. (1986). Intergenerational preference ordering. *Social Choice Welfare*, 3, 151-160.

Gale, D. (1967). On optimal development in a multi-sector economy. *Review of Economic Studies*, 34, 1-18.

Gong, P. (1992). Multi-objective dynamic programming for forest resource management. *Forest Ecology and Management*, 48, 43-54.

Harsanyi, J. (1955). Cardinal welfare, individualistic ethics, and interpersonal comparison of utility. *Journal of Political Economy*, 63, 309-321.

Hartwick, J.M. (1977). Intergenerational equity and the investing of rents from exhaustible resources. *American Economic Review,* 67 (5), 972–74.

Hartwick, J.M. (1978a). Investing returns from depleting renewable resource stocks and intergenerational equity. *Economics Letters,* 1, 85–88.

Hartwick, J.M. (1978b). Substitution among exhaustible resources and intergenerational equity. *Review of Economic Studies* 45:347–54.

Hegel, G.W.F. (1964). *Political writings*, translated by T. M. Knox. Oxford: Oxford University Press.

Hegel, G.W.F. (1967). *Elements of the philosophy of right.* Translated by T. M. Knox. Oxford: Oxford University Press.

Heisenberg, W. (1963). *Physics and Philosophy: The revolution in modern science.* London: George Allen and Unwin.

Howarth, R.B., & Norgaard, R.B. (1990). Intergenerational resource rights, efficiency, and social optimality. *Land Economics,* 66 (1), 1–11.

Howarth, R.B. (1991a). Intertemporal equilibria and exhaustible resources: An overlapping generations approach. *Ecological Economics,* 4 (3), 237–52.

Howarth, R.B. (1991b). Intergenerational competitive equilibria under technological uncertainty and an exhaustible resource constraint. *Journal of Environmental Economics and Management,* 21 (3), 225–43.

Howarth, R.B., & Norgaard, R.B. (1992). Environmental valuation under sustainable development. *American Economic Review,* 82 (2), 473–77.

Howarth, R.B., & Norgaard, R.B. (1993). Intergenerational transfers and the social discount rate. *Environmental and Resource Economics,* 3 (4), 337–58.

Howarth, R.B. (1995). Sustainability under uncertainty: A deontological approach. *Land Economics,* 71 (4), 417–27.

Ikerd, J.E. (1997). *Toward an economics of sustainability*. Available from the website:
http://www.ssu.missouri.edu/faculty/jikerd/papers/econ-sus.htm Accessed on August 1, 2004.

Jevons, S. (1977). The coal question: An inquiry concerning the progress of the nation and the probable exhaustion of our coal mines. *In The Study of the Future*, edited by E. Cornish and others. Washington, DC: World Future Society. (Originally published in 1865.)

Kahneman, D., & Tversky, A. (Eds.) (2000). *Choices, values, and frames.* Cambridge: Cambridge University Press

Kangas, J. (1993). Integrating biodiversity into forest management planning and decision making. *Forest Ecology and Management*, 61, 1-15.
Kant, S. (2003a). Extending the boundaries of forest economics. *Forest Policy and Economics*, 5, 39-56.
Kant, S. (2003b). Choices of ecosystem capital without discounting and prices. *Environmental Monitoring and Assessments*, 86, 105-127.
Kant, S., & Lee, S. (2004). A social choice approach to sustainable forest management: An analysis of multiple forest values in Northwestern Ontario. *Forest Policy and Economics*, 6, 215-227.
Koopmans, T.C. (1960). Stationary ordinal utility and impatience. *Econometrica*, 28, 287–309.
Krautkraemer, J.A. (1985). Optimal growth, resource amenities, and the preservation of natural environments. *Review of Economic Studies,* 52 (1), 153–70.
Krutilla, John V. (1967). Conservation reconsidered. *American Economic Review*, 54 (4), 777–86.
Lauwers, L. (1997). Continuity and equity with infinite horizons. *Social Choice Welfare*, 14, 345-356.
Liu, G., & Davis, L.S. (1995). Interactive resolution of multi-objective forest planning problems with shadow price and parametric analysis. *Forest Science*, 41, 452-469.
Malthus, T.R. (1976). *An essay on the principle of population.* New York: Norton. (Originally published in 1798.)
McFadden, D. (1999). Rationality for economists? *Journal of Risk and Uncertainty*, 19(1-3), 73-105.
McKenzie, L.W. (1983). Turnpike theory, discounted utility, and the von Neumann facet. *Journal of Economic Theory*, 30, 330-352.
McKenzie, L.W. (1986). Optimal economic growth, Turnpike theorems and comparative dynamics. In K.J. Arrow, M. Intrilligator (Eds.), *Handbook of mathematical economics*, Vol. 3 (pp. 1281-1355). New York: North-Holland Publishing Company.
Meadows, Donella H., Dennis L. Meadows., Jorgen R. & William B. III. (1972). *The limits to growth.* New York: Universe Books.
Misra, D., & Kant, S. (2004). Production analysis of collaborative forest management using an example of joint forest management from Gajarat, India. *Forest Policy and Economics*, 6, 301-320.
Mitra, T., & Wan, H. Jr. (1986). On the Faustmann solution to the forest management problem. *Journal of Economic Theory*, 40, 229-249.
Mitra, T., Ray, D., & Roy, R. (1991). The economics of orchards: An exercise in point-input flow-output capital theory. *Journal of Economic Theory*, 53, 12-50.
Page, T. (1997). On the problem of achieving efficiency and equity, intergenerationally. *Land Economics,* 73(4), 580-596.
Pattanayak, S.K., Abt, R.C., Sommer, A.J., Cubbage, F., Murray, B.C., Ynag, J.C., Wear, D., & Ahn, S. (2004). Forest forecasts: does individual heterogeneity matter for market and landscape outcomes? *Forest Policy and Economics*, 6, 243-260.
Pearce, D. (1988). Economics, equity, and sustainable development. *Futures,* 20 (6), 598–605.
Pearce, D. (1994). Panel Discussion . In I. Serageldin., & A. Steer (eds.) *Valuing the environment: Proceedings of the First Annual International Conference on Environmentally Sustainable Development* (pp.69-71), Washington D.C. : The World Bank
Pearce, D.W., Markandya, A., & Barbier, E.B.. (1989). *Blueprint for a green economy.* London, U.K.: Earthscan.
Pearce, D.W., & Giles D. Atkinson. 1993. Capital theory and the measurement of sustainable development: An indicator of "weak" sustainability. *Ecological Economics,* 8 (2),103–8.
Pezzey, John. C.V. (1989). Economic analysis of sustainable growth and sustainable development. Environment Department Working Paper No. 15. Washington, DC: World Bank. Published as *Sustainable Development Concepts: An Economic Analysis. World Bank Environment Paper* No. 2. Washington, DC: World Bank.
Pezzey, John. C.V. (1992). Sustainability: An interdisciplinary guide. *Environmental Values,* 1 (4), 321–62.
Pezzey, John. C.V. (1994). *The optimal sustainable depletion of non-renewable resources.* Paper presented at Association of Environmental and Resource Economists Workshop, Boulder, CO, and European Association of Environmental and Resource Economists Annual Meeting, Dublin, Ireland.
Pezzey, John C.V. (1997). Sustainability constraints versus "optimality" versus intertemporal concern, and axioms versus Data. *Land Economics,* 73 (4), 448–66.
Pezzey, John C.V., & Toman, M.A. 2002. *The economics of sustainability: A review of journal articles.* Discussion Paper 02-03, Washington, D. C.: Resources for the Future.

Pezzey, John C.V., & Toman, M.A. 2003. Progress and problems in the economics of sustainability. In T. Tietenberg and H. Folmer (Eds). *The International Yearbook of Environmental and Resource Economics 2002/2003* (pp. 165-232). Cheltenham: Edward Elgar.

Radner, R. (1961). Paths of economic growth that are optimal with regard to final states. *Review of Economic Studies*, 28, 98-104.

Rawls, J. (1971). *A theory of justice*. Cambridge, MA: Harvard University Press.

Repetto, R., Michael W., Christine B., & Fabrizio R. (1989). *Wasting assets: Natural resources in the national income accounts*. Washington, DC: World Resources Institute.

Suppes, P. (1966). Some formal models of grading principles. *Synthese*, 6, 284–306.

Sen, A.K. (1970). *Collective choice and social welfare*. Edinburgh: Oliver and Boyd.

Sen, A.K. (1973). Behavior and the concept of preference. *Economica*, 40, 241-259.

Sen, A.K. (1977). Rational fools: A critique of the behavioral foundations of economic theory. *Philosophy and Public Affairs*, 6, 317-344.

Solow, Robert M.( 1986). On the intergenerational allocation of natural resources. *Scandinavian Journal of Economics*, 88(1), 141-149.

Solow, R.M. (1974). Intergenerational equity and exhaustible resources. *Review of Economic Studies, Symposium on the Economics of Exhaustible Resources*. Edinburgh, Scotland, Longman Group Ltd.

Stiglitz, J.E. (1974). Growth with exhaustible natural resources: Efficient and optimal growth paths. *Review of Economic Studies, Symposium on the Economics of Exhaustible Resources.* Edinburgh, Scotland, Longman Group Ltd.

Svensson, Lars E. O. (1986). Comments on Solow: On the intergenerational allocation of natural resources. *Scandinavian Journal of Economics* 88 (1), 153–55.

Svensson, L.G. (1980). Equity among generations. *Econometrica*, 48, 1251-1256.

Toman, M.A., Mark, P., & Ashton, S. (1996). Sustainable forest ecosystems and management: A review article. *Forest Science*, 42(3), 366-377.

von Weizsäcker, C.C. (1965). Existence of optimal programs of accumulation for an infinite time horizon. *Review of Economic Studies*, 32, 85-104.

Wang, S. (2004). One hundred faces of sustainable forest management. *Forest Policy and Economics*, 6, 205-213.

WCED (World Commission on Environment and Development). (1987). *Our Common Future*. Oxford, U.K.: Oxford University Press.

# CHAPTER 2

# COMPLEXITY, MUDDLING THROUGH, AND SUSTAINABLE FOREST MANAGEMENT[1]

DAVID COLANDER

*Department of Economics, Middlebury College,*
*Middlebury, Vermont 05753, USA.*
*Email: colander@middlebury.edu*

**Abstract.** I distinguish two stories that economists have in their mind when they think of economics—one is a story of efficiency and control; the other story is of complexity and muddling through. I argue that the new work in economics of sustainable forest management that is being discussed in this volume is part of a broader trend that is occurring in economics—switching from the efficiency and control story to a complexity and muddling through story. As such it is associated with current changes going on at the cutting edge of economics.

## 1. INTRODUCTION

In *The Worldly Philosophers* Robert Heilbroner (1953) tells a story of a dinner John Maynard Keynes had with Max Planck, the physicist who was responsible for the development of quantum mechanics. Planck turned to Keynes and told him that he had once considered going into economics himself, but he decided against it--it was too hard. Keynes repeated this story with relish to a friend back at Cambridge. "Why, that's odd," said the friend. "Bertrand Russell was telling me just the other day that he'd also thought about going into economics. But he decided it was too easy." That story captures two typical reactions that students often have to economics. For some it is too easy; for others it is too hard.

In this paper I argue that both these reactions are reasonable, depending on what economic story one is trying to explain. I distinguish two stories that economists have in their mind when they think of economics—one is a story of efficiency and control that has its foundation in the work of David Ricardo and Leon Walras. The other story is a story of complexity and muddling through, and its roots are in the work of Adam Smith and John Stuart Mill. I argue that the new work in sustainable forest management that is being discussed in this volume is part of a broader trend that is occurring in economics—switching from the efficiency and control story to



*Kant and Berry (Eds.), Economics, Sustainability, and Natural Resources: Economics of Sustainable Forest Management,* 23-37.

the complexity and muddling through story. As such it is associated with current changes going on at the cutting edge of economics.

## 2. TWO ALTERNATIVE STORIES

One of the reason economics can be viewed as both easy and hard is that it is a highly complex subject, which, for pedagogical reasons, has to be simplified to a basic story line. Some tangents are allowed, but ultimately those tangents must interweave with the main story line, or they do not appear. I suspect that Planck and Russell differed because they were referring to different story lines.

Russell was likely thinking of the story line currently used in the micro texts, which is what might be called the *efficiency story line*. The efficiency story is a story about the *state* of competition. It is a static story, which nicely fits into a calculus (especially LaGragrangian multipliers or Euler equations) framework. While few principles of economics students completely understand the full efficiency story line, they generally have a sense of a number of examples of it in a partial equilibrium setting--the effect of taxes, quantity restrictions, price ceilings, and price floors on efficiency--and the way in which the economy adjusts to sudden changes in tastes.

Students are also presented with the general equilibrium efficiency story--that under appropriate conditions individual maximization will lead to social maximization, although, to be honest, few principles of economics students come away from the course with a deep understanding of that broader story. They are usually struggling with the simple individual optimization story. Carrying the analysis through to the aggregate level and understanding the welfare implications about markets of that social optimization story is beyond most students. In fact, most of those welfare implications are negative—the arguments cannot be carried over to social maximization under reasonable assumptions. We tell it nonetheless because it is a useful story in organizing thinking about very complicated policy issues.

One of the reasons this social maximization story makes an acceptable textbook story is that it provides space for economists who prefer government action, and those who oppose it. While, under the "right" set of conditions the market maximizes social welfare, it may not; externalities can upset that market-based social maximization. But, not to fear; the government can offset those externalities through appropriate tax policy. Thus, the efficiency story line has the needed neutrality to sell to a wide market—a necessary attribute of any textbook story —and fits with the reasonable proposition that there are costs and benefits to government regulation. It neither opposes nor favours government action. Moreover the story can be spun in a variety of ways to fit individual instructor's biases.

Many students have a hard time understanding the efficiency story because, even though it is highly simplified, it is still complicated. Since the stories are often told graphically and algebraically, languages that are difficult for many principles of economics students to understand, the language problem makes the story difficult. In fact, many students never get around to learning the ideas of economics; they spend all the time learning math.

This maximization cost/benefit story line, which is a key element of the efficiency story as it relates to policy, is a very useful one for students to learn, and to carry with them for the rest of their lives. Since principles of economics is only one of about 35 courses that make up student's training in college, it seems a reasonable story to teach. But, as with all things, it comes at a cost, and that cost is that many students are never introduced to other important stories that economists could tell. One of those alternative stories involves developments that are currently ongoing in the economics profession--developments to which many of the papers in this volume are contributing. That alternative story line might be called the complexity story line.

The complexity story is a much more complicated story than the efficiency story, and is the story Planck was likely referring to. It is about the *process* of competition, and is based in a dynamic framework. It is an evolutionary story of an economy operating over time--drifting along on a slowly moving river with occasional rapids, none of which are directly controlled, or controllable. The complexity story is an almost magical story, one in which the invisible hand of the market takes what should be chaos, and turns it into an elegantly complex structure that fits together, not perfectly or efficiently, but sustainably. Patterns and pictures develop out of nowhere. The resulting system is admired not for its efficiency, nor for any of its static properties; the resulting system is admired for its very existence. Somehow the process of competition gets the pieces of the economy to fit together and prevents the economy from disintegrating into chaos. Observed existence, not deduced efficiency, is the key to the complexity story line.

While the complexity story line has its origins in the economics of economists such as Mandeville, Smith, and Malthus, its more recent development is to be found in the work of evolutionary biologists, such as Edmond Wilson and John Maynard Smith. Both stories are centered around constrained optimization, but whereas the efficiency story line structures the story so that it comes to an answer, and, in principle, a set of policy recommendations, the complexity story line is the never-ending story in which every answer simply raises new questions, and the hope of control gives way to a realization that the best we can hope for is to muddle through.

## 3. SUSTAINABILITY AND THE TWO STORIES

Recognizing the existence of these two stories helps explain the neglect of issues of sustainability in economics and provides a broader framework within which the emerging work in sustainable forest management can be understood. Traditional work in forest management falls within the efficiency story line. The standard literature in forest management, the tradition started by Faustmann and Ohlin, considers the problem of optimal forest rotation assuming fixed tastes and homogeneous super rational, independent, agents, and shows what would be efficient, and what would not. That work does not deal with the question of whether efficiency is society's goal in forest management, or whether it should be.

Looking at broader issues in social welfare theory, it is very clear that that work is contextual—it can only be understood within a much broader framework of thinking about institutions, social wellbeing, and social welfare. Within the broader contextual framework found in the Classical economics of John Stuart Mill and the grand tradition of liberalism, it is clear that *efficiency is not an end in itself*; it is a means to an end; efficiency only has meaning when one specifies what the goals are, whose goals they are, how the goals are to be weighted, and what method we have of resolving conflict among goals. The textbook presentation of economics avoids this broader discussion, as does the traditional work in forest management. I see the work in sustainable forest management as one of the many movements currently going on in economics that is bringing back these broader issues.

Sustainability in the efficiency story is reduced to a question of aggregate existence. Since that efficiency story is generally told in reference to a unique equilibrium model, the presumption of the model is that markets have a natural way of achieving sustainability. We all know the story: Scarcity leads to price rises, which leads to conservation and substitution of the scarce resource, which leads to sustainability—the system simply changes—as forests decrease, we switch to other means of providing the services that forests provided—plastic trees, photosynthesizing machines, whatever. In the efficiency story substitutability will solve any problem of scarcity, so why even discuss sustainability? To discuss sustainability means you don't truly understand the scarcity story.

The gross substitutability answer to sustainability, such as that presented in Goeller and Weinberg (1976) is a reasonable one, but is not the concept of sustainability that most people have in mind when they discuss sustainability. They have a different idea in mind, an idea that does not fit in a unique equilibrium model. The sustainability literature fits into models with multiple equilibria, with equilibria selection mechanisms, and with some equilibria being preferred to others. The model sustainability advocates have in mind has multiple dimensions, one in which a world of rows of neatly organized trees is not the same as a world of old growth forests where ecological competition has prevailed. Such multiple dimension nonlinear optimization issues quickly go beyond the mathematical abilities of the students, and indeed of even the brightest mathematicians. So, once one expands the models to include such issues, it becomes clear that economists' models no longer provide answers, but instead provide, at best, a heuristic solution, not a formal solution to the problems most individuals are interested in. To avoid getting into such issues, the texts, and much of the research in traditional economics, avoid discussing sustainability.

Another reason that the term, sustainability, is not used in the texts is that it conveys to many economists an integration of normative judgments into the analysis. Such a normative use of the term involves not only an interruption of the efficiency story, but a complete incompatibility with it. The efficiency story has struggled to keep such normative judgments out of the reasoning process being taught, even though almost every economist, if pushed, will accept Hume's Dictum that you cannot derive a should from an is, and that policy necessarily involves normative judgment.

It is for these two reasons that if you look at principles of economics texts you will see very little discussion of sustainability of any type. In fact, among Rust Belt economists (Chicago/Rochester and their satellites) the very mention of the term "sustainable" makes their eyes roll in a signal to other Rust Belt economists that "Here we go again; we are talking with another of those wishy washy environmentalists, who are trying to instill their values on others." For Rust Belt economists, sustainability simply isn't an appropriate topic of discussion for proper economists.[2]

Sustainability fits much better into the complexity story. In models of complex systems one doesn't talk about equilibrium; one talks about basins of attractions. Nonlinearities are accepted, and one can expect phase transition jumps as the system evolves. Sustainability means keeping within the existing basin of attraction, and not going to another that is considered less desirable. Within a complex system a "rational choice" is much harder, and indeed impossible, to specify. It is multiple levels of the system, not only the individual, that are optimizing, so the individual is the result of lower-level optimization at the physiological level, is himself optimizing, and is a component of higher level systems which are themselves optimizing, and competing for existence. Everything, including agents, are coevolving. Within a complex system, even if one can specify what one means by rational choice non-contextually, the systemic forces rewarding "rational choice" are often weaker than they are in simple systems. This means that instead of weaving the textbook story around a predetermined equilibrium that must finally be reached if the system is left to its own devices, as is done in the efficiency story, the complexity story is woven around the dynamic process through which one basin is reached temporarily, but other forces are building up to push it into another basin; it is a never-ending story.

Generally, complex systems will have no single equilibrium; but instead a collection of possible basins of attraction, with some basins more likely than others. One can only discover the likelihood of certain basins of attraction by considering the evolution of the entire system with either a heuristic or formal simulation. Instead of thinking of equilibrium, one thinks of replicator dynamics, which drive the system forward in a variety of possible ways. By the replicator dynamics I mean the way in which the aggregated decisions of the agents in the system have a tendency to lead to certain outcomes often not foreseen by individual agents, and possibly not predictable by any agents in the system. Because of the multiple paths, and the potentially complicated dynamics, complex systems are generally analytically indeterminate. To gain insight into a complex system one must think in an evolutionary framework in which many different paths are possible, some more sustainable than others.

In the complexity story the market isn't desirable because of some grand sense of efficiency, and government isn't seen as an entity that can tweak a market process result in a certain way to achieve efficiency. Because the market is seen as fully integrated with the society, tweaking one aspect of the market process can imply a major change in another aspect—the proverbial butterfly flapping its winds in China can change the weather pattern in the U.S. Sudden shifts of the system from one basin to another become part of the analysis, and thus the sustainability of a particular basin, which in the complexity literature generally goes under the name resilience, becomes an interesting issue.

This complexity story conveys a quite different sense of what is happening to an economy than does the efficiency story. It sees change as an evolutionary process occurring at many levels simultaneously. There are interdependent slow and fast moving variables, and policy is affecting all of them. Since one does not see the effects on the slow moving variables in the short run, short run empirical measures of the effects of policy may be highly misleading. You not only have to look for optima, but you also have to look for early indicators of switch points, such as the level of phosphorous that will fundamentally change the nature of a lake.

The policy problem of complex systems is exponentially complex, and pure theory provides far less guidance than it does in simpler systems. There is no one model, so model uncertainty must be part of analysis. Policy must take account of multiple levels of optimization occurring at different speeds. For example, the selection of a certain policy can change tastes, so any policy built upon current tastes may be less than optimal. Policy that does not take account of the cumulative process of policy change can miss important elements of what is really going on. Moreover, in a complex system optimizing likely involves nonlinearities and kinks, making first order conditions of little use in drawing out robust global policy conclusions.

It is into this complexity world that sustainable forest policy is stepping. Ironically, the concern about stepping into that new world tends to be the reverse of the various sides' concern about sustainability. Traditional economics is not concerned with sustainability of the system but seems to be very concerned with sustainability of traditional economics. Their argument for not dealing with the true complexity of the system is that to do so would threaten the current research environment where researchers are comfortable; it would take them out of the theoretical and methodological terrain that has made economics the queen of the social sciences. If we give up our efficiency model, and start dealing with the complexity model, it will be hard to differentiate us from other sociologists. We might even be mistaken for sociologists!

Sustainable forest management advocates take the opposite position. They argue that our current research terrain is too restrictive, and doesn't allow economists to reach their full potential. They suggest that economists should step into a research world where economists have little training, and where the comfort level of tradition and well-worked tools are gone.

Both sides have a point: Who knows—if we go there, will there be an economics profession left? Will we destroy the good that economics does, as we try to deal with these more complicated questions? Isn't it better if we stick with what we know, and have explored, and reach out ever so tentatively and cautiously? Will the economics profession be sustainable in the new uncharted territory?

The sustainable forest management answer to them is essentially the same one that efficiency advocates give to sustainability concerns: don't worry; extending beyond where we are will make things better; economists will have more to offer; we do not have to give up our current benefits to extend the analysis. Of course, economics will be sustainable; we'll just be doing things better.

By temperament, I find myself very much in sympathy with the brave new world view, which, I suspect, accounts for the invitation to speak to you here. But I think the potential of entering that brave new world for undermining economics must be admitted, and accepted. In academia there may indeed be multiple basins of attractions, and some of them may not include economists as we know them today. Traditional economists lack a spirit of adventure of adventure for themselves even as them embrace a framework that advocates it for the economy as a whole.

But it is that same feeling of adventure that places me in opposition to many of the views of those who argue for sustainability as a key goal of society. What I mean by this is that to say that sustainability becomes a potential concern of the system is not to say that the way in which sustainability is used by researchers is not subject to implicit, unstated value judgments, ambiguity, and assumptions that are not in accord with empirical observations. As I read popular articles on sustainability it is often unclear to me precisely what the authors mean by sustainable. When I look at the empirical and historical evidence, I find that the system has continually adjusted much more than sustainability advocates predicted. But just because not all individuals who use the term have cleared up the definitional ambiguity, and just because the term does not neatly fit into the efficiency story, does not mean that sustainability is not a relevant topic for economists to consider, and a highly relevant topic for public policy. I believe it is.

## 4. DIFFERENCES IN THE TWO STORIES

Let me know return to Planck and Russell's different reactions to economics. Judged from the perspective of a Planck, or a Russell, the efficiency story is a piece of cake; it involves elementary algebra and calculus. To Russell that story was too easy to study. The complexity story, however, is formally untellable, and is far more difficult than particle physics. It requires mathematics that was not yet developed in Keynes' time, and is only today beginning to be developed. For Planck, that story was too hard to even contemplate studying.

The following story told by Brian Arthur of a discussion at the first Santa Fe conference on complexity gives one a sense of why the story is so difficult. At that conference Arthur was discussing the problem of including increasing returns in the economic model with one of the physicists there. The physicist said that increasing returns is like spin rotation and that therefore economics with increasing returns is very much like physics. The physicist went on to say that since there are more atoms than people, physics must be harder than economics. Arthur changed the physicist's view by pointing out that in economics one has an additional complication; to make the analyses comparable one would have to assume that each atom had a will of its own, and that what it was trying to do is to take advantage of the other atom, and thwart any attempt at control. Thus, every time you tried to control them, they modified their spin to make any control more difficult. With that explanation the physicist agreed that economics was much more difficult.

## 5. WHY THE COMPLEXITY STORY ISN'T TOLD IN THE TEXTS

I am both a textbook author and an economist, so I feel the pull between the two stories. As an economist, I direct all my thoughts toward the complexity story, trying to understand the work that is being done on it. But little of that work shows up in my principles text. There are two reasons. The first is the sheer complexity of the complexity story. I believe there is a story there, but I'm not sure I can tell that story in a meaningful way to students, or even to myself. A second reason is that I believe the complexity story and apply it to all my decisions. Applied to textbooks, it makes me, and I suspect other authors of successful texts, reluctant to change something that is working. Currently textbooks are working and serving a useful purpose. I believe that the story we are telling in our teaching of economics—the efficiency story--is a useful one for all students to learn; it is far more useful than the stories they learn in most of their other classes. I want every student to come out of college with a strong understanding that there is no such thing as a free lunch. Telling the efficiency story achieves that end, and thus seems justifiable, so it is only reasonable to be hesitant to change from that story.

The underpinnings for a major change in the story economists see themselves studying, and eventually that they will see themselves telling, are, however, currently taking place. As the complexity story develops, it will become more and more tellable, and, as the current texts die out, new texts that make the change to the complexity story will eventually replace the older texts. But I suspect that because the change involves a totally different story line, the change in stories will be a sudden shift rather than a smooth movement (Colander, 2000c). In the meantime, by which I mean the next 20 to 40 years, the real cutting-edge changes will be made in research in particular fields such as forest management. In most of these field areas researchers have already fully mined the efficiency arguments, and have extracted much of the insight from that model. Thus, they have an incentive to explore alternatives, such as the complexity approach.

## 6. THE CHANGES CURRENTLY GOING ON IN THE PROFESSION

While all the field courses are proceeding on their own path, there is sufficient similarity in the changes that are occurring in the profession to suggest the nature of these changes. It is a change in the allowable assumptions, from the holy trinity of *rationality, greed and equilibrium* to a broader set of allowable assumptions, which might be called a new holy trinity of *purposeful behavior, enlightened self-interest, and sustainability*.

The acceptance of these changes by the profession can be seen in a variety of theoretical work, such as work in behavioral economics, evolutionary game theory, agent based modeling, experimental economics, and new institutional economics. In this new work utility maximization is enriched by insights about the individual from psychology and neuroscience. Behavioral economics is the most developed. It is considering issues such as reference-dependent preferences, the replacement of expected utility with prospect theory that seems to capture individuals' decision process much better than simply utility maximization, the development of hyperbolic discounting arguments, the formalization of cognitive heuristics, the replacement of theories of self-interest by theories of social preference, and the development of adaptive learning models.

Once one accepts that the behavioral foundations of choice are important, one is directed to experiments, and experimental economics is another expanding area. Experimental economics provides an almost endless set of possible dissertation topics using a methodology that is quite outside the efficiency framework. It gives one a method of choosing among assumptions, and an alternative to statistical empirical testing. Economists still have a long way to come in experimental work, but that work has the possibility of changing economics significantly.

The acceptance of behavioral economics also leads to using evolutionary game theory as the setting for a foundational theory of economics. Evolutionary game theory allows one to redefine how institutions are integrated into the analysis and to develop a social dimension of individuals, which was previously lacking in current textbook story of economics. The movement is slow, but it is happening, and is reflected in the recent allocation of awards in economics. For example, Daniel Kahneman and Vernon Smith recently won a Nobel Prize for their work in experimental economics and Matt Rabin won the John Bates Clark medal for work on behavioral economics. Because of these changes today one can no longer describe modern economics as neoclassical economics (Colander, 2000a).

I do not want to overstate how these changes are currently affecting economists. Most economists do variations of what they were taught to do, and so have not changed significantly. "Same economist" research changes only slightly. But the economics profession is not a static group, and so the research also changes with the evolving composition of economists, with younger, newly trained economists coming in, and older economists going out. Thus the evolutionary hiring and retirement process affects research.

As time passes, younger, differently trained, economists replace older economists, and the average image of what economics is and of how one does economics changes. Since the profession replaces itself every 35 years or so, the rate of change is only about 3% per year. However, even that rate may be an over-estimate of the degree of change in the initial stages of a cycle of change, because most students choose to work with established professors in established methodologies; the newer methodologies and techniques are risky. Initially only a few risk-preferrers choose that path. So, at the beginning of a cycle of change, the rate of change toward a new acceptable approach is smaller than that 3%, probably closer to 1%. However, at some point a critical mass of work is accumulated, a shift point occurs, the new approach becomes the hot approach, and students flock toward it. At that time the rate of change increases to greater than 3%.[3]

## 7. WHERE ARE THE CHANGES LEADING US?

Ultimately I see these changes leading to a change in the basic story we are telling in economics from the current efficiency story told in the texts—*the story of infinitely bright agents in information rich environments*—to the complexity story—*the story of reasonably bright individuals in information poor environments.* Another way of describing my thesis is that the vision of the economy will evolve from its previous vision of a highly complex "simple system" to a highly complex "complex system."[4] Simple systems, no matter how complex, are reducible to a low dimensional set of equations, making it possible to model the system analytically. A complex system is not, and must be represented in another fashion—through simulation, or through insights gained with replicator dynamics. One can never have a full analysis of the entire complex system.

As I stated above, the current steps the profession is taking toward a complex systems approach are minimal, but the ultimate result of these steps is a movement from telling the efficiency story to telling the complexity story in their research and eventually in the texts. The acceptance of this complexity vision of the economy involves a shift in economics far more fundamental than anything associated with the movements away from the holy trinity that the profession has made so far.

## 8. WHY NOW?

Heterodox and heuristic economists have long argued that economics should deal with broader issues. So the questions arise: Why is the change occurring now? And: Why didn't it occur previously? My answer to these questions is that what has changed is not the recognition that these broader issues are important; that's always been there; what's changed is the belief that economics may have something to bring to bear on these broader questions. The reason is twofold. First, economists now believe that they have something to bring to these questions because of changes in the analytic and computing technology. Second, the efficiency model, developed in relation to the holy trinity, has been developed, and the "low hanging fruit" has been picked. Thus, theorists have an incentive to branch out. In short, the changes will take place because they offer exciting dissertation topics to graduate students and research possibilities for young researchers, not because of any new insights into what the nature of the problem are. The efficiency model will die because, given current technology, it's too simple to generate that dissertations and articles that are the underbelly of the profession in the current institutional structure.

From a technical standpoint, the mathematics involved in the efficiency model is really quite simple; they assume away path dependency, non-linear dynamics and many similar complicating features that could well characterize real world processes. A unique equilibrium is no longer likely or supportable as an assumption, which undermines the efficiency vision of how markets lead the economy to a social optimum.

Schumpeter (1957) made the assumption of a unique equilibrium as a necessary component of a science of economics. With the higher level of mathematics being taught in graduate school, and with the greater mathematical sophistication of those entering the profession, that restriction is no longer necessary, which is why these more complicated issues are being explored. By understanding the processes that guide the economy in its evolution one can gain insight into the economy and to the future direction of the economy, even if one does not know what it's ultimate equilibrium will be.

But as soon as one moves to these more complicated mathematical approaches, neat analytic solutions are far less likely to be forthcoming. This leads to a third change that is occurring in the profession, and is likely to be the most significant change in the more distant future. That is the movement from analytics to simulations. The reality is that advances in computing power involve a fundamental change in technology that is reducing the value of deductive theory. If one can gain insight through simulation, one has far less need to gain insight through deductive analytic theory. As long as computing power continues to double every 18 months, agent-based simulation will become more and more important in economist's tool kit, and will eventually replace deductive analytic theory, and the supply/demand framework of the current texts.

In these agent-based models the researcher "grows" an economy, letting simple algorithms describing agent behavior (algorithms developed in behavioral work) compete with one another, and see which wins out.[5] Agent-based simulations are fundamentally different than simulations designed to solve equations. In agent-based modeling one analyzes the system without any equations describing the aggregate movement of the economy; one simply defines the range and decision processes of the individual actors. Through multiple simulation runs one can gain insight into the likelihood of certain outcomes and of the self-organized patterns that emerge from the model. As computing power becomes cheaper and cheaper, such modeling will likely take over the profession. Ultimately, I see virtual economies being created in which policies are tested to determine their effectiveness in the same way that virtual designs are currently tested.

Is such agent based modeling still economics? I believe it is; it keeps much of standard economics—it sees individuals as purposeful, although the precise nature of purposeful behavior is derived from the model rather than assumed. It assumes individuals interact and trade, and that successful individuals continue; unsuccessful individuals do not. But to be honest, it is likely that the simulation based economics will be more social science generally, and fall under a general "cognitive science" discipline.

## 9. POLICY IMPLICATIONS OF THE TWO STORIES

What relevant policy lessons for students come out of the complexity story is far less clear to me, and I think to the entire profession, than is the fact that the changes are occurring. In thinking about the policy implications of the complexity story, Hayek, following the ideas of complexity, initially pushed the implications too far, and seemed to be saying that there was no room for policy activism—that the economics system should be left alone.[6] There clearly is some sense of that coming out of the complexity story, but I see a more nuanced policy view coming out of the complexity story, in which the theory is neutral about general policy prescriptions in the same way that the current textbook efficiency story is neutral. There are reasons for government intervention and reasons for laissez faire in the complexity story.

They just are not necessarily the same ones as found in the efficiency story, and they are much harder to pull out of the analysis. Determining a firm foundation for the implications of the complexity story for policy is a long way away.

I have tried to develop that sense of policy nuance in my work on what I call the economics of muddling through (Brock & Colander, 2000, and forthcoming-a, b) which I contrast with the efficiency story's economics of control approach. In the economics of control, one can, at least in principle, state what the optimal action for each agent, and the optimal policy for the policy maker, will be. In the economics of muddling through, specifying the optimal action for the agent and the optimal policy is far beyond the capabilities of the modeler. The best agents can do is to muddle through; similarly, the best policy makers can do is to muddle through.

Instead of controlling the economy, the goal of policy makers is to muddle through as effectively as possible, perhaps improving the workings of the economy in certain specific instances, but with no grand vision that one is going to suggest an optimal policy. In the economics of muddling through there is no such thing as a free lunch, but once in a while you can snitch a sandwich. Policy work is designed to snitch as many sandwiches as one can. I am pleased with this "muddling through" policy story, and believe that eventually it will be the way economists think of themselves and policy. But it is still in development and is not yet ready for prime time.

I do not claim that muddling through is a breakthrough in our understanding of economic policy issues; it simply is recognition of the limitations of our knowledge of the effects of economic policy. In nuanced discussions among good economists, the limitations of the current theory are well known, and the policy implications of any model they develop have always been for more nuanced, and considered in a much broader framework, than in policy discussions found in the texts. However, to make the story simple enough for the texts, the policy presentation has to be simplified, and it is that simplified version that students learn, which reporters present as economist's views, and which economists sometimes fall back on when they are pushing an idea, or simply being lazy.

Muddling through is conducting policy without an ultimate set of plans. So not only are the agents of the new economics operating in an information poor environment, so too are the policy makers. In such a situation policy becomes problem driven, not theory driven. Economics becomes not a single theory that guides policy, but a set of tools—statistical tools, modeling tools, and heuristic tools--that when incorporated with knowledge of the institutional structure can help the policy maker achieve the solution to problems posed by agents in the system.

This muddling through approach is a quite different view of policy economics than the view that is presented in the texts, where economists are the holders of knowledge of what policies will achieve global efficiency. In muddling through global efficiency is beyond what one can hope to achieve. One can still talk about efficiency, but it is defined locally in relation to existing institutions, and means producing what one is currently producing within existing, or only slightly modified institutions, at the lowest cost. It is useful only in analyzing incremental change, where issues of sustainability are minimal. Used in this limited sense its implicit assumption that one's normative ends are little changed from previously, and that they can therefore be left implicit, can be seen as a reasonable simplification.

Broader, less locally defined efficiency, is much more difficult to either define or use in policy discussion. Policy work in muddling through must make ones goals and assumptions clear. Thus, more generally, in muddling through efficiency is not a goal, but a condition imposed by the analysis about the costs of achieving whatever goal has been specified. It is achieving given ends as cheaply as possible, and only has meaning in regard to those ends. In this muddling through framework you hire an economist, tell him or her your goals, and he or she will bring his or her expertise in modeling and data analysis to help achieve those goals at the least possible cost.

The textbooks will not be telling the story of muddling through for a long time; it is too radical a change in vision. Initially, changes that are least challenging to the textbook story will find their way into the texts. The field of behavioral economics that is exploring the meaning of the "purposeful behavior" assumption is offering the type of modifications that will show up in the texts soon. These modifications offer a slight change in the policy prescriptions that flow from the analysis. An example of what I have in mind is Cass Sunstein and Richard Thaler's (forthcoming) concept of "libertarian paternalism." It proposes a set of policies that are consistent with the standard economic policies prescriptions that follow from the efficiency story, but which take into account agent's ill-formed preferences, one of the insights that follows from behavioral economic work.

These ill-formed preferences mean that individual's choices influenced by default rules, and libertarian paternalism is designed to take advantage of this fact. For example, say the policy maker believes that individuals will be better off with more forests in the world, and that a policy allows individuals to direct a part of their taxes to forests. By making that policy option the default option, and requiring the individual to default out of the program, rather than requiring the individual to choose to be in the program, the policy maker can increase participation in the program significantly. Doing so does not take away the individual's choice since the individual has the same choice in both situations, but the behaviors will be quite different.

Applying even this small implication of behavioral economics to policy is a major step. It means that economists must accept that normative judgments become part of the policy process. But a full acceptance of the policy implications is a much larger step. If tastes are endogenous, then normative issues become a central role in economic policy and cannot be escaped or ignored.

## 10. CONCLUSION: ECONOMICS AND SUSTAINABLE FORESTRY

Let me conclude with a few brief comments about the implications I see this shift having for forestry research. As I see it, many people have a sense that it may be a good goal for society to have the economy move to an equilibrium that is characterized by more of our land devoted to forests, than they believe is likely to be the case under existing institutions. While I tend to agree with that normative view, I also believe that what one means by forest is often ambiguous. What can be called a forest, and how to weight different types of forests, are difficult problems and can lead to much confusion in the debate.

In dealing with this debate the efficiency story is not especially helpful, because it excludes many of the issues upon which the debate is based. By being more open to alternative assumptions, the complexity approach to economics brings economists back into the broader theoretical and policy debate. Rather than defining the model and policy questions that can be asked, economic reasoning can be used as an input into broader models. That, in my mind is a plus for everyone involved. Thus, I disagree with those economists who fear this movement; to fear it means that one does not believe that the policy insights of economics will be able to compete with the insights from other disciplines and from other approaches. I believe that economic insights are strong enough to survive, and even prosper, on this expanded terrain. The complexity approach gives up the pillars upon which our welfare economics is built and in doing so it gives up the almost theological sense of what is right that is often associated with that view. In doing so it loses some influence. But by entering the debate, and letting economic ideas procreate with other ideas, it gains, and becomes stronger.

In the complexity approach we will not have theory to rely upon to say what policy is right or wrong. But we will have tools that can add insights about how to create the desired ends. Will certification actually increase the amount of forests, or will it have unintended effects? Are there other ways to achieve that goal? Can trees be made into an "image good" so that individuals can gain pleasure from the existence of trees? Can land trusts be expanded, so that people have a method of changing their notional demand for forests into a real demand that can be revealed in a satisfactory way? Can we structure institutions so that our society is more forest friendly? For example, I have often wondered about the wastefulness of cemeteries and the granite monuments to death that somehow have been built into our culture. Why couldn't we have found a basin of attraction that, whenever a person dies, instead of being buried in a cemetery, that person is buried in a sacred cemetery forest, which will be kept for generations and generations. I'm not sure what the answers to these questions are, but in asking them, and others like them, the research in forestry is moving to the new complexity story approach to economics that will eventually take over the way economics is done.

## NOTES

---

[1] Parts of this paper come from early drafts of a book I am currently working on with William Brock entitled The *Economics of Muddling Through.* (Brock & Colander, forthcoming-b) At this point only I am responsible for the arguments presented here.

[2] Steve Landsburg in *The Armchair Economist* (Landsburg, 1993) is a good representative of an excellent Rust Belt economist.

[3] That is close to happening in behavioral economics in certain fields such as finance. As Richard Thaler has said, once, people asked what was behavioral finance; now people ask what other type of finance is there. A leading indicator of the changes that are occurring, one looks at the hiring priorities of top schools, and the needs their hiring departments see. In the early 2000s behavioral economics is seen as a hiring priority; experimental economics is not yet a totally accepted hiring priority, and agent based

modeling is hardly on the horizon.

[4] For a discussion of what is meant my complex system see Auyang (2000)

[5] For a discussion of agent-based modeling see Robert Axtell and Josh Epstein (1996) and Robert Axelrod (1997).

[6] In his later writings, he modified these views and focused more on the importance of institutions and law. For a discussion see contributions on Hayek in Colander (2000b).

## REFERENCES

Auyang, S.Y. (2000). *Foundations of complex-system theories in economics, evolutionary biology, and statistical physics*. New York: Cambridge University Press.

Axelrod, R. (1997). *The complexity of cooperation, agent based models of competition and collaboration.* Princeton: Princeton University Press.

Axtell, R., & Epstein, J. (1996). *Growing artificial societies: Social science from the bottom up.* Washington, D.C.: Brookings Institution Press.

Brock, W., & Colander, D. (2000). Complexity and policy. In D. Colander (Ed.), *The complexity vision and the teaching of economics (pp.73-93)*. Aldershot, UK: Edward Elgar.

Brock, W., & Colander, D. (2005). Complexity, pedagogy, and the economics of muddling through. In M. Salzano and A. Kirman (Eds.), *Economics: Complex Windows.* Springer-Verlag Italia: Milan, Italy.

Brock, W., & Colander, D. (forthcoming-b). *The Economics of Muddling Through.* Princeton: Princeton University Press. (manuscript in process).

Colander, D. (2000a). The death of neoclassical economics. *Journal of the History of Economic Thought.* 22(2), 127-43.

Colander, D. (Ed.) (2000b). *Complexity and the history of economic thought.* New York: Routledge Publishers.

Colander, D. (Ed.) (2000c). *The complexity vision and the teaching of economics.* Cheltenham, U.K.: Edward Elgar.

Goeller, H., & Weinberg, A. (1976). The age of substitutability. *Science*, 191 (4228), 683-689.

Heilbroner, R. (1953). *The worldly philosophers: the lives, times, and ideas of the great economic thinkers.* New York: Simon and Schuster.

Landsburg, S. E. (1993). *The armchair economist: economics and everyday life.* New York: Free Press.

Schumpeter, J. A. (1957). *History of economic analysis.* New York: Oxford University Press.

Sunstein, C., & Thaler, R. (forthcoming). *Libertarian paternalism.* Chicago Law Review

# CHAPTER 3

# INTER-TEMPORAL ETHICS, MODERN CAPITAL THEORY AND THE ECONOMICS OF SUSTAINABLE FOREST MANAGEMENT

M. ALI KHAN

*Department of Economics, Johns Hopkins University*
*Baltimore, MD 21218, USA.*
*Email: akhan@jhu.edu*

**Abstract**: In this exploratory chapter, I examine how the disciplines of forest economics, capital theory and ethics, insofar as they pertain to decisions taken over time, each provide a lens with which to view the other. More specifically, I read texts of Kant, Laslett, Bourdieu, Cowen-Parfitt and Mitra-Wan-Ray-Roy and attempt to place the general theory of inter-temporal resource allocation within a larger conversation on intergenerational justice taking place in political and sociological theory. I thereby seek to develop a vocabulary for exploring alternative possibilities for social, political and communal bonding by giving meaning to terms such as sustainability, efficiency and equity for the 'optimal' allocation of common or environmental (measurable or non-measurable) resources over time.

> There is no better way ... of compelling us to recognize the character of our subject, its problems and its limitations, than by asking questions of an ethical type. (Laslett, 1987)[1]

> The choices of ecosystem capital are complex and problematic precisely because these entail systems (holistic) aspects that defy reduction to the venerable fiction of commodities and gross substitution along undifferentiated needs. [The] results may be frustrating for those who seek simple answers, but such are not to be found. The decision on the appropriate rate of discount or allocation of ecosystem capital would entail judgements concerning the relevant context and constraints. (Kant, 2003b)[2]

> As everyone knows, priceless things have their price, and the extreme difficulty of converting certain practices and certain objects into money is only due to the fact that the conversion is refused in the very intention that produces them, which is nothing other than the denial of the economy. (Bourdieu, 1983).[3]

> It has never been usual, and it is certainly not easy, to think in terms of duration when considering issues of ethical and political theory. ... Here political theorists encounter a circumstance notorious to mathematicians and statisticians, that infinity is a fundamentally elusive concept (Laslett & Fishkin, 1992)[4]



*Kant and Berry (Eds.), Economics, Sustainability, and Natural Resources: Economics of Sustainable Forest Management,* 39-66.

## 1. INTRODUCTION

Economists do not have a comparative advantage when it comes to ethics, particularly equity and justice across generations, at least by virtue of the expertise that economic science[5] affords them; and the science of forestry economics has been increasingly framed in the last twenty years so that issues of justice and intergenerational equity have been brought to the forefront and given a central role. Thus, in terms of the capital-theoretic issues that I want to present as being relevant to the economics of forestry, I have two options. The first is simply to ignore an imposing and rich critical literature, recently surveyed by Kant (2003a), on the grounds of disciplinary competence, if not of disciplinary relevance. The other option, the one that I do take in this chapter, is present the work in the light of this criticism, coping with it and learning from it as best as it can, and committed to an inter-disciplinarity, however fuzzy, in the belief that "we must do what we can with instruments whose inadequacies and capacity to mislead have been recognized and allowed for."[6] This being said, and especially since I am exercising this option under the rubric (and partial title) of the relevance of modern capital theory to the economics of forestry, I need to underscore my belief also in the importance of the technical work that I report; the danger of disciplinary hubris, an arrogation of a single discipline's voice as the sole and substantive one, is easier to recognize in others than in oneself.

The outline of this chapter, then, is as follows. First, by an exegetical reading of Kant's delineation of the boundaries of the economics of forestry, I spell out what I see to be the basic motivating vocabulary of this side of the subject, the grammar of a language game that explicitly engages inter-temporal ethics, and in its multi-facettedness, goes both inside and outside economic theory bound to its more restricted notions of inter-temporal equity, conventionally fueled by utilitarian impulses and reflexes. Next, using the work of Laslett as a guide, work that deserves to be better known and engaged not only by economists, I place this vocabulary of inter-temporal ethics and sustainability into that of another conversation, one being conducted, for want of a better characterization, in the space of political theory. These larger issues of inter-temporal obligation and submission, when conceived within the relatively narrower frame of environmental economics, and in particular that of the economics of forestry, inevitably revolve around the notions of capital and the rate of discount. Thus, in a subsequent section, I draw on a neglected paper of Bourdieu to give (perhaps a fuller) meaning to the notion of ecosystem capital and use this more capacious view of capital to point to recent work of Cowen-Parfit that surveys and reconsiders the issue of a positive social rate of discount. With these markers and guideposts in place, I turn to capital theory, as conventionally but not universally articulated, and use the Mitra-Wan tree farm and the Mitra-Ray-Roy orchard to cull from the literature a "folk theorem" which, despite an imposing amount of work, has not been given the attention that it perhaps deserves. In a concluding section, I point towards the interstices and lacunae that are identified in the various vocabularies that I have read, recapitulated, recounted.

## 2. ON THE BOUNDARIES OF FOREST ECONOMICS

In his useful survey, Kant (2003a) brings future generations into sharp salience, and uses them to articulate basic principles of the economics of sustainable forest management *(SFM)*.

> The basic idea behind *SFM* is to manage forests in such a way that the needs of the present are met without compromising the ability of future generations to meet their own needs. Under the umbrella of 'both-and' principle, four sub-principles – principles of existence, relativity, uncertainty, and complementarity – will be of paramount importance to guide the evolution of the economics of the *SFM.* [These] five principles may become the foundations of the economics of *SFM* (51).[7]

I defer to the next section a consideration of the terms embodied in the phrase needs and ability of future generations, all of fundamental consequence for my subject, and turn to the explication and translation of the second sentence. The 'both-and' principle is seen as stemming from post-Newtonian physics and to be contrasted with the 'either-or' principle of Newtonian physics and of neo-classical economics, an important marker for several papers in the subject.[8] These difficulties and indeterminacies of translation do not dog the four sub-principles.[9]

> The 'principle of existence' suggests that we cannot ignore the existing situations because these conditions have survived a long time. The 'principle of relativity' suggests that an optimal solution is not an absolute but a relative concept. The 'principle of uncertainty" suggests that due to uncertainties in natural and social systems, a social agent may never be able to maximize his outcomes, but will always search for positive outcomes, and therefore resource allocation will be improved by adaptive efficiency and not by allocative efficiency. The principle of complementarity suggests that human behavior may be selfish as well as altruistic, people can have economic values as well as moral values, and people need forests to satisfy their lower level needs as well as higher level needs (51).

The first principle can be read in two opposing ways: first, to take account of existing conditions so as to change them, and not to avoid facing them simply because they have survived so long into to the present; or secondly, to take account of them in a way that is resistant to change and reads their survival as an equilibrium that is not only stable but desirable, an equilibrium that presumably testifies, in the words of Burke, to a social contract.

> Society is indeed a contract ... a partnership in all science; a partnership in all art; a partnership in every virtue and in all perfection. As the ends of such a partnership cannot be obtained in many generations, it becomes a partnership not only among those who are living, but between those who are living, those who are dead, and those who are to born. Each contract of each particular state is but a clause in the great primeval contract of eternal society, linking the lower with the higher natures, connecting the visible and the invisible world.[10]

It is this identification with Burkean conservatism that leads Kant to argue for "forest rotation based on the annual allowable cut" as opposed to Faustmann's rotation; it has "dominated forestry practices all over the world for centuries against all economic arguments of forest economists (51)," and before advocating changes, "one" needs to be clear that the resulting "new situations would be, in total, better or worse, than existing situations (51)." The important qualification here is the phrase *in total*; it cautions that all costs, including transaction and institutional costs, ought to be taken into account. More specifically, in his discussion of the forester's rotation, Kant draws attention to uncertainties in production, transactions costs and increasing returns. Thus, conservative or radical, the tension in the 'principle of existence' lies in subscription to the way in which the *status quo* has been conceived and formalized, and to the resulting argument that rests on this formalization and thereby validates interferences with it and to it. The confidence that one attaches to proposed changes derives from the confidence that the optimization problem mandating these changes has taken the essentials of the situation into account, on how its initial conditions have been formalized, aspects that it excludes and includes. In asking for the theorems that have been appealed to for the formulation of a theorem, Kant's 'principle of existence' is also gesturing in an important way, it seems to me, towards the ethics of theorizing.[11]

It is through his second principle that Kant reaches out to a binary with an illustrious geneology in both ethics and economics. The relative/absolute terminology forms the basic decomposition of meaning that Wittgenstein draws on in his 1929 lecture on ethics,[12] and when Kant emphasizes, for illustrative purposes, the distinction between Aboriginal and industrial values, and between different "frames of reference" leading to different principles of forest management, he is well within the orbit of Wittgenstein's discussion.

> If we consider (6.422) an ethical law of the form "You ought ..." the first thought is "And what if I don't?" – as though it were a statement of *relative value.* With a judgement of absolute value the question makes no sense. To understand any judgement of [absolute] value we have to know something of the culture, perhaps the religion, within which it is made, as well as the particular circumstances that called it forth; what the man had done, what the question was when I spoke to him, and so on.[13]

Thus, in his second principle, Kant has moved from the relative comfort of the technicalities of the solution of an optimization problem to the absolute difficulties of its "correct" formulation.

It is perhaps here that I also need to mention how Keynes (1930) appropriates Wittgenstein's binary for his own purposes; namely, to distinguish between absolute and relative needs.

> Now it is true that the needs of human beings may seem to be insatiable. But they fall into two classes – those needs which are absolute in the sense that we feel them whatever the situation of our fellow human beings may be, and those which are relative in the sense that we feel them only if their satisfaction lifts us above, makes us feel superior to, our fellows. Needs of the second class ... may indeed be insatiable ... [b]ut this is not so true of the absolute needs – a point may soon be reached ...when these needs are satisfied in the sense that we prefer to devote our further energies to non-economic purposes.[14]

Thus, when Kant deduces from his principle of relativity that "optimal solutions will be situation specific and will in many cases will be beyond market cases (51)," he is referring both to situations that Keynes had in mind as well as to those where the economic problem is pressing to such an extent, the absolute needs so overwhelming, that questions of immediate justice rather than those of efficiency come to the fore. And here as well, the question of what theoretically constitutes absolute needs, and how a collective agreement is to be reached on their precise constitution, goes directly to what I am labelling by the phrase "ethics of theorizing." As such, there is an important overlap, a common orientation if one prefers, between Kant's principle of existence and his principle of relativity.

Kant's third principle, involving as it does the distinction between the *natural* and the *social,* and in particular in emphasizing a *social agent,* rests on the distinction between *adaptive* and *allocative efficiency.* In the review itself, he does not elaborate this distinction, confining himself to a footnote abstracting the work of Douglas North.

> Adaptive efficiency is concerned with the kinds of institutions that shape the way an economy evolves over time. It includes the willingness of a society to acquire knowledge and learning, to induce innovation, to undertake risk and creativity, and to resolve problems of society through time.[15]

The issue here is not the subscription to these laudable objectives – who would disagree with them? – but one of how a *society* and a *social agent* is conceived so as to further the attainment of these objectives? And again, how are such a collective agreements to be reached? In emphasizing institutional design, Kant is clearly emphasizing the formulation of the optimization problem, and the purposes that it embodies, to which the new institutions are to respond. Put another way, institutions have to be designed with respect to a picture of some common objectives, and it is the articulation of this commonality, an agreement as to their outlines, the less fuzzy the better, that is the heart of the issue.

In Kant and Berry (2001), an attempt is made to go beyond the standard economic prescription of solving these issues of the "commons" through either a precise delineation of property rights or through government intervention involving price or quantity directives. The authors focus on what they term *resource regime,* and the dependence of output on such a regime through the *transaction function* as a crucial variable in the formulation of policy. Thus, they conclude their useful paper as follows.

> [I]n developing economies the state regime will frequently not be optimal for management of forest resources located near populated areas. Similarly where forests are leased to private companies but the local communities are heavily dependent on these forests, a joint regime between the company and local communities may be optimal.

The ground is by now a familiar one. There are different stake-holders and any solution that does not takes their interests and leverages into account will be undercut to yield outcomes that can be improved upon. Optimal solutions are only optimal to the extent that they take adequate account of the conditions that go towards determining the problem. This is a rather obvious point but it is indeed surprising as to the extent to which disciplinary imperatives, rather than local conditions, motivate the relevant theorizing. The numerical and geometrical illustrations provided in (Kant & Berry, 2001) show how effectively this point can be made even when the complex notion of a *resource regime* is formulated in the stark simplicity and uni-dimensionality of a real number. Thus, Kant's distinction between *allocative* and *adaptive* efficiency is more a plea for care in the formulation of the optimization problem rather than a distinction between theoretical and institutional economics or between qualitative and quantitative theorizing, neo-classical economics and its "other".[16] Once we focus on efficiency, adaptive or allocative, we are focussing on maximization, on how best to attain our objective given the means that are available to us. The adjectival qualifiers simply alert one to how the objectives and the means are to be conceived and formulated. It is again a question of theorizing, and the ethics and politics that underlie it.

Kant's fourth and final principle, that of complementarity, is, at one level, a succinct summary of the issues that I have already tried to articulate. The crucial marker here, one that takes the place of *society* and *social agent* in the 'principle of uncertainty', is that of the *people.* It is only with reference to it that the binaries of selfish/altruistic, economic/moral, higher/lower are given play. I have already located these binaries in the work of Wittgenstein and Keynes in connection with the 'principle of relativity', and Kant (2003a) gives them further attention in the reading that recent literature, particularly that of Haines (1982), gives to the work of Marshall.[17] While emphasizing the need for this important hierarchical decomposition of the space of unknowns, commodity space if one likes, that is at stake here, I shall not give it further consideration, and turn instead to the formalization of a public agency, the agent on whose behalf the optimization problem is being formulated, and for whom its solutions are being implemented.

In this connection, and given my emphasis on Laslett's work in the sequel, it is perhaps appropriate to begin with a 17th-century thinker who first interrogated the concept of the *people* and the contracts, agreements and arrangements that rest on it.[18]

> The people, to speak truly and properly, is a thing or body in continual alteration and change, it never continues one minute the same, being composed of a multitude of parts, whereof divers continually decay and perish, and others renew and succeed in their places. They which are the people this minute, are not the people the next minute.[19]

Even now these sentences go to the nub of the matter: in their denial of the fact that *people* can have no durational existence and therefore cannot enter into political arrangements with a well-defined representative, they ask whether the understanding of *human behavior* is secured through the aggregation of the social from the individual, or does one, by necessity, have to rely on the social to give meaning and definition to the individual. This basic question regarding methodological individualism can be put another way. Can cooperative outcomes, to be sustainable, be generated only through competition and the pursuit of individual self-interest – the so-called Nash program? Or is a common history or tradition or a supra-individual agency, a collective such as a *state, society* or *community,* necessary for the requisite bonding and trust that is indispensable for the allocation of (common or environmental) resources? And if so, how is such an agency and a basis for commonality to be discerned, formalized and articulated? To approach the matter yet another way, one that gives an adequate emphasis to issues involving planning over horizons of time concerning which no single agent has purview, much less jurisdiction or control, how is one to attain the conceptual and philosophical clarification concerning formalizations of social and community interdependence that is to be incorporated in any proposed optimization exercise?

Just as an illustration of how these two poles – an acceptance of the qualitative and essential difference between micro and macro frameworks versus an aspiration towards giving the latter a foundation through the former – are to be negotiated, I appeal to two thinkers that represent these positions. In the first place, Rawls is a thinker whose work is a sustained theoretical attempt at narrowing divergent interests and delineating positions on which members of a particular polity can reach consensus and agreement. Rawls (2001) writes:

> Our aim is to uncover a public basis for a political conception of justice. In describing the parties we are not describing persons as we find them but rather ... according to how we want to model rational representatives of free and equal citizens. We impose on the parties certain reasonable conditions as seen in the symmetry of their situation with respect to one another and the limits of their knowledge (veil of ignorance).

The question is what does Rawls' theory of justice[20] say about *sustainability, efficiency* and *intergenerational equity?* How does the symmetry of individual parties, especially those not yet born, translate into optimization exercises based on a zero rate of discount? On what basis does one form a sustainable consensus? How does one delineate those considerations which are amenable to agreement and exclude those that are not? In the second place, and as a representative of thinking that is orthogonal to methodological individualism, I turn to Hegel. My interest in his *oeuvre* lies in its singular attempt to develop a tri-partite general equilibrium system based on the *family, civil society* and the *state.*

> In dealing with ethical life, only two views are possible: either we start from the substantiality of the ethical order, or we proceed atomistically and build on the basis of single individuals. This second point of view excludes mind [spirit] because it leads only to juxtaposition [conglomeration, aggregation]. A living relationship exists only in an articulated whole whose parts themselves form particular subordinate spheres. French abstractions of mere numbers and quanta of property must be finally discarded ... Atomistic principles of that sort spell, in science as in politics, death to every rational concept, organization and life.[21]

In conclusion, two points emerge from the 'principle of complementarity': one relates to the character of the resource that is to be allocated, the extent to which *natural* is implicated and imbricated in the *social*; and the other, to the *agency* doing the allocating, the extent to which it is public and thereby divorced from the private.

All in all, my consideration of these four sub-principles draws attention to the broad interdisciplinary framing that the subject demands, and emphasizes, rather than a particular theory, the theoretical principles that go into its theorizing. However, one essential aspect of the situation has been totally neglected.

## 3. ON A CONVERSATION BETWEEN GENERATIONS

In the way that I have read them, Kant's four sub-principles for sustainable forest management *SFM* – existence, relativity, uncertainty and complementarity – as well as the authors I have appealed to illuminate them (with the possible exception of Filmer) do not involve time in any explicit way. They all deal with more classical and timeless problems of political and ethical obligation. This is hardly accidental. In their 1979 introduction, Laslett and Fishkin (1979) stress that an "entirely new moral perspective may have to be worked out now to meet the intellectual demands upon us by environment, population and futurity." Under the rubric of what they term "arithmetic humanity in relation to politics, especially the correct boundaries which should surround any human collection so that a proper political society may appear", they stress the danger of "preoccupation with a small traditional agenda of classical 'problems' in political philosophy and of too much reliance on respected names from the past."

> The issues to do with arithmetic humanity are continuous with those to do with democratic theory, and two-fold in their character. They are geographical, as when Peter Singer talks so urgently about our duties to distant yet contemporary humans in times of famine, and temporal when Peter Laslett addresses the problems of generations past and generations yet to come.[22]

Thus, when Kant locates the basic idea behind *SFM* is to "manage forests in such a way that the needs of the present are met without compromising the ability of future generations to meet their own needs," he is locating his subject, at least in part, in precisely the terrain that Laslett investigates in two famous essays. In the remainder of this section, I try to bring Kant and Laslett together.

Laslett (1979) unpackages the term "generation" along three dimensions: a procreative one, as in all fathers or grandfathers; a temporal one, as in a group born within a particular interval of time; and finally, an attitudinal one, as a "unity capable of a attitude or of a responsibility," as in a post-war generation. Not unlike Filmer's criticism of the term *people,* or Burke's metaphysical use of an *eternal society,* Laslett shows how the word *generation,* in shuttling between the three meanings that he has identified, does not stand up to a rigorous analysis. Unlike static general equilibrium theory, say as articulated in (Debreu, 1959), it is not the difficulty of one agent Pareto-optimally appropriating for herself all of the societal resources that have been bequeathed to her; the problem lies in delineating a boundary to the term, in the recognition that one generation is intertwined in another, in giving meaning to the assertion that one generation consumes everything and leaves nothing for its successor.

> Since the concept of a generation is elusive and confusing, it is difficult to see how one can talk at all convincingly about rights and duties in respect of such an unmanageable entity (39). ... [T]he concept bristles with ambiguities and difficulties of a logical and empirical kind. ... [P]alpable consequences for all of us seem to flow from the uses we make of the word 'generation'. These consequences are practical and moral (39).[23]

Laslett asks whether he is obliged by the actions of his predecessors, and if so, what is it that obliges him?[24] This question clearly goes to the heart of inter-temporal ethics; and if the term *sustainability* is to be given a determinate and coherent meaning in this context, it clearly must be located in the domain that these questions open. Laslett proceeds by exploiting a slippage between the procreational and temporal usages of the term *generation.* He identifies an asymmetry in the former and building on it, applies it to the latter.

> Parents have duties towards their children but the fact of procreation gives parents no rights in them. Children have rights in their parents, but no duties towards them, not, that is to say, duties towards them as progenitors. The duties in respect of procreation are owed their own offspring. In the ethical exchange between procreational generations, then, duties do meet rights; but not in respect of the same persons (48).

Laslett's proposal is that the "ethical reciprocity characteristic of procreational generations ... can by inexact analogy, be held to apply to temporal generations, that is to generational relationships in society at large, though within one collectivity only (49)." By exploiting an indeterminacy of meaning, he has extracted an operational principle of inter-temporal ethics. Just as Filmer appealed to the father-son relation to understand the monarch-subject relation, Lalsett appeals to principles of equity within the family to articulate a principle of justice within society. The elaboration is worth quoting in full.

> In the same way as Children within the family can expect nurture from their parents as a right, conferring no obligation upon them, so can the members of any generation of Englishmen take for granted the material, technical, cultural, social and political benefits which accrue to them from their predecessors. Their 'debt to the past' is to be satisfied by their duties to the future, and 'future' in this last phrase must be construed as 'the foreseeable future' (49).

The limitation to one collectivity, to Englishmen for example, is of fundamental importance because it is precisely that very collectivity which is to secure adherence to the principle. It is here that we come up against the concept of *arithmetic humanity* and the need to give boundaries to the society which is being theorized for. In terms of my earlier discussion, it is the non-procreational changes in the collectivity that overturn provisional solutions, and run through all of Kant's four sub-principles: of existence, relativity, uncertainty and complementarity.

It is important to understand that Laslett has secured a space of generational obligation that is distinct from political and social obligation. It is to the former, and to the former alone, that his basic principle is addressed. He adduces two sets of considerations for exempting from his theory the (material) support that children give their parents. He refers to such support as "predominantly social or even political in character – as an instance, in fact, of the universal obligation we all have towards contemporaries in need – rather than as generational (51)." The first of these is the conception of "nature of affection, familial and otherwise," as a commodity.[25] Parental love is not a commodity that asks for repayment, the relationship is not based on a *quid pro quo*.[26] Laslett's second consideration is based on the past; in particular, on an appeal to the Poor Law in English history whereby destitute parents were supported by the state irrespective of whether they had "grown-up independent children at the time." He emphasizes that the transfer of a right from society to one's progeny is a deliberate, and presumably, political decision. In summary, generational justice, Burke's eternal intergenerational contract, simply consists in these "unidirectional, hook-eye linkages."[27]

In his reconsideration of the subject a decade later, Laslett's earlier essay is summarized as the following principle of inter-temporal justice.

> It consists in an obligation on all present persons to conduct themselves in recognition of the rights of all future persons, regardless of geographical location and temporal

> position. No generation is at liberty to ransack the environment, or to overload the earth with more people than can be supported, or even, though this is more debatable, to act in such a way as to ensure that the human race will disappear. The duty goes beyond beneficence, the idea that it would be better to act this way and magnanimous to our successors (15).[28]

This is a deontological principle that stands on its own, which is to say, receives no warrant from some prior Rawlsian conversation, or a Hegelian conception of *geist* or some utilitarian pleasure-pain principle based on aggregation or a Nash program. But while it recognizes the *rights of the unborn,* it clearly does not go far enough in giving operational precision to *conduct,* to the words *ransack* and *overload.* What is to be noted, however, that Laslett accepts the ambiguity of the term *generation,* the fact that he cannot "give the generational contract a local habitation and a name, any more than could be done for the social contract itself," and moves on to a free use of "generational images, generational language, and the association of generational relationships (25)." Through two additional metaphors to complement his earlier one of a chain, that of a procession and a rope, he works his earlier ideas through the concepts of an *intergenerational tricontract* and the *intercohort trust relationship.*

It is through a picture of a procession that Laslett and Fishkin propose to get at the open-endedness, locality and irreversibility of inter-temporal justice. An individual in a procession cannot see where the procession begins and where it ends, communicate only with people immediately preceding and succeeding him, influence the progress of only those in it who follow him, and has only a rough idea as to whether the procession is headed in the direction that has been determined for it.

> [T]he processional image is particularly useful because a visible segment of the procession apparent from a single point of observation does provide an intelligent sample of the whole, while preserving the condition that people should be perpetually entering from one side and leaving from the other. If the ideal observer tries to get up as high as possible above the procession to see as much as can be seen of it, however, he or she runs into the difficulty about length in relation to infinity. The observer would do best to stay metaphorically on the ground, but to seek a vantage point with the widest possible view of the procession as it passes (13).

There is a trade-off; the more of a procession that an observer attempts to see, the less of its character she sees. In the limit, she may see all of the procession, obtain a view *sub specie aeternitatis,*[29] but it is not clear what meaning she is to extract from such a view.

> A deathless collectivity, identified with the political purposes of the state, and not itself subject to the limitations of duration imposed on political cooperation and exchange, might make dealing with the problems and puzzlements much easier. But the image of a eternal, all-inclusive collectivity embracing everyone alive, scarcely belongs in the arena of individual rights, government by consent of the governed, and the rule of law. Awkward as the processional image may be, awkward because the reality to be signified is itself so elusive, we are required to accept its logic in preference to the Hegelian march of the state through history (14).

It is through such pictures that Laslett imbibes the lessons from Samuelson's (1958) overlapping generations model, and rather than seeing the work's principal contribution as a counterexample to the fundamental theorems of welfare economics, on the explication of the fact that these theorems hinge crucially on their underlying assumption of a fixed finite number of commodities, makes it a basis of an *intergenerational tripartite contract.* In answer to the reformulation of Wittgenstein's question as to "Why should I do anything for future generations [when] they have done nothing for me (28)," Laslett substitutes a tripartite arrangement for the two-generational procreative contract. Such an arrangement concerns *removed generations,* where the term refers to "those who do not overlap but stretch backward and forward from the present generation, itself thought of as a removed generation with respect to the others (25-26)."

> Obligations between removed generations cannot in consequence be addressed at all under the two-generational contract. [T]he intergenerational tricontract ... seems particularly well-suited to securing justice between removed generations. [It] gives formal expression to the widespread conviction about the obligations of generations to those coming after them, not only removed generations but also those which overlap in the same time space and which under another aspect can be regarded as contemporaneous, successive cohorts or age-groups. This is the conviction that each generational entity must deliver the world to its successor in the condition in which it was received (29).

As in the 1979 analysis, it is the analogy, inexact analogy, to relations within the family that forms the basis for the tricontract, the principle that generational obligation is unidirectional, that it always moves forward in time.

> To look upon the symmetrical interchange between parents and children as having anything to do with an agreement or a contract between them seems to me to lack all power to convince. It does so even as a metaphorical construct, a simile, or an analogy (29). Everyone, therefore, has rights to what he or she receives from his precursors or hers, rights that are or will be met by duties they perform to their successors. But they do not have ... any duties anterior to them, or any rights to those posterior to them (31).

Laslett's principle is important enough that it can be underscored: "the child generation receives transfers being made to it by the parental generation as of right under the contract, but that tricontract gives no title to transfers from the child generation, or to the grandparental generation from the parental generation (31-32)." But there is another aspect to Laslett's discussion in 1992 that was not there in 1979, and this concerns the contribution that children do make towards the support of their parents, that the flow of resources is not only, cannot only, be in one generational direction. Here he relies on the metaphor of a rope to supplement, intertwine if one prefers, the intergenerational tricontract with intercohort trust.

> These cohorts, nearly all of which are based in practice on a year in time rather than a moment, continuously intertwine with each other over the whole length of human history. They do so like the strands that wind round one another to create a piece of thread, each strand being shorter than the piece of thread itself, which unlike the strands, is capable of indefinite extension (46).

When Laslett refers to *trust,* he does not conceive of it either as a commodity or as a grammar of relationships, but as an institution based on an object of value and constituted by three types of agents: trustor, trustee and a beneficiary.

> In the trust, a trustor makes items of value over to trustees, not for the good of the trustees, but solely for the good of the beneficiaries. If the trust is ... discretionary, the trustees can and should vary the distribution of the assets, without necessarily referring to the trustor if available, provided always that the object of the trust, that is, the welfare of the beneficiaries, is enhanced. In selecting the trustees and specifying their duties the trustor must be presumed to know that they are sufficiently well-informed of the relevant circumstances and of what would be the best for the beneficiaries under changing conditions, and in view of how conditions might change in the future. If the beneficiaries should need counseling in the extent and character of their justifiable expectations from the trust, it is for the trustees to supply it (33).

The point is that the trustor cannot deliver the trust, now conceived in Laslett's usage, directly to the beneficiary for reasons having to do with time but whose further articulation is neither necessary nor relevant, but has to resort to a trustee, with *well-specified duties, sufficiently well-informed* both about the beneficiary and about the circumstances in which such a beneficiary may possibly find himself, and who could be called on for *counselling* of the beneficiary. It is a lot to ask, but an application to inter-temporal equity and justice, and in particular the economics of forestry, asks even more. The reason has to do with the confounding of roles stemming from the fact that the trustee and the trustor are also beneficiaries. For the trust, again in Laslett's usage, to be workable, it must be enveloped in an atmosphere of trust that deals with incompleteness, not of contract, but of the underlying implicit agreement[30] and of the vulnerabilities of each of the parties.

It is precisely to make all of this manageable that Laslett grounds the trust within a cohort. Just as a single collectivity localizes the intergenerational tricontract, so does a single cohort furnishes a necessary limitation that makes it feasible for the actors to fulfill all of the roles that have been assigned to them. Members of a productive generation is concretely, rather than abstractly, acquainted with its non-productive elders, and can influence through the political machinery, a concrete set of politicians, rather than an abstract deathless collectivity, to cope with the requirements of inter-temporal distributive justice. Laslett emphasizes these limitations by questioning a "picture of the natural world being entrusted to humanity" and asking "how the entrustment of the world is to be conceived." It is again a question of the proper boundaries within which a theory is conceived, the collectivity for which the theorizing has been done.

> The difficulties with the concept of the world itself as a trust to humanity serve to direct attention to the fact that ethical principles other than those informing contract and trust might be invoked for environmental purposes. This might be done in combination with versions of either of these two, or both of them, perhaps otherwise (45).

And here Laslett joins with Kant in the invocation of aesthetic and religious principles.[31] We are back to a view of the world *sub specie aeternitatis,*[32] Kant's reminder that the 1992 Earth Summit "acknowledged the social, cultural, recreational and spiritual values of forests, and viewed these benefits as fundamental to *SFM*."[33]

In conclusion, it is difficult to deny that Laslett's notions of *inter-temporal tricontract* and *intercohort trust* go to the heart of the economics of forestry, but they are to be used without *hubris,* as a basis for a theoretical opening of a conversation rather than a closing of it, for examining the consequence of perturbations of a model rather than the model itself, not for flaunting expertise but for drawing attention to the fact that the final word is simply not available, and that therefore the distance between the theorist and the theorized needs to be minimized rather than maximized.[34]

> Theorists, social theorists, political theorists, and ethical theorists have yet to get an effective hold on the realities that would have to enter into any truly adequate account of justice over time (46).

## 4. ON ECOSYSTEM CAPITAL AND ON AN ECONOMY OF PRACTICES

Kant (2003b) observes that "aesthetic, spiritual, religious and cultural attributes are not subject to commoditization," and attributes of a system that can be commodified are orthogonal to those which cannot, and the "orthogonal attachment – incongruous nature – will restrict the aggregation of all attributes to a single economic (monetary) measure (119)." Kant proposes a notion of capital, *ecosystem capital,* to grapple with, and subdue, tendencies in theorizing in which such commodification is rampant and unchecked.

> The ecosystem capital is valuable to human society not only for the products which may be thought of as commodities, that it contributes to the economic system but also for its functional contributions to the well-being of humanity. ... Thus, most of the contributions of ecosystem capital are derived by keeping its different components working in their existing functional relationship as a fully functionalized system [in which] each part is as 'valuable' as the whole and hence the value of any single component cannot be understood separately from its contribution to the whole (117-118).

In the previous section, I considered the difficulty in giving rigor to phrases such as the *well-being of humanity,* and in the one previous to it, the importance of theorizing that is self-aware in what it includes and excludes, in being transparent in what, and how, a theorist, as theorist, sees as a *fully functionalized system.* So when Kant asserts the holistic characteristics of *ecosystem capital,* he clearly does not have in mind static general equilibrium theory, as articulated in (Debreu, 1959) for example, but is trying to reach and say something beyond the standard theory. In this section, I try to understand his concept through Bourdieu's 1983 notion of *symbolic capital* and the conceptual schema with which Bourdieu gives it meaning.

As far as the noun *capital* is concerned, Bourdieu's conception seems entirely conventional.

> Capital, which, in its objectified or embodied forms, takes time to accumulate and which, as a potential capacity to produce profits and to reproduce itself in identical or expanded form, contains a tendency to persist in its being, is a force inscribed in the objectivity of things so that everything is not equally possible or impossible (241).[35]

In this definition, ideas basic to capital theory such as durability and irreversibility are incorporated,[36] but nuanced in that the transformation of resources over time, is qualified by the term *potential,* and rather than simply as a "stock of tangible, solid, often durable things such as buildings, machinery and inventories," as in recent definition in Solow (2000), capital is seen as a *force inscribed in the objectivity of things.* It is this that gives a singularity to Bourdieu's vision, in its holistic thrust in an "economy of practices which would treat mercantile exchange as a particular case of exchange in all its forms (242)."

> A general science of the economy of practices, capable of reappropriating the totality of the practices which, although objectively economic, are not and cannot be recognized socially as economic, and which can be performed only at the cost of a whole labor of dissimulation (*euphemization*), must endeavour to grasp capital and profit in all their forms, and to establish laws whereby the different types of capital (or power, which amounts to the same thing) change into one another (242-243).

We have already seen this emphasis on *totality* in Kant's 'principle of existence,' but what is additionally involved here, it seems to me, is the 'principle of complementarity' whereby any water-tight distinction between the *social* and the *natural* is denied. By seeing the non-economic as economic, and complementarily, by embedding the economic within what is seen to be the non-economic, Bourdieu goes beyond Solow's definition and the questions that follow from it: how do you measure its stock? what is its rate of return? what is its rate of depreciation? To what quantitative extent do the rapidly growing East Asian economies owe their success to it? There is a reliance, perhaps even a crucial dependence on the language and apparatus of capital theory, along with a denial, perhaps even an assertion of impossibility, of aggregation to single number. There is an optimism that anything that *persists* over time, and directs circumstances into one channel rather than another, as any "special proclivity or talent that exemplify the value of some specifically Asian virtues of character and social organization: diligence, teamwork, compromise and so on," is amenable to the insights of capital theory.[37]

An *economy of practices* then rests on the notion of "*symbolic capital,* that is to say capital – in whatever form – in so far as it represented, i.e. apprehended symbolically, in a relationship of knowledge or, more precisely, of misrecognition and recognition, presupposes the intervention of the habitus, as a socially constituted cognitive capacity (255)." This simultaneous *misrecognition and recognition* is simply a restatement of *euphemization,* and a reach to practices that are capital theoretic even though nonquantifiable, subject to economic laws even though treated as non-economic.

> Economic theory ... by reducing the universe of exchanges to mercantile exchange, which is objectively and subjectively oriented toward the maximization of profit i.e., (economically) *self-interested,* it has implicitly defined the other forms of exchange as noneconomic, and therefore, disinterested (242).

With an invocation to the term *habitus,* I am in the very vortex of Bourdieu's *oeuvre,* and given the scope of this chapter, shall constrain myself only to observe that *habitus, field* and *capital* constitute concepts that "have no definition other than systemic ones, and are designed to be *put to work empirically in systemic fashion.* Such notions as habitus, field and capital cannot be defined, but only within the theoretical system they constitute, not in isolation."[38] It is in this kind of advocacy of holism that Bourdieu's work naturally dovetails into that of Kant.

With this background, I can move relatively quickly and observe that Bourdieu decomposes *symbolic capital* into three forms: *economic capital,* when it is "immediately and directly convertible into money and may be institutionalized in the form of property rights"; *cultural capital,* when it is "convertible, under certain conditions, into economic capital and may be institutionalized in the form of educational qualifications"; and *social capital,* "made up of social obligations ("connections"), and when it is "convertible, under certain conditions, into economic capital and may be institutionalized in the form of a title of nobility." After noting that *cultural capital* is further trichotomized into *embodied, objective* and *institutionalized* forms, I turn to *social capital.*

> Social capital is the aggregate of the actual or potential resources which are linked to possession of a durable network of more or less institutionalized relationships of mutual acquaintance and recognition – or in other words, to membership in a group – which provides each of its members with the backing of the collectively-owned capital, a "credential" which entitles them to credit, in the various senses of the word (248-249).

The capital-theoretic emphasis is on durability and on fungibility, and both rest on purpose. Categories such as *obligations, rights* and *duties* that we earlier encountered under the rubric of political theory have now been appropriated by social theory, and given an economic basis for reciprocity.

> Network of relationships is the product of investment strategies, individual and/or collective, consciously or unconsciously aimed at establishing or reproducing social relationships that are at once necessary and elective and that are directly usable in the short or long term. Creation of durable obligations subjectively felt (gratitude, respect, friendship) or institutionally guaranteed (rights) (249-250).

I am now finally in a position to give meaning to Kant's 'both-and principle', possibly in a way different from his.[39] Arrow (2000) counters the "widespread consensus on the plausibility of the hypothesis that social networks can affect economic performance" by the "considerable consensus that much of the reward for social interactions is intrinsic" which is to say, individuals interacting for the sake of interacting, non-purposive and without any other discernible objective. "Indeed, this is what gives them their value in monitoring." At the same time, there is no denying that "networks and other social links may also form for economic reasons" – to "guard against market failure" and to "exploit monitoring devices not otherwise available." It is important to be clear that this simultaneous presence and absence of purpose – yes/no and both/and rather than either/or – is not necessarily contradictory. The point is that one pertains to the individual while the other pertains to the relevant group of which the individual is part. What is a given unchosen parameter from the individual's perspective is an unknown choice variable from the perspective of the group. As we saw in Laslett's emphasis on a single collectivity and in the limitation to a single cohort, the "process of *consecration* of the group presupposes and produces mutual knowledge and recognition, affirms the limit of the group, and is constituted by exchange of material and symbolic resources (250)."

> Each member of the group is thus instituted as a custodian of the limits of the group: because the definition of the criteria of entry is at stake in each new entry, he can modify the group by modifying the limits of legitimate exchange through some form of misalliance. Through the introduction of new members into a family, a clan, or a club, the whole definition of the group, i.e., its fines, its boundaries, its identity, is put at stake, exposed to redefinition, alteration, adulteration (250).

This point can be made another way. Once the sovereignty and integrity of the realm of the market is accepted, and "existing social relations [taken] as a preexisting network into which new parts of the economy (for example, development projects) have to be fitted," one can focus on efficiency, however broadly interpreted, and "exploit complementarity relations and avoid rivalries." At the same time, there is no denying that "new projects will create their own unintended social relations, possibly destroying existing ones," and as such, one has to keep track of these variations, of how the parameterized background is changing. However, here again, it is important to be clear that in this treatment of something as being simultaneously fixed and changing – both/and rather than either/or – involves no contradiction. The point is not that one pertains to the moment while the other pertains to the evolution of that moment, and consequently what is a parameter from the perspective of static analysis becomes a variable to be tracked from the standpoint of dynamics, but rather that an entity which is a variable for sociological, or perhaps anthropological, analysis is, and ought to so remain, a parameter for economic analysis. "The market needs supplementation (for efficiency) by nonmarket relations", and the question as to whether the "market (or, for that matter, the large efficient bureaucratic state) destroy social links that have positive implications for efficiency" is important and long standing.[40] We are back to the Kant-Berry emphasis on working within particular resource regimes and being sensitive to their evolution.

I conclude my discussion of *ecosystem capital* with Solow's (2000) words regarding *social capital,* with his saying that "those who write and talk about social capital are trying to get at something difficult, complicated and important: the way a society's institutions and shared attitudes interact with the way its economy works. It is a dirty job, but someone has to do it; and mainstream economics has puristically shied away from the task."[41]

## 5. ON THE 'CORRECT' SOCIAL RATE OF DISCOUNT

So far, my consideration of the work of Kant, Laslett, and Bourdieu has revolved around the following questions: does a particular collectivity need a supra-collective agency to guide it? and if so, what is nature of the agency that is to assume the responsibility for such guidance? and in what domains does its guidance lie? Simply put, given my concern with issues of inter-temporal ethics and justice, does a society or a state need a supplementation of individual (generational) decisions by a public policy-prescriber? who does he represent? and how does the answer to these questions rebound on the validity and implementation of all that he proposes?[42] And from all of the texts that I have read so far, it is clear that however we give meaning to the notion of optimality, and however we formulate commonality of resources, the optimal inter-temporal allocation of such (environmental) resources available to any collective (society, community, regional or ethnic group, nation state, a particular unity of nations) cannot, ought not to, be based on criteria that discount the weight that is attached to future generations or cohorts of that collective just because they are temporally removed. If the words *sustainability* and *inter-*

*temporal equity* are to have any analytical thrust, sustainable policies cannot be rejected, or decided upon, on criteria that have already incorporated in them some form of inter-generational *myopia* or *impatience.* The benefits of a forest, or of public investment in clean air, or of a project such as the Hoover dam, or of the preservation of Yellowstone National Park are hardly limited to the generation that commits the resources to secure them.

This observation is well-understood. Economists know well the relevant quotations from Ramsey, Harrod, and others,[43] and are aware that that in his 1935 work on economic growth, von Neumann (1935-36) also did away with time-preference and confined himself to an investigation of maximal *balanced* growth paths. This earlier work is now complemented by more recent surveys of Cowen (1992), Cowen-Parfit (1992), Kant (1999; 2000; 2003a; 2003b) and Peart (2000). And once the issues are considered within the broader rubric of inter-temporal justice and ecosystem capital, as in Kant, the arguments are only underscored. However, even though simple and well-understood, mainstream economic research has bypassed and ignored this basic observation on two grounds: (i) analytical tractability, and (ii) a recommittment to methodological individualism as typified by the analytical construct of the *representative agent.*

In terms of (i), the analytical difficulties of optimizing models with a zero rate of time-preference are easily communicated.[44] Any plan for the inter-temporal allocation of resources, if limited to a finite period and embodying a particular time-horizon, has to take as given – arbitrarily and prior to the determination of unknowns of the plan – the amount of resources that are to be left for generations beyond that arbitrarily chosen horizon. The only analytically viable option, then, is to work with an infinite time-horizon – to plan from now into an *indefinite* future, with the expectation that each successive year these plans will be reworked with additional information. But the problem with an infinite time-horizon is that the time-stream of benefits may not sum (integrate) to a finite number, and therefore the objective may not even be defined!

Ironically, the reasons for (ii) follow from the overwhelming influence that the Ramsey growth model (with an infinite time-horizon but with a positive discount rate) has gained in mainstream economic research. As this model became a central conceptual framework for the discussion of macroeconomic policy, issues of *time-consistency* and *incentive compatibility* led to a devaluation of a (national) policy maker to one of the many participants of the policy game, and to a reinterpretation of his objective function as that of an *infinitely-lived* representative agent. Under this blurring of the individual and the social, a planning model, with the planner formalizing the bonds of the collective, is recast as a positive model[45] of a decentralized economy in which each of a continuum of identical individuals is seen to be pursuing their own individual interests. With this important conceptual manoeuvre, a preoccupation with the short-run – with impatience reorienting an agent towards immediate benefits and profits – is rendered more "rational" and thereby more defensible. Thus, it is not surprising that research on models with a zero time-preference, analytically difficult to begin with, abruptly ceases in the eighties. The current conventional wisdom is to see it as "dispensable and misdirected." The effects of this conventional wisdom are pervasive. The bibliographies of standard textbooks in the field such as those of Arrow-Kurz (1970), Stokey and Lucas (2000), Aghion and Howitt (1998), or Majumdar, Mitra, and Nishimura (2000) simply ignore the earlier literature on the extension of Ramsey's undiscounted setting.[46]

In subsequent sections, I continue the discussion of this point of view. For the moment, I conclude this section by noting the Cowen-Parfit distinction between the *probabilistic discount rate* and the *social discount rate,* a rate that is used simply because of the remoteness of the future. Cowen and Parfit (1992) write as follows.

> Remoteness in time roughly correlates with a whole range of morally important facts. So does remoteness in space. ... But no one suggests that because there are such correlations, we should adopt a spatial discount rate. No one thinks that we would be morally justified if we cared less about the long-range effects of our acts, at some rate of n per cent per yard. The temporal discount rate is, we believe, as little justified.

## 6. ON THE ECONOMICS OF FORESTS AND OF ORCHARDS

In his surveys Kant (2003a) does not give any space to the work of Mitra-Wan-Ray-Roy, and I shall argue in the next section why this omission is not accidental. Here, I shall try to substantiate why the reformulation in this work is of fundamental analytical consequence for the subject.[47]

In their seminal paper on the economics of sustainable forest management, Mitra and Wan (1986) shift their perspective from the number of trees of a particular age in a 'given' forest to the proportion of the acreage of that forest that is devoted to trees of a particular age.[48] Coupled with their assumption that there exists a particular age beyond which a tree rots and yields no timber, it allows them to recast a difficult infinite-dimensional functional-analytic problem into a finite-dimensional one of (albeit infinitely) repeated choice from a finite-dimensional simplex. Toward this end, consider a unit plot of land which, without any replenishment and with costless planting, can support forever trees of ages ranging from one to $n$ years. A tree of age $i, (i = 1, 2, \cdots)$, when chopped down, yields $b_i$ units of timber, and an $n$-year old tree yields nothing if grown beyond $n$ years. Thus, time is measured in discrete (unit) intervals, and necessarily entails the obvious, but nevertheless crucially important, observation that the amount of acreage $x_{i+1}$ devoted to trees of age $i+1$ tomorrow cannot be *more* than the amount of acreage $x_i$ devoted to trees of age $i$ today. Let me refer to this observation as a *feasibility* constraint; it surely distinguishes a problem in the economics of forestry from a problem of more general capital theory. Thus, in this today-tomorrow world, a given value of these two acreages $(x_i, x_{i+1})$ directly translates, through subtraction, into a proxy for the number of tress chopped down today, and hence the amount of timber consumed today.

Now, suppose there is agreement that the forest must be managed to maximize the aggregate amount of timber over the foreseeable future. The question then how is the forest to be managed? Even in such a simple setting, the word *sustainability* does not have an unambiguous meaning. One obvious candidate is a forest that does not change its composition from year to year – the composition of the forest would be identical across all time periods. But an alternative would be a time-profile that is repeated – a forest would be sustainable if its changing composition over a period of time remains unchanged for subsequent, identically long periods of time. Mitra-Wan commit themselves to the former interpretation, and thereby reduce the problem of determining the composition of a maximal sustainable forest from an infinitely varying sequence, albeit with a determinate pattern, to a constant sequence, and thereby to a real number! Under this reduction, the first step involves a static optimization problem, trivially amenable to Kuhn-Tucker theory.

The interest, however, is not primarily in the composition of maximally sustainable forest but one that is optimal in terms of its timber yield, optimal especially if we prohibit discounting. This is to say that we work under the constraint that timber available today has the same *social* value as timber available a hundred, or indeed a thousand, years from today. As discussed above, we circumvent the problem of an undefined sum of timber by an appeal to the overtaking criterion of Atsumi (1965) and von Weizsäcker (1965), and it is the execution of this appeal that is my primary interest here. The point is a fascinating one. Consider the accounting or shadow prices churned up by the Kuhn-Tucker solution to the static optimization problem directed to determining the composition $\hat{x}$ of the sustainable forest with a maximal yield, and given any other (infinite) sequence characterizing the possible evolution of the forest, compute the value-loss of timber in terms of these prices relative to $\hat{x}$. Consider, as a third step of the argument, the infimum (the greatest lower bound) of the value-losses of all possible paths of evolution of the forest. Under the feasibility constraint, these (uncountable since the possible paths are uncountable)[49] numbers are not all infinite, and hence the problem is well-defined and implementable. The third step is to show that for any given initial composition of the forest, there exists a path that attains this minimal value-loss. And now under an assumption that has guided this analysis, and has been a subtext of my description, can be made explicit. It is simply that the composition of the maximally sustainable forest is unique! Under this assumption, we have the *coup de grace,* the final fourth step of the argument, that any given initial composition of the forest, this minimal value-loss path is optimal.

The question then is what assumption on the primitive data of the problem, the $n$ non-negative numbers $(b_1,\cdots,b_n)$, guarantees that the composition of the maximally sustainable forest is unique? This is simply the assertion that the average productivity of the tree of a particular age, $b_i/i$, is maximized at a unique age, say $\sigma, 1 \leq \sigma \leq n$. And with this assumption in place, it is all a matter of easy harvesting. Mitra-Wan consider two situations. The first concerns plot of land that is initially barren. The solution, the Faustman solution, is simply to chop down all trees of age $\sigma$, resulting is a path that minimizes the value-loss of all paths that can be generated from this (barren) plot of land and is therefore, by the Mitra-Wan theorem, optimal. The second concerns a plot of land which is already parcelled out among trees of varying ages. Now the solution is simply to chop down all trees of age equal to, or greater than, $\sigma$, resulting is a path that minimizes the value-loss of all paths that can be generated from this plot of land, and is therefore optimal. The fact that Mitra-Wan prove this theorem by working with a set of prices that are different from the ones used to obtain the first solution need not concern us here; the principal analytical point is that in either case, the paths that are proved to be optimal minimize the period-by period (and therefore the aggregate) value-losses of any other path starting from the given plot of land, and it is this that allows us to show the optimality of the Faustman and other policies.

This being said, the question arises as to the extent to which the composition of a well-managed forest eventually resembles, which is to say converges, to that of the maximally sustainable forest. The answer here hinges on whether it is the aggregate of the timber or the aggregate of a (strictly concave) function of the timber in each period that is being maximized. In either case, there is a non-degenerate interval containing the maximally sustainable forest composition to which the optimal path converges. However, it is only in the latter (strictly concave) case, that this interval is degenerate in the sense that it reduces to a singleton. In other words, when the period-by-period utility function of timber levels is strictly concave, the optimal path forest composition converges to that of the composition of the maximally sustainable forest. In all other cases, and in particular the case of a linear period-by-period utility function, the optimal path may be periodic.

As I discussed in the previous section, it is the undiscounted theory that has been neglected in applications, and that the discounted theory – recursive dynamic programming – has received extended treatment. Thus, once the Mitra-Wan formulation is well-understood, the tracks for developing its discounted version are all well-laid out and well-understood; see Mitra (2000) for example. This is not to say that the results are not surprising. Mitra and Wan (1985) conclude the introductory section of their paper with the following two sentences.

> In fact, this study together with Mitra and Wan (1981)[50] show[s] that the asymptotic properties of optimal programs are similar when the utility function is linear, regardless of whether there is positive or zero discounting. But these properties may be quite dissimilar, when the utility function is strictly concave, depending on whether future utilities are undiscounted (in which case we have the "turnpike property", with the unique OSP as the "turnpike"), or positively discounted (in which case, a "turnpike property" need not hold, and periodic optimal solutions are definitely possible).

Moving beyond the economics of forestry, as developed by Mitra and Wan (1986; and 1985), to the economics of sustainable orchard management, a seminal paper of Mitra, Ray, and Roy (1991) views the earlier work as a contribution to "point-input, point-output" capital theory and extends it to "point-input, flow-output" capital theory. This is to say, it allows the trees in the forest to yield fruits in each year of their life possibly in addition to the timber when they are cut down. Thus, in addition to the output, point-output, when a tree of a particular age is cut down, it yields a flow-output at every prior year. The basic outline of the model and the benchmarks of the analysis remain broadly similar, but now the determination of the flow-output requires a determination of the number of trees of each age, and thereby leads to a veritable thicket of mathematical difficulties. It is of course outside the scope of this chapter to discuss these difficulties in any detail other than to point out that the authors consider a special case of the problem – one in which there is positive discounting and no utility from the timber that is available once a tree is cut down. It is indeed an analysis of orchards! The authors' conclusions are summarized in the following two sentences.

> Under a mild condition on the flow-output vector, we establish that optimal programs for every discount factor and every initial state (other than a unique stationery optimal state) will exhibit non-convergence. Furthermore, we provide a necessary and sufficient condition on the flow-output vector for which a neighborhood turnpike theorem; that is long-run fluctuations on an optimal program are "small" when the discount factor are "close" to unity.

## 7. ON A 'FOLK THEOREM' OF CAPITAL THEORY AND ON A DIRECTION FOR FUTURE WORK

Kant (2003a) begins the first substantive section of his survey by referring to Faustman's solution, and notes that "through 1999, 278 identifiable works have been published [and] 85% of these have been published since 1979."[51] In this context, he sights the papers of Samuelson (1976), Anderson (1976), and Reed (1984) as the pioneering ones. In his own "dynamic approach to forest regimes in developing economies", Kant (2000) formulates and solves an optimal control problem. And since optimal control theory in its most accessible form relies on Pontryagin's principle, applications to the economics of forestry have relied most heavily on techniques developed for continuous time. And it is not a theorem that is applied but rather a principle, a presupposition and subscription to professional identity that requires the three hallowed steps: determination of the optimal controls from the maximization of the Hamiltonian, the determination of the auxiliary differential equations and the satisfaction of the transversality conditions. The early warning of Aumann (1965) retains its cautionary significance, and rigorous determination of the policy function is difficult[52].

Mitra-Wan are clear that their work is simply an application of the principles of the general theory of inter-temporal resource allocation, as developed by Radner (1961), Gale (1967), Brock (1970) and McKenzie (1968, 1976, 1983, and 1986) to a setting that models salient features of the economics of forestry. It is then somewhat of an irony that in their recent work on a canonical model in capital theory, Khan-Mitra follow the guidelines laid out in the 1981 analysis of the economics of forestry. The following sentences from McKenzie's (1983) introduction lay out the setting for a fuller appreciation of this point.

> Asymptotic theory for optimal paths of capital accumulation is more difficult when the utility function for the single period is concave, but not strictly concave. However, in the case of stationary models where future utility is not discounted, the theory is rather fully developed.[53] There is convergence to a subset of processes which span a flat on the epigraph of the utility function. This flat is often referred to as the von Neumann facet.
>
> In the case of discounted utility and quasi-stationary models ... we must use the convergence of the von Neumann facets associated with discount factors to the von Neumann facet of the undiscounted model as the discount factor approaches 1. Then, as before, it is possible to appeal to the stability properties of the optimal paths for the undiscounted case that lie on the von Neumann facet. We may prove that optimal paths

> are confined to smaller neighbourhoods of an optimal stationary path as the discount factor approaches 1 if the von Neumann facet for the undiscounted utility is stable, that is, contains no infinite cyclic paths.

What is remarkable in this statement is the reliance on the undiscounted case to yield insight into the properties of the discounted case. Thus even if one is interested primarily, if not only, in the discounted setting, and considers the undiscounted problem misguided for philosophical or ideological reasons, the analysis demands that attention nevertheless be paid to it. McKenzie puts the analysis of the discounted and undiscounted cases on the same table, so to speak, and this is of particular import, it seems to me, for a field which takes the scepticism of the magnitude of the discount factor as its cardinal tenet and an as an important marker of identity.[54] Indeed, a "folk theorem" for the general theory of inter-temporal resource allocation can be culled from McKenzie's (1983) statement. It is simply that for any dynamic problem falling within the rubric of the theory, there is a threshold discount factor such that the stability properties of thc optimal paths are qualitatively the same as those obtained for the undiscounted case for all discount factors above that threshold, and that complicated and rich dynamics, possibly including chaos, obtain for all discount factors below that threshold[55].

But now the direction for future work for the theory that I would like to indicate can be spelt out. It is not only a subscription to discrete time and to work within the rubric of the general theory of optimal resource allocation over time, as developed by Gale, Brock and McKenzie, but to be especially alert to the synthesis around the undiscounted case that the theory offers. Thus, in the specific context of the results obtained by Mitra and his coworkers, the next order of business is to integrate the undiscounted and discounted cases for both the economics of forestry as well as that of orchards; namely, to integrate Mitra and Wan (1985) and Mitra and Wan (1986), and to develop the undiscounted analysis for Mitra, Ray, and Roy (1991).

## 8. CONCLUSION

The concluding question then is what has been gained by putting my readers (and myself) through Kant's four principles of forestry economics, Laslett's intergenerational tricontract and intercohort trust, Bourdieu's social capital and an economy of practices, the Cowen-Parfit reworking of the argument against social discounting and the Mitra-Wan-Ray-Roy work in capital theory – all under one set of covers? More specifically, how does the direction for further work identified in the previous section depend on the preceding ones? The fact that each of these texts, and the subjects they textualize, can be usefully pursued in isolation is incontestable; what is at issue is the possibility of other productive directions in the interstices that become evident when these texts are read *together.* To put it another way, given that reasons of efficiency demand that an economist work with particular presuppositions in keeping with her comparative advantage, conform to a particular disciplinary idiom and standard of rigour, project her work into a particular subspace; are there other considerations, perhaps of efficiency in the much longer run, perhaps even of inter-temporal ethics, that demand the development of the subject in which its various facets and factors are examined not only in isolation, but rather also in a way that has the potential of mutual reinforcement and global insight? After specialization, when do the disciplines begin trade? This chapter, and the narrative it attempts to forge, is obviously tilted towards a particular answer to this question, and to the extent that this is justified, which is to say that this joint reading has provided, at least in part, a coherent and useful chapter, it is Section 5 on the 'correct' discount factor that provides the hinge between its two parts, between forest economics and inter-temporal ethics on the one hand, and between inter-temporal equity and capital theory on the other.

**Acknowledgements:** A preliminary version of this paper was prepared for presentation at an *International Conference on the Economics of Sustainable Forest Management* held on May 20-22, 2004 at the *University of Toronto.* I am grateful to Shashi Kant for his invitation, and for all his help in initiating a beginner into the intricacies of his subject. I am also grateful to Ellen Silbergeld for references and encouragement, and to *The Center for a Livable Future* for research support. Arnab Basu and Chris Metcalf both (independently) wanted a deeper comparative analysis, but they will have to wait for future work. Finally, my dependence on (continuing) invaluable discussions with Tapan Mitra goes beyond mere acknowledgement.

## NOTES

[1] See the second paragraph of Laslett (1987). My interest in Laslett also lies in ascertaining the meaning he gives to "questions of an ethical type."

[2] See the last section on summary and conclusions in Kant (2003b)

[3] See the introduction of Bourdieu (1983, pp.242-243); this defense of economics by a professional sociologist/anthropologist is of interest in itself.

[4] See the first paragraph of the section titled "Obstacles to the Understanding and Analysis of Justice over Time" in Laslett and Fishkin (1992, p 6). For a view from the community of mathematics, see Derbyshire (2004, Chapter 6) and the discussion of infinitesimals and of the 'infinitely large' in Halmos (1990).

[5] The uneasy relation between science and ethics, at least in the meaning that is conventionally given to both of these terms, is outside the scope of this essay. For a contribution around the time of the "founding" of so-called neoclassical economics, the reader can do worse than see Huxley's 1886 essay in Huxley (1894). For the author's subscription to Wittgenstein's views, see Khan (2003).

[6] This is a programmatic assertion in a 1992 volume devoted to "justice between age groups and generations"; see Laslett and Fishkin (1992, p. 11).

[7] In this section, all numbers in brackets refer to page numbers in Kant (2003a).

[8] In Kant (2003a, Footnote 18), there is a justification for this dichotomy that I leave to the expertise of others, though not without wondering how Heisenberg's uncertainty principle fits into the 'both-and' principle. Also see Kant (2003a, Footnote 1) and the references therein to the work of Dugger and Hamilton.

[9] However, I shall return to the 'both-and' principle in the sequel in the context of the discussion of Bourdieu's work in Section 4.

[10] This passage is quoted in Laslett (1979) and in (Laslett and Fishkin, 1992), and is of obvious importance to this essay. I shall keep returning to it in the form of allusions to a "deathless collectivity".

[11] For a more detailed elaboration of this phrase, see Khan (2003).

[12] For a reading of Wittgenstein's lecture from this point of view, see Khan (2003).

[13] These two sentences are taken from pages 18 and 21 of Rhees (1965). The number 6.422 refers to a particular paragraph in Wittgenstein's *Tractatus Logico-Philosophicus.* For a more detailed explication of Wittgenstein's absolute/relative distinction, see Khan (2003).

[14] See Keynes (1930, p. 365). For connecting this distinction to Wittgenstein, and to Hirsch's notion of "positional goods," see Khan (2004a).
[15] See Kant (2003a, Footnote 20). Since no quotes are given, I am assuming that this is Kant's own paraphrase.
[16] For a delineation of the "other" of neo-classical economics, and thereby its own delineation, see Endnote 8 and the references cited in the endnote.
[17] I leave an investigation of this connection to future work.
[18] For a more 'up-to-date' discussion, see Bird (1999). My motivation here, as in the rest of this essay, is to bring the issues into sharp relief without conceiving them to have been conclusively resolved one way or another.
[19] See Laslett and Fishkin (1992, p.8) for this quotation from Filmer's *Patriarchia.*
[20] In addition to Rawls (1999 and 2001), see Sandel (1998) and Mouffe 1993) and their references.
[21] See Hegel (1964, §156A) and Hegel(1964, p.263). For the reader without even a passing acquaintance with this work, a cursory perusal of its contents may be enlightening for its emphasis on ethics, and for its 'general equilibrium' sweep.
[22] See Laslett and Fishkin (1979, pp.3-4).
[23] Until indicated, all numbers in brackets refer to page numbers in Laslett (1979).
[24] I cannot help returning at this point to the quotation from Wittgenstein furnished in Section 2 above.
[25] For the dangers of reckless commodification, see, for example, Kant (2003b, Section 4.1) and Khan (2002).
[26] In Laslett and Fishkin (1992, pp.28-29), Laslett writes, "It is an absurdity to construe the attitudes and behavior of children, the procreated, with respect to their parents, the procreators, in the mode if-you-do-something-for-me-now-I-will-in-due-course-do-the-equivalent-for-you. This is particularly so for procreation itself, which is surely the greatest of the goodies generators offer to the generated."
[27] These refer to Laslett's metaphor of a "chain made out of hooks and eyes, where hooks all have to lie one way, and at the point where the chain stops a hook without an eye is always hanging forward (48). It sis permissible also to look upon these hook-eye linkages extending indefinitely into the future (49)."
[28] All numbers in brackets from now till the end of the section refer to Laslett and Fishkin (1992).
[29] The phrase is Wittgenstein's; see Rhees (1965, p.20). I shall refer to it further on in the sequel.
[30] For Laslett's difference between contract and agreement, see Laslett and Fishkin (1992, pp.32-33).
[31] Laslett writes "Aesthetic and religious principles might be invoked and the issues construed in different ways than expounded here (45)." He also mentions the possibility of a theory along sociobiological lines.
[32] See Endnotes 29 and 13 and the text they endnote.
[33] See Kant (2003a, p.48). In Section 2.2 of the same paper, he notes that "Ecological, aesthetic and spiritual values do not lend themselves to economic measurements (43)." For this connection to the dangers of commodification, also see Endnote 25 above.
[34] For this distinction, see Khan (2004b) and its references.
[35] In the rest of this section, all numbers in brackets refer to page numbers in Bourdieu (1983). I might note here that I find the neglect of this fundamental paper in current discussions of the "social capital" and of "social networks", as for example in Dasgupta-Sirageldin (2000), particularly egregious.
[36] Arrow (2000) singles out three aspects in any substantive discussion of capital: extension in time, deliberate sacrifice in the present for future benefits and alienability.
[37] This paragraph draws heavily on Solow (2000); and quotations are all Solow's words though used in ways different from his.
[38] See Bourdieu and Wacquant (1992, p.96). In this book, Bourdieu observes that "The question of the limits of a field is a very difficult one, if only because it is *always at stake in the field itself* and therefore admits of no *a priori* answer (100). *A capital does not exist and function except in relation to a field.* It confers a power over the field, over the materialized or embodied instruments of production or reproduction whose distribution constitutes the very structure of the field, and over the regularities and the rules which define the ordinary functioning of the field, and thereby over the profits engendered in it (101)." For a discussion of *habitus,* see pages 133-137.
[39] In this connection, see Endnotes 8 and 9 and the text they endnote.
[40] This paragraph draws heavily on Arrow (2000), and quotations are all his.
[41] Even though I read Solow's text as strongly complementary to that Arrow's, I do not see in it the same doubts as to a possible integration of sociology and economics. Whereas Arrow urges the "abandonment

of the metaphor of capital and the term, “social capital," seeing the measurement of “social interaction", and presumably thereby of the concept itself, as “a snare and a delusion", Solow’s only requirement seems to be the avoidance of “vague ideas and casual empiricism".

[42] These are of course basic problems of political theory; for one discussion of public agency, see Bird (1999, Chapter 3).

[43] In his pioneering 1928 paper on optimal economic growth, Ramsey (1928) emphasized that “we do not discount later enjoyments in comparison with earlier ones, a practice which is ethically indefensible and arises merely from the weakness of the imagination." In 1948, (Harrod (1948) went further than Ramsey: “A government ... capable of planning what is best for its subjects ... will pay no attention to pure time preference, a polite expression for rapacity and the conquest of reason by passion." For an extended discussion, see Koopmans (1965 and 1967).

[44] In essence, these go back to the quotation from Laslett that constitutes my fourth epigraph. The paragraph to follow can be complemented by Derbyshire (2004, Chapter 1) for a more intuitive understanding.

[45] The “other" of a positive model is what economic theory sees as a normative model. However, there are obvious presuppositions, if not difficulties, underlying this positive/normative distinction.

[46] For a close discussion and extension of Ramsey’s precise model, the classic references are Samuelson and Solow (1956), Samuelson (1965), and Koopmans (1965 and 1967).

[47] Its bears underscoring that the responsibility for this presentation lies solely with the author.

[48] The extent to which the work of Mitra and Wan goes beyond the pioneering work of Wan (1978 and 1989) and his references, deserves an essay on its own. In this connection, the reader may also want to see Kemp and Moore (1979) and Wan (1993, 1994).

[49] I mean an uncountable infinity as opposed to a countable infinity which, colloquially speaking, is also uncountable. This is related to Derbyshire’s illuminating distinction between counting logic versus measuring logic in Derbyshire (2004, pp.82-86).

[50] This reference is now Mitra and Wan (1986), and would presumably lead purists fixated on the subject of priority and acknowledgement to antedate Mitra and Wan (1986) to 1981.

[51] See his Footnote 3 in which he also gives a reference to Newman’s review of these papers.

[52] See Dasgupta-Mitra (undated), and Khan-Mitra (2002b, 2003b).

[53] In this connection, McKenzie references his papers McKenzie (1968 and 1976).

[54] Even though it has no sharp formulation of this issue, I think it important not to overlook Koopmans’work in this connection, as in (1965 and 1967).

[55] The substantiation of this program for the particular case of the RSS model is being conducted by Professor T. Mitra of Cornell and the author: for preliminary and partial results, the reader is referred to all of the papers referenced under Khan and Mitra. These are available on request.

# REFERENCES

Aghion, P., & Howitt, P. (1998). *Endogenous growth theory*, Cambridge: MIT Press.

Anderson, F. J. (1976). Control theory and the optimal forest rotation. *Forest Science*, 22, 242-246.

Arrow, K. J. (2000). Observations on social capital and economic performance. In P. Dasgupta, & I. Sirageldin. (Eds.). *Social capital: A multifaceted perspective* (pp. 6 -10). Washington, D. C.: The World Bank.

Arrow, K. J., & Kurz, M. (1970). *Public investment, the rate of return and optimal fiscal policy. Baltimore:* The Johns Hopkins Press.

Atsumi, H. (1965). Neoclassical growth and the efficient program of capital accumulation. *Review of Economic Studies*, 32, 127-136.

Aumann, R. J. (1965). Review of ‘optimal capital adjustment’ by K. J. Arrow. *Mathematical Reviews*, 28(3876)

Bird, C. (1999). *The myth of liberal individualism.* Cambridge: Cambridge University Press.

Bourdieu, P. (1983). Ökonomisches kapital, kulturelles kapital, soziales kapital. In K. Reinhard (Ed.), *Soziale ungleichheiten* (pp. 183-198). (Soziale Welt, Sonderheft 2) (Goettingen: Otto Schartz & Co.). English translation available as, Forms of capital, in J. G. Richardson (Ed.), *Handbook of theory and research for the sociology of education.* New York: Greenwood.

Bourdieu, P., & Wacquant, L.J.D. (1992). *An invitation to reflexive sociology.* Chicago: Chicago University Press.
Brock, W.A. (1970). On existence of weakly maximal programmes in a multi-sector economy. *Review of Economic Studies*, 37, 275-280.
Cowen, T. (1992). Consequentialism implies a zero discount rate. In P. Laslett, J.S., Fishkin (Eds.), *Justice between age groups and generation* (pp. 162-168). New Haven: Yale University Press.
Cowen, T., & Parfit, D. (1992). Against the social discount rate. In P. Laslett, J.S., Fishkin (Eds.), *Justice between age groups and generation* (pp. 144-161). New Haven: Yale University Press.
Dasgupta, P., & Sirageldin, I. (2000). *Social capital: A multifaceted perspective.* Washington, D. C.: The World Bank.
Dasgupta, S., & Mitra, T. (n. d.) *Optimal policy function in the forestry model: An example*. mimeo.
Debreu, G. (1959). *Theory of Value*, New York: John Wiley.
Derbyshire, J. (2004). *Prime obsession.* London: Penguin Books Ltd.
Faustmann, M. (1849). Translated into English as "Calculation of the value which forestry lands and immature stands possess for forestry. In M. Gane (Ed.), *Martin Faustmann and the evolution of discounted cash flow* (pp. 27-55). Oxford: Comonwealth Forestry Institute..
Gale, D. (1967). On optimal development in a multi-sector economy. *Review of Economic Studies*, 34, 1-18.
Halmos, P.R. (1990). Has progress in mathematics slowed down? *American Mathematical Monthly*, 97, 561-589.
Harrod, R.F. (1948). *Towards a dynamic economics*. London: MacMillan.
Haines, W.M. (1982). The psychoeconomics of human needs. *Journal of Behavioral Economics*, 11, 97-121.
Hegel, G.W.F. (1964). *Political writings*, translated by T. M. Knox. Oxford: Oxford University Press.
Hegel, G.W.F. (1967). *Elements of the philosophy of right.* Translated by T. M. Knox. Oxford: Oxford University Press.
Huxley, T.H. (1894). *Evolution and ethics.* London: Macmillan Publishing Company.
Kant, S. (1999). Endogenous rate of time preference, traditional communities and sustainable forest management. *Journal of Social and Economic Development*, II, 65-87.
Kant, S. (2000). A dynamic approach to forest regimes in developing economics. *Ecological Economics*, 32, 287-300.
Kant, S. (2003a). Extending the boundaries of forest economics. *Forest Policy and Economics*, 5, 39-56.
Kant, S. (2003b). Choices of ecosystem capital without discounting and prices. *Environmental Monitoring and Assessments,* 86, 105-127.
Kant, S., & Berry, A. (2001). A theoretical model of optimal forest resource regimes in developing economics. *Journal of Institutional and Theoretical Economics*, 157, 331-355.
Kemp, M.C., & Moore, E.G. (1979). Biological capital theory. *Economics Letters,* 4, 141-144.
Keynes, J.M. (1930). Economic possibilities for our grandchildren. In J.M. Keynes, (Ed.), *Essays in Persuasion* (pp. 358-373). New York: W. W. Norton and Co.
Khan, M.A. (2002). Trust as a commodity and on the grammar of trust. *Journal of Banking and Finance*, 26, 1719-1766. An erratum (publisher's mistakes) in 27 (2003) 773.
Khan, M.A. (2003). *On the ethics of (economic) theorizing.* Presented at an American Statistical Association panel on ethics at the JSM Meetings held on August 3-7, 2003 in San Francisco, California.
Khan, M.A. (2004a). Self-interest, self-deception and the ethics of commerce. *The Journal of Business Ethics,* 52, 189-206.
Khan, M.A. (2004b). Composite photography and statistical prejudice: Levy-Peart and Marshall on the theorist and the theorized. *European Journal of Political Economy*, 20, 23-30.
Khan, M.A., & Mitra, T. (2002a). On choice of technique in the Robinson-Solow-Srinivasan model. *International Journal of Economic Theory*, forthcoming.
Khan, M.A., & Mitra, T. (2002b). *Optimal growth in a two-sector model without discounting: A geometric investigation.* Institute of Economic Research, Kyoto University.
Khan, M.A., & Mitra, T. (2002c). *Optimal growth under irreversible investment: The von Neumann and Mckenzie facets.* Mimeo, Cornell University.
Khan, M.A., & Mitra, T. (2003a). *Optimal growth under irreversible investment: A continuous time analysis.* Mimeo, Cornell University.

Khan, M.A., & Mitra, T. (2003b). *Undiscounted optimal growth under irreversible investment: A synthesis of dynamic programing and the value-loss approach.* Mimeo, Cornell University.
Khan, M.A., & Mitra, T. (2004). *Topological chaos in the Robinson-Solow-Srinivasan model.* Mimeo, Cornell University.
Koopmans, T.C. (1965). On the concept of optimal economic growth. *Pontificae Academiae Scientiarum Scripta Varia*, 28, 225-300.
Koopmans, T.C. (1967). Inter-temporal distribution and "optimal" aggregate economic growth. In W. Fellner et al. (Eds.), *Ten economic studies in the tradition of Irving Fisher* (pp. 95-126). New York: John Wiley and Sons.
Laslett, P. (1979). The conversation between Generations. In P. Laslett, J.S. Fishkin (Eds.), *Philosophy, politics and society*, New Haven: Yale University Press.
Laslett, P. (1987). The character of family history, its limitations and the conditions for its proper pursuit. *Journal of Family History*, 12, 263-287.
Laslett, P. (1992). Is there a generational contract? In P. Laslett, J.S. Fishkin (Eds.), *Justice between age groups and generations* (pp 24-47). New Haven: Yale University Press.
Laslett, P., & Fishkin, J. S. (Eds.) (1979) *Philosophy, politics and society.* New Haven: Yale University Press.
Laslett, P., & Fishkin, J. S. (Eds.) (1992) *Justice between age groups and generations.* New Haven: Yale University Press.
Mouffe, C. (1993). *The return of the political.* London: Verso Press.
Majumdar, M., Mitra, T., & Nishimura, K. (2000). *Optimization and chaos.* Berlin: Springer-Verlag.
McKenzie, L.W. (1968). Accumulation programs of maximum utility and the von Neumann facet. In J.N. Wolfe (Ed.), *Value, capital and growth* (pp. 353-383). Edinburgh: Edinburgh University Press.
McKenzie, L.W. (1976). Turnpike theory, the 1974 Fisher-Schultz lecture. *Econometrica*, 43, 841-865.
McKenzie, L.W. (1983). Turnpike theory, discounted utility, and the von Neumann facet. *Journal of Economic Theory*, 30, 330-352.
McKenzie, L.W. (1986). Optimal economic growth, Turnpike theorems and comparative dynamics. In K.J. Arrow, M. Intrilligator (Eds.), *Handbook of mathematical economics*, Vol. 3 (pp. 1281-1355). New York: North-Holland Publishing Company.
Mitra, T., & Wan, H. Jr. (1985). Some theoretical results on the economics of forestry. *Review of Economic Studies*, 52, 263-282.
Mitra, T., & Wan, H. Jr. (1986). On the Faustmann solution to the forest management problem. *Journal of Economic Theory*, 40, 229-249.
Mitra, T., Ray, D., & Roy, R. (1991). The economics of orchards: An exercise in point-input flow-output capital theory. *Journal of Economic Theory*, 53, 12-50.
Mitra, T. (2000). Introduction to dynamic optimization theory. In M. Majumdar, T. Mitra, Nishimura, K. (Eds.), *Optimization and Chaos* (pp. 31-108). Berlin: Springer-Verlag.
Peart, S. J. (2000). Irrationality and inter-temporal choice in early neoclassical thought. *Canadian Journal of Economics,* 33, 175-189.
Radner, R. (1961). Paths of economic growth that are optimal with regard to final states. *Review of Economic Studies*, 28, 98-104.
Ramsey, F. (1928). A mathematical theory of savings. *Economic Journal*, 38, 543-559.
Rawls, J. (1999). *A theory of justice*, Cambridge: The Belknap Press.
Rawls, J. (2001). *Justice as fairness: A restatement.* Cambridge: The Belknap Press.
Reed, W.J. (1984). The effects of the risk of fire on the optimal rotation of a forest. *Journal of Environmental Economics and Management,* 11, 180-190.
Rhees, R. (1965). Some developments in Wittgenstein's view of ethics. *Philosophical Review*, 74, 17-26.
Samuelson, P.A. (1958). An exact consumption-loan model of interest, with or without the social contrivance of money. *Journal of Political Economy*, 66, 467-482.
Samuelson, P.A. (1965). A catenary turnpike theorem involving consumption and the golden rule. *American Economic Review*, 55, 486-496.
Samuelson, P.A. (1976). Economics of forestry in an evolving society. *Economic Inquiry*, 14, 466-492.
Samuelson, P.A., & Solow, R.M. (1956). A complete capital model involving heterogeneous capital goods. *Quarterly Journal of Economics*, 70, 537-562.
Sandel, M.J. (1998). *Liberalism and the limits of justice* (2nd Edition). Cambridge: Cambridge University Press.

Solow, R.M. (1974). Intergenerational equity and exhaustible resources. *Review of Economic Studies*, 41, Supplement, 29-35.

Solow, R.M. (2000). Notes on social capital and economic performance. In P. Dasgupta, I. Sirageldin (Eds.), *Social capital: A multifaceted perspective* (pp. 6-10). Washington, D. C.: The World Bank.

Stokey, N.L., & Lucas, R.E. Jr. (1989). *Recursive methods in economic dynamics*. Cambridge: Harvard University Press.

von Neumann, J. (1935-36). Über ein Ökonomisches Gleichungs-System und eine Verallgemeinerung des Brouwerschen Fixpunktsatzes. In K. Menger (Ed.), *Ergebnisse eines Mathematischen Kolloquiums*, No. 8. Translated as "A model of general economic equilibrium. *Review of Economic Studies*, 13, 1-9.

von Weizsäcker, C. C. (1965). Existence of optimal programs of accumulation for an infinite time horizon. *Review of Economic Studies,* 32, 85-104.

Wan Jr., H.Y. (1978). A generalized Wicksellian capital model: An application to forestry. In V. Smith (Ed.), *Economics of natural and environmental resources* (pp. 141-153). New York: Gordon-Breach.

Wan Jr., H.Y. (1989). Optimal evolution of tree-age distribution for a tree farm. In C. Castillo-Chavez et al. (Eds.), *Mathematical approaches to ecological and environmental problems* (pp. 82-99). Berlin: Springer-Verlag.

Wan Jr., H.Y. (1993). A note on boundary paths. In R. Becker et al. (Eds.), *General equilibrium, growth and trade II*, (pp. 411-426). New York: Academic Press.

Wan Jr., H.Y. (1994). Revisiting the Mitra-Wan tree farm. *International Economic Review*, 35, 193-198.

# CHAPTER 4

# POST-KEYNESIAN CONSUMER CHOICE THEORY FOR THE ECONOMICS OF SUSTAINABLE FOREST MANAGEMENT

MARC LAVOIE

*Department of Economics, University of Ottawa,*
*200 Wilbrod street, Ottawa Ontario, Canada, K1N 6N5*
*Email: mlavoie@uottawa.ca*

**Abstract.** Post-Keynesian economics is one of the many heterodox schools of thought in economics, such as the Marxist, Institutionalist and neo-Ricardian schools. Its members mainly deal with macroeconomic issues, but post-Keynesian economics also has a theory of the firm and a theory of consumer choice. As with most other heterodox variants of economics, post-Keynesian economics is based on four presuppositions: its epistemology is based on realism, its ontology is based on organicism, rationality is procedural, and the focus of analysis is production and growth issues. By contrast the symmetric presuppositions of neoclassical theory are: instrumentalism, atomism, hyper rationality, and the focus of analysis is exchange and the optimal allocation of existing resources.

Post-Keynesian consumer theory arises from a multitude of influences, including those of socio-economists, psychologists, marketing specialists, and individuals such as Herbert Simon and Georgescu-Roegen, who are or were fully aware of the complexity of our environment, as well as the disparate clues that were left by the founders of post-Keynesian theory, clues that turn out to be surprisingly consistent with each other. Post-Keynesian consumer theory can be said to be made up of seven principles: procedural rationality; the principle of satiation, the principle of separability, the principle of subordination, the principle of the growth of needs, the principle of non-independence and the heredity principle. These seven principles will be explained in the paper. A key consequence of these seven principles, in particular the principle of subordination, is that the utility index cannot be represented by a scalar anymore, but rather by a vector, and that the notions of gross substitution and trade-offs, which are so important for neoclassical economics, are brought down to a minor phenomenon, which only operate within narrow boundaries. This consumer theory does not rely on the Archimedes principle that "everything has a price". In particular, it is presumed that the principle of subordination, or hierarchy, is particularly relevant when dealing with moral issues, for instance questions of integrity, religion, or ecological issues.

Past work in ecological economics has shown indeed that a substantial proportion of individuals refuse to make trade-offs with material goods when biodiversity, wildlife, or forests are concerned. This has implications for contingency value analyses, based on willingness to pay or willingness to accept compensation, that attempt to take into account the non-market value of ecology or forestry preservation. The claim made here is that post-Keynesian consumer choice theory is highly relevant to forest economics, and could be used as a basis for consumer choice models in the economics of sustainable forest management.



*Kant and Berry (Eds.), Economics, Sustainability, and Natural Resources: Economics of Sustainable Forest Management*, 67-90.

*Messerchmann: combien voulez-vous pour partir sans le revoir?*
*Isabelle: Rien, Monsieur. Je ne comptais pas le revoir ....*
*Messerchmann: Je n'aime pas quand les choses sont gratuites, Mademoiselle.*
*Isabelle: Cela vous inquiète?*
*Messerchmann: Cela me paraît hors de prix .... Vous m'êtes très sympathique et je suis disposé à être très généreux avec vous. Combien voulez-vous?*
*Isabelle: Rien, Monsieur.*
*Messerchmann: C'est trop cher.*

*[Jean Anouilh, L'invitation au château, Éditions de la Table Ronde (Folio), Paris, 1972 [1951], acte IV, pp. 325-328]*

## 1. INTRODUCTION: POST-KEYNESIAN ECONOMICS AT LARGE

Post-Keynesian economics is one of the many heterodox schools of thought in economics, which stand in opposition to the mainstream in economics, the so-called neoclassical school. These heterodox schools include, among others, Marxists, Sraffians, Institutionalists, Evolutionarists, the Regulation school, Humanist and Social economists, and Feminist economists.

As with most other heterodox variants of economics, post-Keynesian economics is based on four presuppositions: its epistemology is based on realism, its ontology is based on organicism (or the possibility of the fallacy of composition), rationality is procedural, and the focus of analysis is production and growth issues. By contrast the symmetric presuppositions of the neoclassical school are: instrumentalism, atomism, hyper rationality, and the focus of analysis is exchange and the optimal allocation of existing resources. Some heterodox economists claim that the all-powerful presupposition is that of realistic abstraction.

It is sometimes said that the purpose of vision of post-Keynesian economics is that economics ought "to make the world a better place for ordinary men and women, to produce a more just and equitable society" (G.C. Harcourt, as cited by Dow 1990, p. 354). This being said, what are the distinguishing features of post-Keynesian economics, compared to those of other heterodox schools? I would argue that there are three: true uncertainty, historical time, and the importance of aggregate demand (See the recent surveys of Arestis (1996) and Palley (1996). Still relevant is the survey by Eichner and Kregel (1975)).

While all economists, or nearly all of them, would agree that economic systems are demand-led in the short-run, few of them would extend the preponderance of demand-led phenomena to the long-run. Rather, neoclassical authors, and Marxist ones for instance, would claim that in the long run supply-side factors dominate. This is not the point of view of post-Keynesians, who believe that effective demand rules even in the long-run, while supply-side factors will adjust. There is thus an infinity of possible growth rates, not just one, determined by population growth and exogenous technical progress, as in the Solow neoclassical growth model.

The multiplicity of equilibria, or the belief that models must be open-ended, is a characteristic feature of post-Keynesian economics. This is linked in particular to the importance which is given to true or fundamental uncertainty, also called Keynesian or Knightian uncertainty. Along with representatives from the Austrian school, post-Keynesians have long emphasized the need to distinguish between fundamental uncertainty and probabilistic risk. The future is uncertain, not only because we lack the ability to predict it, which is tied to epistemological uncertainty and procedural rationality, but also because of ontological uncertainty – the future itself is in the making and the decisions that we are to take will modify its course (Rosser, 2001). When private agents take decisions that affect them directly, fundamental uncertainty leads them to adopt a course of action that will generate safety. In public matters, such a course should also be followed, but is less likely to be followed because incentives are lacking. The precautionary principle in environment is clearly tied to fundamental uncertainty.

All this is also tied to historical time, in contrast to logical time. In historical time, time is not reversible, and actions cannot be easily reversed. The sequence of events is of fundamental importance. There is no optimal equilibrium out there, waiting to be achieved, whatever the route being taken. Post-Keynesian economics is very much influenced by the notion of irreversibility, path-dependence, lock-in effects, hysteresis, non-linearities. These notions, before they became fashionable and formalized, were very much described by earlier post-Keynesian economists such as Keynes, Kaldor or Minsky, under the term of non-equilibrium economics, and they can also be associated, as we shall see, with what Georgescu-Roegen called dynamic time.

In simpler post-Keynesian models, with only two equilibria rather than a multiplicity of them, the best equilibrium cannot usually be achieved through the usual market forces (Lavoie, 2001, p. 455). Indeed, market forces usually lead the model towards the low equilibrium. As a general statement, it could be said that post-Keynesian economists, in contrast to their neoclassical colleagues, are very much distrustful of the ability of the market price mechanism to solve most contemporary economic problems. This suspicion is based on a rejection, or at least a questioning, of the allocation role which is being attributed to prices by mainstream authors. Post-Keynesian economists doubt the general validity of the principle of substitution. These doubts arise in particular from the results achieved with the Cambridge capital controversies, but also from the observation that factors of production are generally complementary, rather than substitutable, and also from the observation that most successful activities are being pursued on the basis of cooperation and trust.

Following on the footsteps of their past leaders, Keynes and Kalecki, most post-Keynesians work on macroeconomic theories and issues, but post-Keynesian economics also has a theory of the firm and a theory of consumer choice, which will be the focus of this chapter. In the next section, I recall that the main post-Keynesian authors have a common vision of what an alternative consumer theory ought to look like, and I list the seven principles that should underwrite such a theory. The third section develops these seven principles. In the fourth section, I purport to show that these principles could be a basis for forest sustainable management, and that indeed several researchers working in the field of ecological economics have adopted these principles in the past. The fifth section illustrates this claim, by examining the issue of contingency valuation studies in light of the new principles of consumer behavior.

## 2. THE COHERENCE OF VIEWS ON POST-KEYNESIAN CONSUMER CHOICE THEORY

While post-Keynesians have spent a great deal of effort on macroeconomics and monetary issues as well as methodological issues, they have devoted less attention to microeconomics, seemingly avoiding in particular the subject of consumer choice. For instance, in the two guides on post-Keynesian economics, published at a twenty-year interval, there is no chapter devoted to consumer theory (Eichner, 1979; Holt & Pressman, 2001). However, despite its apparent neglect, there exists a Post-Keynesian theory of consumer choice, based on the indications left by the best-known and most productive post-Keynesian authors, such as Joan Robinson (1956, p. 251), Luigi Pasinctti (1981, p. 73), Edward Nell (1992, p. 396), Philip Arestis (1992, p. 124), Bertram Schefold (1997, p. 327). These indications on consumer choice show a great degree of coherence, and in my opinion they fit tightly with the rest of post-Keynesian theory. Indeed, Drakopoulos (1992b) goes so far as to argue that Keynes himself had in mind such a heterodox consumer choice theory.

Post-Keynesian consumer theory arises from a multitude of influences, including those of socio-economists, psychologists, marketing specialists, and individuals such as Herbert Simon and Georgescu-Roegen, who are or were fully aware of the complexity of our environment, as well as the disparate clues that were left by the founders of post-Keynesian theory, clues that turn out to be surprisingly consistent with each other.

The most detailed examination of a possible post-Keynesian consumer theory can be found in two books by Peter Earl (1983, 1986), and the motivations supplied above are quite apparent there. Other specific contributions to post-Keynesian consumer choice can be found in the works of Arrous (1978), Eichner (1987, ch. 9), Drakopoulos (1990, 1992a, 1994), Lavoie (1992, ch. 2), Lah and Sušjan (1999), and Gualerzi (2001), and a substantial amount of overlap with Earl's initial attempt at defining a specific post-Keynesian consumer choice vision is obvious. However, I consider that the first article that ought to be read to get a sense of what a post-Keynesian consumer theory implies is the one written in French by René Roy (1943) – an author unrelated to heterodox economics, although he was teaching in the same school as Nobel prize-winner Maurice Allais. As far as I know, Roy's contribution is first noted by Encarnaciòn (1964) in his short formalization of lexicographic choice.

In his little-known article, Roy (1943) puts forward several propositions that would seem to constitute the core of a post-Keynesian theory of consumer choice and that are quite compatible with the rest of post-Keynesian theory. For instance, Roy denies that the preferences of consumers (demand) explain the prices of consumer goods, thus rejecting the neoclassical view of value based on scarcity. He rejects the generalized use of indifference curves, for he believes that such a representation is not a realistic representation of human needs. He argues that goods can be, to some extent, separated into groups of goods with common features, substitution effects playing an important role within a group, but not in-between groups. In addition, he believes that these groups can be ordered in a hierarchy, with consumers moving from one group of goods to another, as their most urgent needs get progressively satiated and as their incomes rise. Variations in the prices of goods located in core groups (the basic needs) will have an impact on the demand for the goods of peripheral or discretionary groups ("luxury goods"), but by contrast variations in the prices of goods located in discretionary groups will have no impact on the demand for the goods of core groups.

As will become obvious, there is a tight link between Roy's views and the common ground of post-Keynesian consumer theory, as represented below under the form of seven principles (Lavoie, 1994; Drakopoulos, 1999). Most of the names of these heterodox principles of consumer behavior arise from the terms used by Georgescu-Roegen (1954, p. 514-5). Separability is taken from Lancaster (1991), while non-independence is taken from Galbraith (1958). The seven principles are:

i. The principle of procedural rationality;

ii. The principle of satiable needs;

iii. The principle of separability of needs;

iv. The principle of subordination of needs;

v. The principle of the growth of needs;

vi. The principle of non-independence.

vii. The principle of heredity.

The above seven principles will be briefly explained in the next section, although their names are by themselves evocative. In the meantime, the following quote by a leading post-Keynesian author clearly illustrate four of the above seven principles. The numbers in square brackets inserted within the quote refer to the above list.

> [v] ...Post-Keynesians generally assume that, in an economy that is expanding over time, it is the income effect that will predominate over the relative price, or substitution, effects.... [iii] Substitution can take place only within fairly narrow subcategories. [iv] Consumer preferences are, in this sense, lexicographically ordered... . [vi] A household's consumption pattern, at any given point in time, thus reflects the lifestyle of the households that constitute its social reference group. [Eichner 1986, p. 159-60].

A quick survey of the literature on environmental economics demonstrates that the more radical environmental economists have used the principles mentioned above in their effort to present a consumer choice theory that would be an alternative to the standard neoclassical model. The claim made here is that post-Keynesian consumer choice theory is highly relevant to forest economics, and could be used as a basis for consumer choice models in the economics of sustainable forest management.

## 3. THE PRINCIPLES OF POST-KEYNESIAN CONSUMER CHOICE

### *3.1 Procedural Rationality*

The principle of *procedural rationality*, as proposed by Herbert Simon (1976), asserts that agents have designed rules and procedures that allow them to reach decisions quickly and efficiently, despite an environment of imperfect knowledge and an overload of information. A lot of these rules are based on non-compensatory procedures, where only some elements or possibly a single one, are taken into consideration, provided they reach a certain threshold, so that they *satisfice* a given target, as in the so-called conjunctive and disjunctive rules, or in elimination by aspects. Rules are often based on a hierarchic design.

The fact that procedurally rational agents often do use compensatory procedures or do not behave as if they were approximating regression analysis or expected utility theory to arrive at their decisions does not mean that these agents are error-prone or suffer from some biases. Rather, as Gigerenzer (2000, ch. 8) has demonstrated, non-compensatory procedural rules can arrive at the right decision just as often, when such a decision exists, and much more efficiently than compensatory ones. In addition, as recalled by Dhar (1999), non-compensatory rules lead to more decisions than compensatory ones; when there is no decisive advantage, there is no choosing at all and the purchase decision is postponed.

These means and procedures include rules of thumb, the acceptance of social conventions, and reliance on the hopefully better informed opinion of others. Seen from the perspective of neoclassical substantive rationality, procedural rationality may seem to be *ad hocery*, but procedural responses are the only sensible answer to an environment characterized by bounded knowledge and computational capabilities, time constraints and fundamental uncertainty (in the sense of true, ontological, uncertainty). It could also be called the principle of *reasonable rationality* and is sometimes called *ecological rationality*. This involves "heuristics that are matched to particular environments [and that] allow agents to be ecologically rational, making adaptive decisions that combine accuracy with speed and frugality" (Todd & Gigerenzer, 2003, p. 148). The purpose of economics ought not to define an ideal consumer that would have all the nice mathematical properties that are required by an elegant theory; rather the purpose should be to define realistic behavior.

Non-compensatory rules are also sometimes followed when choices involve either certain outcomes or uncertain or probabilistic ones. Some agents will refuse under any circumstance to select an alternative that carries some unknown outcome when a certain alternative is available. "A sure alternative and a risk proposition, being relatively heterogeneous, can in no case be indifferent" (Georgescu-Roegen, 1954, p. 525). In particular, many business people may decide not to go ahead with any project, whatever the potential gains, if they fear that these projects might bring about bankruptcy with a high enough probability. These entrepreneurs will run for safety, in a non-compensatory fashion, despite the fact that expected utility theorists would declare them irrational, since their preference orderings could not be represented by any standard utility function (Blatt, 1979-80).

Instead of trying to demonstrate that x percent of subjects fail to behave in accordance with the standard neoclassical axioms of rationality, one should provide evidence describing actual behavior. Also, when designing policies, the behavior of agents should be modeled as is, rather than as it should be if the world were devoid of information limits and fundamental uncertainty.

### *3.2 Satiable Needs*

The second principle, that of *satiable needs*, can be likened to the neoclassical principle of diminishing marginal utility (or its non-satiable principle), but it takes a particular meaning in the Post-Keynesian theory of the consumer. Here satiation arises with positive prices and finite income. There are *threshold* levels of consumption beyond which a good, or its characteristics, may bring no satisfaction to its consumer. Beyond the threshold, no more of the good will be purchased, regardless of its price.

One has to carefully distinguish wants from needs, as do Lutz and Lux (1979). Following the pyramid of needs proposed by psychologist Abraham Maslow, and in line with the group classification proposed by Roy (1943), they argue that there is a hierarchy of needs, where some are more basic than others, which implies that they must be fulfilled in priority. In that sense all needs are not equal. Some needs are bound to be satiated much earlier than others. Needs are subject to a hierarchic classification and are the motor of consumer behavior. By contrast, wants evolve from needs. They can be substituted for each other and constitute "the various preferences within a common category or level of need" (Lutz & Lux, 1979, p. 21). This distinction will be useful in defining the next two principles of a post-Keynesian consumer theory.

### *3.3 Separability of Needs*

The principle of the *separability of needs* asserts that categories of needs or of expenditures can be distinguished from each other. In the case discussed by Lancaster (1972; 1991), with goods described by a matrix of consumption technology with various characteristics, a separate need will be associated with a sub-matrix of goods and characteristics arising out of a decomposable matrix. The principle of the separability of needs is illustrated by the widely-used econometric models of consumer demand, which assume that broad categories of expenditures enter separately into the overall utility function. In the utility-tree approach of Strotz (1957), the principle of separability is pushed one step further, since these broad categories of expenditures are further subdivided into several branches.

The separability of needs allows the consumer to divide the decision-making process into a series of smaller multi-stage decisions, and is consistent with the *hierarchic principle* designed by Simon (1962) to deal with complex systems or complex issues, and with mental accounts and *categorization*, as described by decision specialists (Henderson & Peterson, 1992). The consumer first makes an allocation of his budget among needs, and then spends that allocation among the various wants or subgroups of each need, independently of what happens for the other needs. Changes in the relative prices of goods within a given category of wants will have no effect on the budget allocation between various needs, while a fall in the overall price of a group of goods corresponding to a given need will have repercussions on the budget allocation of all needs. The principle of the separability of needs imposes substantial restrictions on the neoclassical principle of price substitution (without dismissing it), since separability limits severely the degree of substitutability between goods in different groups.

Indeed, a substantial amount of empirical evidence shows that general categories of consumption expenditures have low own-price elasticities and cross-elasticities. Eichner (1987, 656) points out that most of these elasticities are not significantly different from zero, and he argues on technical grounds that all coefficients (their absolute values) probably are an overestimate of the actual values.

### *3.4 The Subordination of Needs*

Further restraints may be added if one goes beyond the principle of separability of needs, by introducing a fourth principle, the principle of the *subordination of needs*. With this principle, utility cannot be represented by a unique catch-all utility measure; it can only be represented by a vector, and there is no continuity of preferences anymore. Gone is the so-called axiom of indifference, or what in modern microeconomics is known as the axiom of continuity. The principle of the subordination of needs is often associated with the notion of a *pyramid* of needs – a *hierarchy* of needs – as described by the humanistic school of psychology (Lutz & Lux, 1979), and as could be found in the works of Menger and Alfred Marshall. The integration of the principles of separability and subordination leads to Nicholas Georgescu-Roegen's (1954) principle of *irreducibility*. Needs can be ordered, but they are irreducible.

In the case of utility-tree analysis, the first-stage budgeting problem is now resolved by assuming that money is allocated first to necessities and then to discretionary needs. There is no substitution between the budget categories apportioned to necessary needs and discretionary ones. All the principles previously invoked culminate in this hierarchy: needs are separable and the most basic needs are first taken care of in their order of priority, until they are satiated at some threshold level. Several studies seem to offer some support to this principle of irreducibility, even when only material goods are concerned. For instance Johnson (1988) has shown that goods that have a small number of common attributes are more likely to be ranked lexicographically when purchasing decisions are taken, regardless of price changes; consumers were eliminating products on the basis of basic expenditures categories. Similarly, Sippel (1997, p. 1439) has found that "every subject showed a marked preference for some of the goods, while other goods were not chosen at all, even at low prices". Frequent substitution occurred, as one would expect, mainly in the case of goods fulfilling similar wants, such as Coke and orange juice. In general, subjects "violated the axioms of revealed preference", those based on the neoclassical theory of the utility-maximizing consumer subject to a budget constraint. Besides arguing that these consumers are error-prone or irrational, a way out of these results is to suppose that many of these consumers acted on the basis of the principles of the separability and the subordination of needs. Thus non-compensatory choices can be said to exist at two levels: within the principle of procedural rationality, as a convenient mean to simplify decisions; and within the principle of subordination, in a more fundamental way.

While strict lexicographic ordering is unlikely, more sophisticated lexicographic approaches have been suggested, with consumers setting targets and threshold, i.e., with the addition of the second principle of post-Keynesian consumer theory, that of satiation (Earl, 1986). These non-compensatory ordering schemes are not only reasonable but also compatible with procedural rationality, since a complete utility map is not required. Decisions about the most basic needs can be taken quite independently of the informational requirements of the higher needs. Consumers need know nothing whatsoever about the prices of the goods that are part of the higher needs, and they need not rank alternatives which they cannot attain or which are beyond their satiation levels (Drakopoulos, 1994).

Neoclassical authors deny that needs are subject to the principle of subordination. This, it must be presumed, is mainly due to the devastating consequences of the irreducibility of needs for neoclassical theory and its substitution principle. Irreducible needs imply that they are incommensurable and therefore that "everything does not have a price". A trade-off is not always possible. The axiom of Archimedes, so popular with choice theorists, does not hold any more (Earl, 1986, 249), and nor does the axiom of gross substitution (Eichner, 1987, p. 632), so often invoked among general equilibrium theorists.

A large number of economists have been tempted to associate Georgescu-Roegen's combined principle of the irreducibility of needs with Lancaster's approach to characteristics. Lancaster (1972, p. 154; 1991) himself has suggested such a move, which he calls *dominance*. Such a combination was in fact already provided by Ironmonger (1972). Among post-Keynesians, Arrous (1978, p. 277) and Lavoie (1992, p. 78-85) have provided an analysis of irreducible needs tied to groups of characteristics. Other post-Keynesians have proposed to adopt the radical form of Lancaster's analysis of characteristics, such as Pasinetti (1981, p. 75) and Nell (1992, p. 392).

A key consequence of the principle of subordination is that the utility index cannot be represented by a scalar anymore, but rather by a vector (Encarnaciòn, 1964; Fishburn, 1974). The notions of gross substitution and trade-offs, which are so important for neoclassical economics, are brought down to a minor phenomenon, which only operate within narrow boundaries. This consumer theory does not rely on the Archimedes principle that "everything has a price". In particular, it is presumed that the principle of subordination, or hierarchy, is particularly relevant when dealing with moral issues, for instance questions of integrity, religion, or ecological issues. Past work in ecological economics has shown indeed that a substantial proportion of individuals refuse to make trade-offs with material goods when biodiversity, wildlife, or forests are concerned. This has implications for cost-benefit analyses, based on willingness to pay or willingness to accept compensation, that attempt to take into account the non-market value of ecology or forestry preservation.

### *3.5 The Growth of Needs*

Having assumed that indeed there exists a hierarchy of needs, how do consumers move up the steps of the pyramid, from the core basic needs to the higher but more peripheral needs? The basic answer is that individuals move upwards in the hierarchy due to income effects (Joan Robinson says that consumers 'step down the hierarchy', meaning that basic or subsistence needs have top priority, while discretionary needs have lower priority!). Beyond the principle of satiation, lies the principle of the *growth of needs* – the fifth principle of post-Keynesian consumer choice.

When a need has been fulfilled, or more precisely when a threshold level for that need has been attained, individuals start attending to the needs which are situated on a higher plane. There are always new needs to be fulfilled. If they do not yet exist in the minds of consumers, they will be acquired through a learning process and industry will promote them through the creation of new goods (Gualerzi, 2001). Needs however often require income to be satisfied. To go from one level of need to another dictates an increase in the real income level of the individual. The fulfilment of new needs, and therefore the purchase of new goods or new services, is thus related to income effects. This is the microeconomic counterpart of the post-Keynesian focus on effective demand, that is, on macroeconomic income effects.

What is being asserted is that income effects are much more important in explaining the evolution of expenditure on goods than are substitution effects. The latter play only a minor role in a static analysis of consumer behavior, when similar goods or goods fulfilling the same wants are being considered. Indeed, changes in relative prices have an impact on budget allocation between needs only in so far as they have an impact on real income. This reflects the general lack of confidence in the principle of substitution. Post-Keynesian authors doubt that price is often a key determinant of purchasing decisions. As claimed by Arestis (1992, p. 124), "The post-Keynesian theory of household demand begins with the fundamental assumption that in an economic system it is the income effects rather than the substitution effects which are most important".

Another issue related to the growth of needs principle is that of material versus moral needs. This is emphasized by Lux and Lutz (1999) in their entry on the dual self, and in the work of Etzioni (1988). While many would still doubt the possibility of a lexicographic ordering in the realm of material goods, a large number of authors seem to agree that, unless one is a rational fool as Sen (1977) puts it, people will entertain lexicographic ordering when moral issues are at stake, and hence when moral issues are incommensurable with material goods. This is more likely when incomes have reached a high enough threshold.

### *3.6 Non-independence and Heredity*

The last two principles are the principle of *non-independence* and the *heredity* principle. The emphasis of traditional theory on substitution effects also has led to the neglect of the learning process in consumption theory. Do past choices modify preferences? How do consumers rank their new spending opportunities? How do they learn to spend their additional spending power?

Preferences are not all innate; they are acquired by experience and by imitation of the consumption pattern of friends or of people of higher ranks in the consumers' hierarchy. Consumers watch and copy other consumers. Fads leading to large sales of specific products reaction are thus explained by the informational content of consumption by neighbors, relatives, friends or acquaintances. The impact of socio-economic contact on purchases reinforces the belief that the composition of demand depends on socio-economic classes. Decisions and preferences are not made independently of those of other agents. A household's pattern of consumption will reflect the lifestyle of the other households that constitute its social reference group.

In addition, marketing officers, through publicity, will attempt to make sure that households follow the appropriate lifestyle (Hanson & Kysar, 1999a, 1999b). It has long been argued that private consumption is being inflated by salesmanship and advertising, and the same could be said in many countries about military expenditures for security reasons. Thus, for instance, taste creation through advertising could be prejudicial when this generates a damaged environment.

The term "principle of non-independence" must be attributed to John Kenneth Galbraith (1958), who called it the dependence effect. But of course it must be related to the large amount of socio-economic studies on conspicuous consumption and lifestyles inspired by Veblen, Duesenberry and Leibenstein, and their snob and bandwagon effects (Mason, 1998). Choices are dependent of the choice of others, and wealth and consumption relative to that of others is a key component of the degree of our satisfaction.

The principle of non-independence can also be interpreted in a second way. This would be what Georgescu-Roegen (1966, p. 176) calls the *heredity* principle. One of the better established fact in modern behavioral economics is that choices are not independent of the order in which they are made (or the method by which they are made), in contrast to the standard neoclassical assumption. The heredity principle incorporates historical time into choice theory. This implies that there is a kind of path dependence. Choices are reference dependent. There is hysteresis in choice, which might be inborn (not caused by advertising), but which arises out of experience.

## 4. THE PRINCIPLES APPLIED TO ECOLOGICAL ECONOMICS AND SUSTAINABLE FOREST MANAGEMENT

Despite the fact that post-Keynesian economists claim that their main purpose is to make the world a better place to live for the ordinary person, there have been few intrusions by post-Keynesian authors into the environmental forum, with the exception of authors such as Gowdy, Spash, or Rosser. This may be linked to the ideas of the Master himself, John Maynard Keynes, since Keynes (1973, p. 117) thought that our conclusions are often influenced "by a psychological trait ... a certain hoarding instinct, a readiness to be alarmed and excited by the idea of the exhaustion of resources". Keynes gave as an example the writings of Jevons, who, in a Malthusian mood, predicted the breakdown of the industrial revolution for lack of coal. Jevons himself hoarded such huge quantities of writing paper and brown packing paper that his children, more than fifty years after his death, still held piles of them. Keynes believed that these fears "omitted to make allowance for the progress of technical methods". The issue of coal, more than one hundred years ago, resembles that of oil today, and it may turn out to have a similar outcome, with substitutes eventually coming in. As to the issue of paper, as pointed out by Behan (1990, p. 13), "a scarcity of wood products has failed to materialized". Thus what should be of concern is not so much renewable products, or products that can be reproduced or for which substitutes can be found. Rather the concern must be about issues for which no substitute is possible: fresh air, landscapes, scenic beauty, wildlife habitat, the multiplicity of species, ethical concerns, and so on. Our concern is tied to the *intrinsic* value of these. As Bengston (1994, p. 524) points out, "they have a good of their own; they are not substitutable". For instance, some people value forests intrinsically. This, I believe, is what Kant (2003) means with sustainable forest management.

A quick survey of the literature on environmental economics demonstrates that the more radical environmental economists – ecological economists – have used all six principles mentioned above in their effort to present a consumer choice theory that would be an alternative to the standard neoclassical model. The claim made here is that post-Keynesian consumer choice theory is highly relevant to forest economics, and could be used as a basis for consumer choice models in the economics of sustainable forest management. Indeed the claim has been made by Kant (2003, 50) that sustainable forest management requires two elements, "multiple equilibria and new consumer choice theory". This new consumer choice theory would incorporate "context specific and dynamic preferences, heterogeneous agents, distinction between needs and wants, and subordination of needs" (ibid, p. 52). For this purpose, Kant (2003, p. 47) mentions ecological economists, socio-economists, humanistic economics and post-Keynesian economics.

Many of the themes evoked by post-Keynesian economists are ranked highly by ecological economists or those favoring sustainable forest management. First and foremost, there is the *precautionary* principle associated with *fundamental* uncertainty. When information is lacking, business people act prudently. They usually postpone taking decisions that might increase the probability of bankruptcy of their institution. The same principle should be applied to environmental issues. In doubt, no decision that increases the probability of an environment catastrophe should be taken. This concern with true uncertainty has been underlined by Vatn and Bromley (1995, p. 18), van den Bergh et al. (2000, p. 57), and Ravetz (1994-95).

Second, there is the *heredity* principle, or a variant of what I have called the principle of non-independence. Preferences are endogenous and context specific, as Kant (2003) would like them to be described by consumer theory. "Utility depends on past experience, the duration and intensity of past experience, and the length of time that has passed since the relevant experience took place" (Gowdy, 1993, p. 235). Habit formation can be seen as a particular case of path dependency (Zamagni, 1999, p. 117). In this framework, the theory of choice reflects the complexities of human nature rather than the mathematical requirements of tractability. As Crivelli (1993, p. 119) points out, "the longest standing invocation of hysteresis seems to be in the context of the theory of choice", and Georgescu Roegen's heredity principle is a case example of hysteresis. The path taken by consumers will have permanent effects on future choices. This linked to the other feature of the principle of non-independence, i.e., the fact that advertising and fads have an impact on the choices made by individual consumers, reinforces the arbitrary nature of consumer choices and the possibility of intransitive preferences and multiple equilibria. Indeed, Gowdy (1993, p. 235) claims that the heredity principle is tied to the large discrepancies that have been observed between willingness to pay and willingness to accept in contingent valuation studies. Gowdy argues that agents will be less likely to give up some environmental landscape that they have had the opportunity to experience.

A third theme which is common to both ecological economists and post-Keynesian economists is that of multidimensional choice. This point was made very early on by Bird (1982, p. 592), who argued that in contrast to neoclassical economics, "the choice between alternative environmental policies must necessarily therefore be made in more than one dimension". This theme is a recurrent one among the proponents of sustainable forest management. Bengston (1994, p. 523-5) for one claims that "the multidimensional or pluralist perspective maintains that held values cannot be reduced to a single dimension and that all objects cannot be assigned value on a single scale – values are inherently multidimensional". This is certainly an important feature of socio-economics (Etzioni, 1988), and it has been endorsed by several ecological economists such as Vatn and Bromley (1995, p. 9), who have called it the *incongruity* problem.

Martinez-Alier et al. (1998) make an interesting distinction. They represent the idea of multiple dimensions under the name of *weak comparability*. When there exists a common unit of measurement across plural values, usually a monetary one, one can speak of strong comparability. When such a common rod does not exist, they speak of weak comparability. The latter implies *incommensurability*. Martinez-Alier et al. (1998) claim that no value is superior to another. We are in a zone where no preference can be ascertained beforehand. Multi-criteria evaluation techniques must be brought in to disentangle conflictual and multidimensional elements of choice. The value that will actually appear to be dominant will depend on the characteristics of each individual case, where each possible alternative will be assessed on the basis of a multiplicity of criteria (cf. Bengston, 1994, p. 525). No algorithm or axiomatization of choice is possible under these conditions. All that matters is that the decision process itself be well defined. In fact, as argued by Gowdy and Mayumi (2001), choices over multiple dimensions are conducive to states where the agent is unable to choose. The so-called indifferent states of neoclassical analysis, when it comes to multidimensional issues, are better interpreted as alternatives that agents "cannot order without a great deal of hesitation or without some inconsistency" (Gowdy & Mayumi, 2001, p. 233; cf. Georgescu-Roegen, 1954, p. 522). Such choices generally will not be transitive. We may also wish to argue that a single individual may have several conflicting "souls", and hence may reach different judgments, depending on the point of view, or dimension, which is being favoured (Steedman & Krause, 1986).

It seems to me that the weak comparability criterion advocated by Martinez-Alier et al. (1998) and Gowdy and Mayumi (2001) is very similar to the post-Keynesian principle of the separability of needs that was presented in section 3. This principle severely restricts the substitution effects that could arise between elements that belong to different groups of needs, but it does not totally eliminate them. One could presume that multi-criteria decision techniques that rely on weak comparability would still entertain substitution effects. If monetary compensations are high enough, they will win the day. Consumers will be swayed by a high enough monetary tradeoff, even if they hesitate to do so. But several ecological economists have denied any role for substitution effects, at least in some circumstances for some categories of households. This is what Vatn and Bromley (1995) call "choices without prices without apologies".

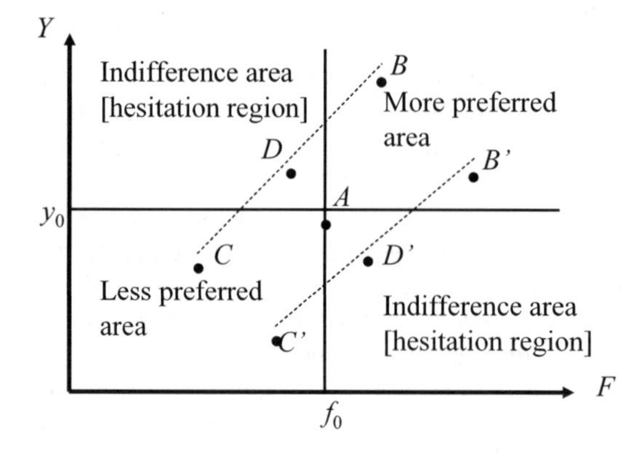


***Figure 4.1.*** *The Neoclassical Indifference Approach and the Hesitation Region*

Substitution effects are totally wiped out when lexicographic choices or choices of a lexicographic nature are entertained. This is tied to the post-Keynesian principle of the subordination of needs, or the irreducibility principle of Georgescu-Roegen. Lexicographic choices in the field of environment have been explicitly put forward by Edwards (1986), Stevens et al. (1991), Lockwood (1996), Spash and Hanley (1995), Spash (1998), van den Bergh et al. (2000), Gowdy and Mayumi (2001), and Kant (2003). The first five of these authors present a graphical representation of lexicographic choice, pointing out that it dismisses the neoclassical axiom of indifference, also called the Archimedes axiom or the axiom of gross substitution, which is so essential to price-based neoclassical environmental policies. These authors do not claim that all agents exhibit behavior based on choices of a lexicographic nature. Rather they argue that a substantial proportion of the population – sometimes called ethicists or altruists – exhibit such a behavior on matters tied to environment, and that neoclassical representations of these consumers are misleading, and lead to inadequate interpretation of surveys on the opinions of people about their environment. This applies in particular to the contingent valuation surveys.

The difference between standard neoclassical consumer analysis and the heterodox approach based on the separability and the subordination of needs, within the context of environmental issues, can be shown most clearly with the help of the following two diagrams, inspired by Spash (1998). Figure 4.1 illustrates standard neoclassical analysis, and possibly the principle of separability of needs with its associated region of hesitation. Income devoted to private goods is on the vertical axis, while the size of an old growth forest is represented on the horizontal axis, as in the example provided by Lockwood (1996, p. 88). Consumers are assumed to be choosing between up keeping a certain provision level of old growth forest on the one hand and the income amount which they can devote to private good consumption on the other hand. The former is called F and the latter is Y. Suppose that the starting situation is one where the size of the forest is $f_0$, while income level is $y_0$, which corresponds to point A. The plane can thus be divided into four quadrants, divided by the vertical and horizontal lines passing through the starting endowment. The north-east quadrant, including the two horizontal and vertical lines defining it, is an area that represents combinations of private consumption and forest size which are preferred, compared to bundle A. Symmetrically, the south-west quadrant, with its two line frontiers, represents an area of less preferred combinations, relative to A. On the other hand, the two remaining zones, the north-west and south-east quadrants, are areas of indifference. These are areas where some trade-off is assumed to be possible. It is possible to have more private consumption in exchange of a smaller forest, or some larger forest in exchange for a lesser amount of private consumption. The consumer is willing to make the trade-off, at some price, because, if the terms of the trade-off are high enough, the trade-off will keep constant the satisfaction (the utility) of the consumer.

In each of the two areas of indifference, there will be a multiplicity of combinations that will keep constant the utility of the consumer. This locus of points, along with combination A, will define the neoclassical indifference curve. What the neoclassical axiom of indifference says, now called the axiom of continuity, is that if there exists a combination B which is preferred to the starting bundle A, while there is another combination C which is less preferred to A, as shown in Figure 4.1, then there must exist a combination D on the segment linking B to C which is indifferent to the initial bundle A. This segment is shown by the dashed line in Figure 4.1. Another such dashed line illustrates the axiom of continuity in the other area of indifference, in the south-east quadrant, with bundles B', C', and D'. The neoclassical indifference curve would then go through the three points A, D and D'.

A first criticism of this indifference curve construction is that of Gowdy and Mayumi (2001, p. 232-4), as already outlined above. They assert that the two areas of indifference, when environmental issues are at stake, are instead areas of *hesitation*, which are likely to carry inconsistent and hence intransitive choices.

These are caused by the high level of fundamental uncertainty associated with environmental issues. Inconsistency is the symptom of the lack of information about the future, and it also reflects the inability and the reluctance of consumers to compare bundles that include weakly comparable components.

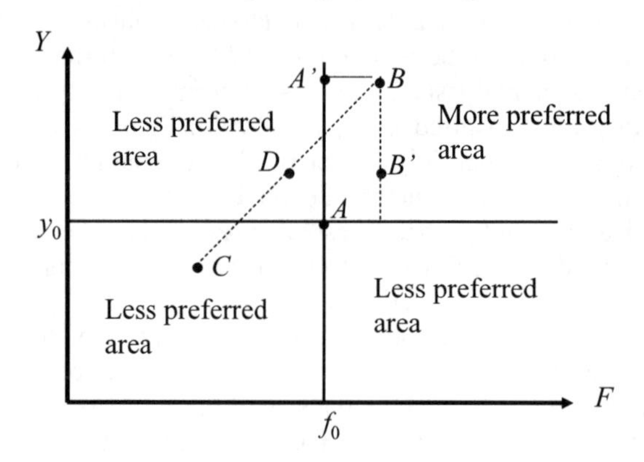


***Figure 4.2.*** *Choices of a Lexicographic Nature with Thresholds*

The second critique of this neoclassical indifference curve construction is that based on the principle of subordination, and its associated choices of a lexicographic nature. This is illustrated with Figure 4.2, inspired by Spash (1998, p. 53). Once more, the individual consumer is assumed to start from bundle A. Let us suppose that the achieved bundle constitutes the thresholds levels that must be minimally obtained for the individual to retain the present level of satisfaction. Any combination that provides an income inferior to $y_0$ would bring about a lower level of satisfaction, whatever the size of old growth forests. Symmetrically, any combination that would reduce the size of forests below $f_0$, whatever the amount of private consumption, would also lead to a lower level of satisfaction. On the other hand, provided the threshold level of income $y_0$ is attained, we presume that the primary determinant of the satisfaction of the consumer is the size of forest $f$. For instance, bundles B and B' on Figure 4.2 would always be preferred to bundle A or A'. Only with bundles providing equal forest size $f$ would the income level $y$ become a (secondary) determinant of the combination choice. For instance B would be preferred to B'. The plane is thus divided into two zones (plus point A). The north-east quadrant, with its horizontal and vertical frontiers, is the area of more preferred combinations relative to A. The other three quadrants are all areas of less preferred combinations relative to the initial bundle A.

Such an alternative consumer behavior does not fulfill the conditions of the axiom of continuity. As was done in Figure 4.1, we may draw in Figure 4.2 a dashed segment line connecting bundle B, which is preferred to A, and bundle C, which is less preferred than bundle A. However there does not exist any point D on this segment which corresponds to a bundle providing an amount of satisfaction which is equal to that of combination A. No combination of forest size and income level is indifferent to that of combination A. The axiom of continuity, or of indifference, does not hold anymore, because of the lexicographic nature of choices. This implies that the Archimedes axiom, according to which everything has a price, does not hold anymore either.

## 5. IMPLICATIONS FOR CONTINGENCY EVALUATION

As is well-known, within the standard neoclassical choice theory framework, the willingness to pay (WTP) and the willingness to accept (WTA) (or willingness to sell, WTS) are well defined measures of the Hicksian consumer surplus, which should be equal to each other (small income effects aside). Still, numerous studies have shown that WTA assessments largely exceed those of WTP. The discrepancy is easily a factor of three to ten (Knetsch, 1990, p. 228), and even a factor of 3 to 50 when environment issues are considered (Gowdy, 1993, p. 236). Lockwood (1996, p. 91) points out that these discrepancies are particularly large when there exists few substitutes for the good being valued, which is line with the distinction that we have made about the separability of needs.

Various explanations have been offered for this phenomenon. The first obvious one is the non-independence principle, more precisely the heredity principle, according to which we hold on more dearly to something which we already have than to something which we never got (Knetsch, 1990; Gowdy, 1993). The second explanation has to do with lexicographic ordering. Consumers might be willing to give up a limited amount of money to improve their environment; but they would demand an unlimited amount of compensation to accept a reduction of the same environment. In fact, they might be unwilling to trade for any reduction in the quality of their environment.

This brings to the fore the large number of zero or infinite bids, as well as refusals to bid, that are encountered in contingency valuation studies. Zero bids or refusals to bid are often interpreted as signaling no interest in improving or preserving the quality of environment. On the other hand, bids that appear absurdly high are waved off, on the basis that they cannot fit the neoclassical theory of the consumer surplus. These anomalous responses, however, are anomalous only within the strict neoclassical framework. As was first pointed out by Edwards (1986, p. 149), the willingness to sell will be undefined for agents that hold preferences of a lexicographic nature whenever their income exceeds their minimum standard of living. In that case, "an altruist committed to the welfare of wildlife and future generations is expected to protest against contingent markets when asked for minimum WTS by either refusing to bid, bidding zero dollars, or bidding an extremely high amount".

Some researchers have investigated these possibilities. Lockwood (1996, p. 99) concludes that his pilot study shows "that some individuals do have complex preference maps which include regions of lexicographic preference for the protection of native forests from logging". Stevens et al. (1991, p. 398) claim that most respondents gave answers that were inconsistent with both the neoclassical trade-off approach or the lexicographic theory. "However, 80 percent of the remainder gave responses that were consistent with lexicographic preference orderings". Spash and Hanley (1995) have investigated the motives behind zero bids. They found that nearly none of the zero bids were given for reasons of zero value. Rather, some participants to the study said that they could not afford to pay anything, while most zero-bidders claimed that ecosystem rights ought to be protected at all costs, and hence should be protected by law. This is consistent with Kahneman & Knetsch, 1992, p. 69), who claim that participants to contingency evaluation are bound to respond with indignation to questions about accepting more pollution over existing pristine landscapes, this indignation being expressed by "the rejection of the transaction as illegitimate, or by absurdly high bids".

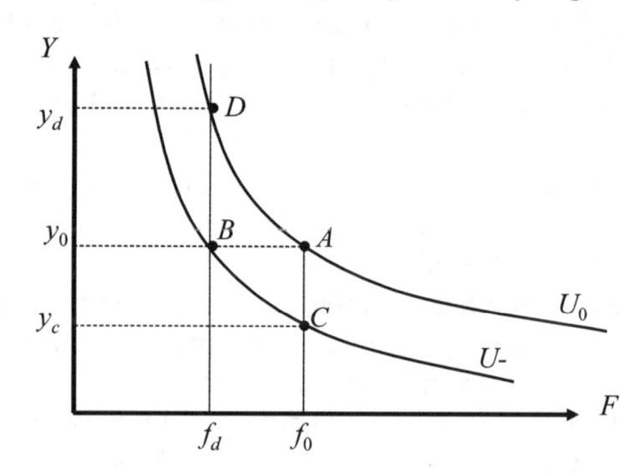


***Figure 4.3.*** *Neoclassical Contingency Value Assessment, with Indifference Curves*

Once again we can give a graphical illustration of these difficulties for neoclassical choice theory. As a basis for comparison, let us start with the illustration of the standard neoclassical case, with indifference curves. Let us assume once again that consumers are concerned with the income level that they can devote to private consumption as well as the size of old growth forest. Figure 4.3 is inspired by the graph provided by Edwards's (1986) pioneer article. Assume the existence of two well-behaved indifference curves, with the consumer being initially located at combination A on the $U_0$ utility indifference curve. Suppose the size of the forest is projected to be reduced from $f_0$ to $f_d$. As is well known, willingness to accept (WTA) is measured by the distance ($y_d$ - $y_0$). The consumer will be indifferent to combinations A and D. As a trade-off for the reduction ($f_0$ - $f_d$ ) in the size of a (presumably public) forest, the consumer is willing to accept a monetary compensation of ($y_d$ - $y_0$). Alternatively, if consumers need to pay to preserve the size of the forest, the consumer may either forsake part of the forest, in which case the person moves horizontally from combination A to combination B – on the lower indifference curve U- –, or the consumer may be willing to pay (WTP) an amount ($y_0$ - $y_c$ ) to retain the size of the forest at $f_0$, in which case consumers move down vertically from point A to point C, on the same lower indifference curve U-. With well-behaved indifference curves, WTP and WTA would be approximately equal, save for the decreasing marginal rate of substitution, as they are drawn in Figure 4.3.

Let us now examine the case of choices of a lexicographic nature. Let us take the simplest case, beyond pure lexicographic choice. Assume the primary element of choice, until income level $y^*$ is achieved, is the level of income. This means that, for any income level below $y^*$, the combination with the highest level of income will be preferred, regardless of the size of the forest. The secondary element of choice, the size of the forest $f$, plays a role only with combinations that feature equal levels of income. By contrast, once the threshold level of income $y^*$ is achieved (cf. Stevens et al. 1991, p. 398), the primary element of choice becomes the size of the forest, while private income reverts to a secondary element of choice, which plays a role only when combinations that feature equal forest sizes are being compared. This is the algebraic example proposed by Lockwood (1996, p. 89), and it corresponds to the graphical example provided by Edwards (1986, p. 148). Figure 4.4 illustrates this case.

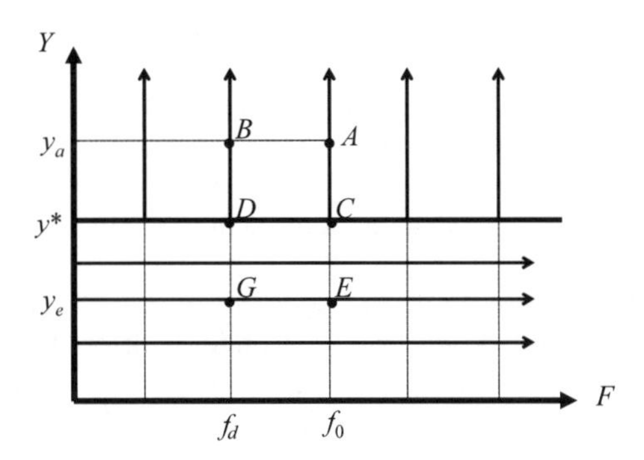


***Figure 4.4.*** *Contingency Value Assessment with Choices of a Lexicographic Nature: Quasi-indifference Curves*

In Figure 4.4, which illustrates the above preference framework of a lexicographic nature, there is not a single indifference curve. No two combinations carry equal satisfaction. Each point on this two-dimensional plane is ordered. The continuous lines with the arrows represent *quasi-indifference* lines, sometimes called *behavioral* curves (Lutz & Lux, 1979, p. 318). Below the level of income $y^*$, these quasi-indifference lines are horizontal, implying that the consumer prefers higher private consumption to lower private consumption, regardless of the size of old growth forests (D is preferred to E). The higher the horizontal quasi-indifference curve, the happier the consumer. However, for a given level of income, say $y_e$, the person prefers a bigger forest to a smaller one (E is preferred to G). This is what the arrows are meant to represent, and this is how these quasi-indifference curves are differentiated from the standard flat or vertical indifference curves that would represent addictive behavior.

When the threshold level of income $y^*$ has been attained, the size of forests becomes the primary ordering criterion. The quasi-indifference curves become vertical. The further to the right the quasi-indifference curve, the better off the consumer is (bundle C is preferred to B). But for a given level of forest size, say $f_0$, the higher the income level the higher the satisfaction of the consumer (bundle A is preferred to C), which is what the arrows on each vertical quasi-indifference curve once again are meant to indicate.

What are the implications of such a preference set for contingency evaluation studies? Assume the consumer starts with combination A, with an income exceeding the minimum threshold. Suppose this consumer is being asked about a possible reduction in forest size from $f_0$ to $f_d$. The likely willingness to pay (WTP) of this person will be $(y_a - y^*)$, that is the entire discretionary income of the consumer, beyond the threshold income level. The consumer would wind up at combination C. Note however that the consumer is not indifferent between combination C and combination B, as was presumed in the neoclassical analysis of Figure 4.3. In Figure 4.4, the consumer still prefers combination C to combination B. The measured WTP thus underestimates the true value of the forest in the consumer mind. Note in addition that whatever the proposed reduction in forest size, the income that can be given up remains the same, unless the reduction in forest size is so small that it does not trigger any negative feeling on the part of the consumer. On the other hand, if the consumer were to start with combination E, below the threshold level of income, WTP would be zero, or near zero, since more income is always preferred to less in this region.

What about the willingness to accept compensation (WTA)? Starting from the above-threshold combination A, the WTA is undefined, or it is infinite, since no amount of money will compensate for the loss of forest (Edwards, 1986, p. 148; Spash & Hanley, 1995, p. 193). Even an infinite amount of additional income would not procure enough compensation for the forest loss to keep constant the consumer's level of satisfaction. Any reduction in forest size causes a reduction in the satisfaction of the consumer, since forest size is the primary criterion of choice.

Choices of a lexicographic nature thus demonstrate that contingency valuation studies that solicit WTP and WTA estimates can arrive at widely different estimates. The use of one method, when the other should be more appropriate, is not a matter of indifference. In addition, the WTP estimate does not correctly reflect the willingness to trade of the consumer. As Lockwood (1996, p. 92) points out, "this sacrifice may not be regarded by the respondent as a transaction based on a free exchange, but as the payment of a ransom for recovery of a valued item. Ransom demands cannot be considered as Hicksian measures of economic welfare, because the person can never be indifferent between the value of the ransom paid and the value of the ransomed entity. The magnitude of the ransom is independent of the value of the entity, so the same payment may be offered for different quantity changes even though each increment in provision is valued". Given all this, it is not surprising that several people surveyed in contingency valuation studies "either refuse to participate in the survey, offer a protest response, try to play the game by inflating their response in an attempt to introduce their non-compensatory value into the process, or offer a WTP which is not a Hicksian measure of welfare change" (Lockwood, 1991, p. 91).

It would be possible to draw a wide variety of choices of a lexicographic nature. Lockwood (1996, p. 90-92) presents an algebraic example where consumers revert to choices based on indifference curves when thresholds for income level and forest size are achieved. This could be represented graphically, with the help of the apparatus developed in Figures 4.3 and 4.4. We could also assume, reciprocally, that compensating choices are made until thresholds are reached, at which point, consumers move on to non-compensatory choices. The principle of heredity could also easily be introduced, by assuming that consumers take as their new forest threshold the most recently experienced forest size.

The quote that was put at the beginning of the present paper illustrates these lexicographic choices. Many ethicists put in zero bids when what they really mean is that the environment is priceless. Any destruction of it should require an infinite amount of compensation. When, in Anouilh's theater play, Isabelle is asked by Messerchmann to leave the room and to express the amount of money that would be sufficient for her to do so, her bid response is zero. But this is a protest bid. Her preferences are lexicographic. Her departure cannot be purchased. The price that would really be required is an infinite amount of money. Messerchmann is not fooled. He understands lexicographic preferences. This is why he interjects, "It's too dear", when Isabelle answers that her willingness- to-accept price is zero – nothing.

> M: How much do you want to leave without seeing him again?
> I: Nothing, Sir. I did not intend to see him again.
> M: Miss, I don't like it when things are free.
> I: Do free things worry you?
> M: They seem priceless to me.... I find you very likeable and I am willing to be very generous to you. How much do you want?
> I: Nothing, Sir.
> M: It's too dear.

## 6. CONCLUSION

The work of Georgescu-Roegen very much inspired Post-Keynesian renditions of consumer theory. Georgescu-Roegen himself was very much concerned with environmental issues, and was one of the earlier economics writers on the topic. Georgescu-Roegen criticized neoclassical choice theory and consumer theory because he felt it lacked realism. He did not want a theory to be based on axioms that provided "a more convenient approach or lead to a simpler scheme"; rather he provided an alternative choice theory because he believed it offered "a more adequate interpretation of the structure of our wants" (Georgescu-Roegen, 1954, p. 519). This was certainly the case of his heredity principle, or the non-independence principle. Improved realism of consumer theory is also the justification for the rejection of the postulate of indifference, and its replacement by the principle of irreducibility, which we associated with a combination of the principles of the separability and the subordination of needs (Georgescu-Roegen, 1968, p. 263). Indeed, all seven principles of consumer choice that have been put forward in section 3 are designed to provide more realism into consumer theory. To develop a more realistic foundation for consumer choice also seems to be the goal of ecological economics (Gowdy & Mayumi, 2001, p. 234; van den Bergh et al., 2000, p. 44) and sustainable forest management (Kant, 2003, p. 40).

All theories require some degree of abstraction by necessity. Still there is a need for realism, specially in the realm of economics of sustainable forest management and environmental economics. I have shown that the principles of consumer behavior which have been put forward by post-Keynesian economists have already been endorsed or put to use by some specialists of ecological economics or sustainable forest management. These principles help to explain some conundrums in empirical work, notably in the contingency valuation studies, and they help to question the relevance or the adequacy of these studies. They also offer a way forward to make future choices on difficult public issues. Environmental or forest management policy theory must be based on proper consumer foundations, that will provide an appropriate agenda for environmental regulation.

## REFERENCES

Arestis, P. (1992). *The Post-Keynesian approach to economics*. Aldershot: Edward Elgar.

Arestis, P. (1996). Post-Keynesian economics: towards coherence, *Cambridge Journal of Economics, 20*, 111-135.

Arrous, J. (1978). *Imperfection de l'information, incertitude et concurrence*. Doctoral dissertation, Université de Strasbourg et Université des Sciences Sociales de Grenoble.

Behan, R.W. (1990). Multiresource forest management: a paradigmatic challenge to professional forestry, *Journal of Forestry*, April, 12-18.

Bengston, D.N. (1994). Changing forest values and ecosystem management, *Society and Natural Resources, 7*, 515-533.

Bird, J.W.N. (1982). Neoclassical and Post Keynesian environmental economics, *Journal of Post Keynesian Economics*, *4*, 586-593.

Blatt, J.M. (1979-80). The utility of being hanged on the gallows, *Journal of Post Keynesian Economics*, *2*, 231-239.

Crivelli, R. (1993). Hysteresis in the work of Nicholas Georgescu-Roegen. In J.C. Dragan, E.K. Seifert & M.C. Demetrescu (Eds.), *Entropy and Bioeconomics* (pp.107-129). Nagard: Milan.

Dhar, R. (1999). Choice deferral. In P.E. Earl and S. Kemp (Eds.), *The Elgar Companion to Consumer Research and Economic Psychology* (pp.75-81). Cheltenham: Edward Elgar.

Dow, S. (1990). Post-Keynesianism as political economy: a methodological discussion, *Review of Political Economy*, *2*, 345-358.

Drakopoulos, S.A. (1990). The implicit psychology of the theory of the rational consumer. *Australian Economic Papers*, *29*, 182-198.

Drakopoulos, S.A. (1992a). Psychological thresholds, demand and price rigidity. *Manchester School of Economics and Social Studies*, *40*, 152-168.

Drakopoulos, S.A. (1992b). Keynes' economic thought and the theory of consumer behaviour. *Scottish Journal of Political Economy*, *39*, 318-336.

Drakopoulos, S.A. (1994). Hierarchical choice in economics. *Journal of Economic Surveys*, *8* (2), 133-153.

Drakopoulos, S.A. (1999). Post-Keynesian choice theory. In P.A. O'Hara (Ed.), *Encyclopedia of Political Economy*, volume 2 (pp. 887-889). London: Routledge.

Earl, P.E.(1983). *The Economic Imagination: Towards a Behavioural Analysis of Choice*. Armonk: M.E. Sharpe.

Earl, P.E. (1986). *Lifestyle Economics: Consumer Behaviour in a Turbulent World*. Brighton, U.K.: Wheatsheaf.

Edwards, S. F. (1986). Ethical preferences and the assessment of existence values: does the neoclassical model fit? *Northeastern Journal of Agricultural and Resource Economics*, *15*, 145-150.

Eichner, A.S. (Ed.) (1979). *A Guide to Post-Keynesian Economics*. M.E. Sharpe: White Plains (N.Y.).

Eichner, A.S. (1986). *Toward a New Economics: Essays in Post-Keynesian and Institutionalist Theory*. London: Macmillan Publishing Company.

Eichner, A.S. (1987).*The Macrodynamics of Advanced Market Economies*. Armonk: M.E. Sharpe.

Eichner, A.S. & Kregel, J. (1975). An essay on post-Keynesian theory: a new paradigm in economics, *Journal of Economic Literature, 13*, 1293-1311.

Encarnaciòn, J. (1964). A note on lexicographic preferences, *Econometrica, 14*, 215-217.

Etzioni, A. (1988). *The Moral Dimension: Toward a New Economics*. New York: The Free Press.

Fishburn, P.C. (1974). Lexicographic orders, utilities and decision rules: a survey, *Management Science, 20*, 1442-1471.

Galbraith, J.K. (1958). *The Affluent Society*. Hamish Hamilton: London.

Georgescu-Roegen, N. (1954). Choice, expectations and measurability, *Quarterly Journal of Economics, 48*, 503-534.

Georgescu-Roegen, N. (1966). *Analytical Economics*. Cambridge: Harvard University Press.

Georgescu-Roegen, N. (1968). Utility. In D.L. Sills (Ed.), *International Encyclopedia of the Social Sciences*, volume 16 (pp. 236-267). London: Macmillan Publishing Company.

Gigerenzer, G. (2000). *Adaptive Thinking: Rationality in the Real World*. Oxford: Oxford University Press.

Gowdy, J.M. (1993). Georgescu-Roegen's utility theory applied to environmental economics. In J.C. Dragan, E.K. Seifert & M.C. Demetrescu (Eds.), *Entropy and Bioeconomics (*pp. 230-240). Nagard: Milan.

Gowdy, J.M. & Mayumi, K. (2001).Reformulating the foundation of consumer choice theory and environmental valuation. *Ecological Economics, 39* (2), 223-237.

Gualerzi, D. (2001). *Consumption and Growth: Recovery and Structural Change in the US Economy*. Cheltenham: Edward Elgar.

Hanson, J.D., & Kysar, D.A. (1999a). Taking behavioralism seriously: the problem of market manipulation, *New York University Law Review*, 74, 630-749.

Hanson, J.D., & Kysar, D.A. (1999b). Taking behavioralism seriously: some evidence of market manipulation, *Harvard Law Review*, 112, 1420-1572.

Henderson, P.W., & Peterson, R.A. (1992). Mental accounting and categorization, *Organizational Behavior and Human Decision Processes,* 51, 92-117.

Holt, R.P.F., & Pressman, S. (Eds.) (2001). *A New Guide to Post Keynesian Economics*. London: Routledge.

Ironmonger, D.S. (1972). *New Commodities and Consumer Behaviour*. Cambridge: Cambridge University Press.

Johnson, M.D. (1988). Comparability and hierarchical processing in multialternative choice. *Journal of Consumer Research, 15* (3), 303-314.

Kahneman, D., & Knetsch, JL. (1992). Valuing public goods: the purchase of moral satisfaction, *Journal of Environmental Economics and Management, 22*, 57-70.

Kant, S. (2003). Extending the boundaries of forest economics, *Forest Policy and Economics, 5*, 39-56.

Keynes, J.M. (1973). *The Collected Writings of John Maynard Keynes: Essays in Biography,* volume 10. London: Macmillan Publishing Company.

Knetsch, J.L. (1990). Environmental policy implications of disparities between willingness to pay and compensation demanded measures of values, *Journal of Environmental Economics and Management, 18*, 227-237.

Lah, M., & Sušjan, A. (1999). Rationality of transitional consumers: a Post Keynesian view. *Journal of Post Keynesian Economics, 21* (4), 589-602.

Lancaster, K. (1972). *Consumer Demand: A New Approach*. New York: Columbia University Press.

Lancaster, K. (1991). Hierarchies in goods-characteristics analysis. In K. Lancaster, *Modern Consumer Theory* (pp.69-80). Aldershot: Edward Elgar

Lavoie, M. (1992). *Foundations of Post-Keynesian Economic Analysis*. Aldershot: Edward Elgar.

Lavoie, M. (1994). A Post Keynesian theory of consumer choice. *Journal of Post Keynesian Economics, 16* (4), 539-562.

Lavoie, M. (2001). Efficiency wages in Kaleckian models of employment. *Journal of Post Keynesian Economics, 23*, 449-464.

Lockwood, M. (1996). Non-compensatory preference structures in non-market valuation of natural area policy. *Australian Journal of Agricultural Economics, 40*, 2, 85-101.

Lutz, M.A., & Lux, K. (1979). *The Challenge of Humanistic Economics*. Menlo Park: Benjamin/ Cummings.

Lux, K., & Lutz, M.A. (1999). Dual self. In P.E. Earl and S. Kemp (Eds.), *The Elgar Companion to Consumer Research and Economic Psychology* (pp. 164-169). Cheltenham: Edward Elgar

Martinez-Alier, J., Munda, G., & O'Neill, J. (1998). Weak comparability of values as a foundation for ecological economics, *Ecological Economics*, *26*, 277-286.

Mason, R. (1998). *The Economics of Conspicuous Consumption*. Cheltenham: Edward Elgar.

Nell, E.J. (1992). Demand, pricing and investment. In E.J. Nell, *Transformational Growth and Effective Demand: Economics After the Capital Critique* (pp. 381-451). London: Macmillan Publishing Company.

Palley, T.I. (1996). *Post Keynesian Economics: Debt, Distribution and the Macro Economy*. New York: St. Martin's Press.

Pasinetti, L.L. (1981). *Structural Change and Economic Growth*. Cambridge: Cambridge University Press.

Ravetz, J. (1994-95). Economics as an elite folk science. *Journal of Post Keynesian Economics*, *17*, 165-184.

Robinson, J. (1956). *The Accumulation of Capital*. London: Macmillan Publishing Company.

Rosser, J.B. (2001). Uncertainty and expectations. In R.P.F. Holt and S. Pressman (Eds.), *A New Guide to Post Keynesian Economics (pp.52-64)*. London: Routledge.

Roy, R. (1943). La hiérarchie des besoins et la notion de groupes dans l'économie de choix, *Econometrica*, *11*, 13-24.

Schefold, B. (1997). *Normal prices, technical change and accumulation.* London: Macmillan Publishing Company.

Sen, A.K. (1977). Rational fools: A critique of the behavioral foundations of economic theory. *Philosophy and Public Affairs*, *6* (4), 317-344.

Simon, H.A. (1962). The architecture of complexity. *Proceedings of the American Philosophical Society*, *106* (6), 467-482.

Simon, H.A. (1976). From substantive to procedural rationality. In S.J. Latsis (Ed.), *Method and Appraisal in Economics*. Cambridge: Cambridge University Press.

Sippel, R. (1997). An experiment on the pure theory of consumer's behaviour. *Economic Journal*, *107*, 1431-1444.

Spash, C.L. (1998). Investigating individual motives for environmental action: lexicographic preferences, beliefs, attitudes. In J. Lemons, L. Westra & R. Goodland (Eds.), *Ecological Sustainability and Integrity: Concepts and Approaches* (pp. 46-62). Dordrecht/Boston/London.: Kluwer Academic Publishers.

Spash, C.L., & Hanley, N. (1995). Preferences, information and biodiversity preservation. *Ecological Economics*, *12*, 191-208.

Steedman, I., & Krause, U. (1986) Goethe's *Faust*, Arrow's possibility theorem and the individual decision-taker. In J. Elster (Ed.), *The Multiple Self* (pp. 197-231). Cambridge: Cambridge University Press.

Stevens, T.H., Echeverria, J., Glass, R.J., Hager, T., & More, T.A. (1991). Measuring the existence value of wildlife: what do CVM estimates really show?, *Land Economics*, *67*, 390-400.

Strotz, R.H. (1957). The empirical implications of a utility tree. *Econometrica*, *25*, 269-280.

Todd, P.M., & Gigerenzer, G. (2003). Bounding rationality to the world. *Journal of Economic Psychology*, *24*, 143-165.

van den Bergh, J.C.J.M., Ferrer-i-Carbonell, A., & Munda, G. (2000). Alternative models of individual behaviour and implications for environmental policy, *Ecological Economics*, *32*, 43-61.

Vatn, A., & Bromley, D.W. (1995). Choices without prices without apologies. In D.W. Bromley (Ed.), *The Handbook of Environmental Economics (pp.1-25)*. Oxford: Basil Blackwell

Zamagni, S. (1999). Georgescu-Roegen on consumer theory: an assessment. In Mayumi, K. and Gowdy, J.M. (Eds.), *Bioeconomics and Sustainability: Essays in Honor of Nicholas Georgescu-Roegen* (pp. 103-124). Cheltenham: Edward Elgar.

# CHAPTER 5

# BEHAVIORAL ECONOMICS AND SUSTAINABLE FOREST MANAGEMENT

JACK L. KNETSCH

*Department of Economics, Simon Fraser University,*
*8888 University Drive, Burnaby, B. C., Canada, V5A 1S6.*
*Email: knetsch@sfu.ca*

**Abstract:** Taking account of recent findings that, for example, people value losses more than otherwise commensurate gains, discount future losses at lower rates than future gains, and tend to make choices on the basis of mental accounts, could markedly improve the guidance offered by economic analyses of forest management options. Asymmetrical incentives and restraints facing individuals and organizations favour continued use of earlier views of standard economic assumptions and such evidence is now largely ignored as are issues such as the appropriate choice of measure to use in valuing the various gains and losses being traded off in managing forest lands.

## 1. INTRODUCTION

Forest management decisions overwhelmingly involve tradeoffs – output A vs. output B, gains to some vs. losses to others, consumption in the near term vs. consumption later. In an effort to make more informed decisions, individuals weigh the alternatives – more formally in benefit-cost analyses, quantification of damages, and impact assessments; and less formally in the ways people think about problems that generate resistance to changes or support for interventions to bring them about.

The analyses of problems, the design of policies to deal with them, and, in particular, the valuations of alternatives are, in practice, based on the dictates of standard economic theory – the assumptions and principles displayed in textbooks and reflected in organization manuals and method and procedural guidelines. However, recent findings from behavior economics research are providing a more informed view of both the principles and assumptions underlying valuations and the values often at issue in resource management decisions. While still largely ignored, these findings offer potentials for better understanding current tradeoffs and greatly improving management decisions.



*Kant and Berry (Eds.), Economics, Sustainability, and Natural Resources: Economics of Sustainable Forest Management,* 91-103.

## 2. WEIGHING TRADEOFFS

Sustainable forest management calls, in large part, for taking account of a wider array of values and uses in forest land management decisions by paying attention not just to timber production, or even to maintaining timber production at some sustained level into the indefinite future, but to other resource values as well: wildlife habitat, soil conservation, forest foods, water retention, carbon repository, biological diversity, aesthetic qualities, recreational opportunities, employment creation, and sense of place.

However, any moves towards taking account of a wider array of uses and values in management and policy decisions will create more than proportionate demands for tradeoffs among them – increases in some may lead to increases in others, but the eventual rule is one of compromises and tradeoffs. This raises problems of identification and quantification, and of weighing or valuation.

All too often little is known about the joint production functions for multiple uses of forest land, so that identification and quantification of the consequences of alternative management practices is not an easy matter (for example, Nautiyal & Rezende, 1985). Further, many of the costs and benefits stemming from forest land management are non-pecuniary in nature, making comparable valuations more difficult, and in some cases problematic at best.

Many, and probably most, of the issues of sustainability are ones for which economic analyses can provide useful guidance and insight, although some are clearly not the exclusive concern of economics. To the extent that economics has been, and continues to be, used in the analysis of sustainable forest management issues, it is economics of a traditional kind. Analysts, and people writing manuals for the guidance of analysts, continue to be regularly admonished – usually by economists – to follow the maxims of standard economic theory: "A core set of economic assumptions should be used in calculating benefits and costs" (Arrow, et al., 1996, p. 222).; "A failure to satisfy the requirements of economic theory would suggest that the appropriate preferences were not being measured" (Diamond, 1996, p. 346). However well-intentioned, and however appropriate this insistence might be for discouraging some of the more egregious misrepresentations of costs and benefits of management options, these exhortations generally ill-serve the cause of more informed decisions in their implicit dismissal of the wealth of empirical findings from recent, and not so recent, behavior studies.

## 3. BEHAVIOR ECONOMICS

The award of a share of the 2002 Nobel Prize for Economics to Daniel Kahneman, a psychologist, for "having integrated insights from psychological research into economic science…", is recognition of both the progress made in this sub-field and the potential benefits of the findings. In many cases these findings provide a more realistic view of people's choices and economic behavior than is available from the standard theory that forms the basis for current economic practice and analyses. The often observed differences between behavior findings and standard theory are far more than random deviations from an expected outcome; they are, instead, systematic and often large. Some are the result of the bounded rationality due to human computational and cognitive limitations, but many – and those of most interest – reflect real preferences that are not well modeled by the axioms of standard theory.

For example, rather than treating their monetary wealth as perfectly fungible, or substitutable between different holdings, people often organize their decisions and choices in terms of separate mental accounts or budgets (Thaler, 1999). Even though they plan to draw on both during their retirement years, they treat investments in their retirement fund differently than they do those in other investment accounts. Another interesting example of the strong motivation provided by mental accounting (Camerer, Babcock, Loewenstein, & Thaler, 1997) is the economically curious, and costly, behavior of New York taxi drivers who quit early on busy days and work longer hours on slow days. The reverse would, of course, allow them to earn more in less time over the year. Rather than the maximizing behavior presumably prescribed by standard theory, the cab drivers appear to set daily income targets and drive until they reach them, even though this results in more unproductive time and less productive time. It also imposes a social cost by having a smaller number of taxis available when demand is high and a larger number when demand is low.

Such mental accounting is likely to give rise to much greater restrictions on people's willingness to substitute and trade off one forest output for another than is anticipated by the postulates of standard theory, or by many forest management proposals. It may also account for at least some of the implied preferences for so-called hard sustainability, a strategy that calls for less substitutability among resource outputs, over a course of soft sustainability which allows for a greater accommodation of substituting gains in the productivity of one resource for losses in the productivity of another.

People also commonly give greater weight to changes that insure certainty, than they do to equal probability changes that do not offer this assurance – the difference between probabilities of 0.99 and 1.00, or between 0.01 and 0.00 are much more important than between, say, 0.45 and 0.46 (Kahneman & Tversky, 1995). While usually not taken into account in traditional risk analysis, or in management decisions, this certainty effect often exerts a strong hold over people's preferences and choices. There is often a great demand for certainty even when certainty is not, nor can it be, on offer, and individuals will go to great lengths to avoid otherwise beneficial actions that carry what are seemingly even the most remote possibilities of downside risks.

## 4. THE VALUATION OF GAINS AND LOSSES

The divergent views of people's valuation of positive and negative changes probably best exemplifies the difference between the directions and suggestions based on standard theory and those based on the empirical evidence from behavior studies. This valuation disparity is also likely the most important and greatest cause for concern, and is therefore used here to illustrate the potential benefits of a greater acknowledgment and consideration of behavior economic findings.

### *4.1 Measures of Value*

A weighing of alternatives and of gains and losses is at the heart of much of the contribution of economics to policy design and management decisions. To deal with such issues, economists and policy analysts focus much of their attention on how much people are willing to sacrifice to secure gains, to mitigate losses, and to avoid present and future problems. This has led to a fairly vast literature on valuation methods and estimates, and to a continuing supply of numbers for benefit-cost analyses and feasibility studies for all manner of proposals. Much good has come of this, not the least of which is a far wider appreciation that economic values include non-market, or non-pecuniary, as well as market returns – that many environmental and preservation returns are equally of economic value as those from timber production..

However, in much of this activity the choice of the particular measure used to assess the economic value of gains and losses has been largely overlooked in favour of easy assumptions and conformity with what has gone before.

There has long been, and continues to be, general agreement among economic analysts that an action or change is considered to be socially beneficial if the gains to those made better off exceed the losses to those adversely affected. This is normally taken to imply that the sums gainers are willing to pay for the gains are sufficient to compensate the losers for their losses – the common interpretation of the potential-Pareto criterion. Accordingly, economists have suggested that the economic value of gains and losses needs to be assessed with different and particular measures: "benefits are measured by the total number of dollars which prospective gainers would be willing to pay to secure adoption, and losses are measured by the total number of dollars which prospective losers would insist on as the price of adoption" (Michelman, 1967, p. 1214).

While valuations of gains and losses call for different measures, the assumption of standard theory, and consequently of economic practice, is that the maximum sum people would be willing to pay (WTP) to gain an entitlement is, except for a normally trivial difference due to an income effect, equal to the minimum sum they would be willing to accept (WTA) to give it up – "…there is no basis consistent with economic assumptions and empirical income effects for WTP and WTA to exhibit sizable differences" (Diamond, Hausman, Leonard, & Denning 1993, p. 66). This remains the empirical assertion of choice, and is seldom questioned by economic analysts.

The empirical evidence is, of course, sharply at variance with the conventional assumption of equivalence between the WTP and WTA measures of economic value. The findings – which have been reported in *all* of the leading economics journals, and those of related fields, for over two decades – suggest that people value losses from two to over four times more than otherwise fully commensurate gains.

Consistent evidence of this reference, or endowment, effect has come from a wide range of studies: surveys, replicated real exchange experiments, and recordings of the choices made by individuals in non-experimental decisions (reviewed in, for example, Samuelson & Zeckhauser, 1988; Kahneman, Knetsch, & Thaler, 1991; and Rabin, 1998). In one experimental group, for example, individuals were willing to pay, on average, $5.60 for a 50 percent chance to win $20.When asked to give up the identical chance to win the same $20 prize, however, those *same* individuals demanded an average of $10.87 (Kachelmeier & Shehata, 1992).

Investors making real portfolio choices also demonstrate a greater sensitivity to losses through their reluctance to realize a loss by selling. This reluctance not only leads to smaller volumes of sales of securities that have declined in price relative to those for which prices have increased (Shefrin & Statman, 1985), but to investors earning substantially lower returns because they replace their winning stocks more often than they do ones with current prices below acquisition prices (Odean, 1998).

In another study of people's actual economic behavior, a significant difference was found in their reactions to price changes. As people value losses more than gains, they were also more sensitive to price increases, which impose a loss, than to price decreases, which provide a benefit. This asymmetry was tested for egg purchases, and resulted in a price elasticity of –1.10 for price increases and only –0.45 for price decreases (Putler, 1992).

In yet another persuasive demonstration of the valuation disparity, employees increased their private retirement savings rates from 3.5 percent to 11.6 percent when their contributions were changed from payments out of current earnings to the less valued foregoing of a portion of future increases in their wages (Thaler & Benartzi, 2004). A number of other studies provide further examples of the difference in people's valuations of gains and losses: in one, participants demonstrated a strong reluctance to give up a default automobile insurance option when an otherwise more attractive choice was available (Johnson, Hershey, Mesaros, & Kunreuther, 1993); in another, people showed a greater sensitivity to losses in judgments of fairness (Kahneman, Knetsch, & Thaler, 1986); and another revealed that stronger legal protection was accorded to losses over foregone gains in judicial choices (Cohen & Knetsch, 1992).

Many other studies have demonstrated that the valuation disparity is pervasive, usually large (though variable depending on the entitlements at issue and the further particulars of the context of the valuation), and not merely the result of income effects, wealth constraints, or transaction costs (for example, Kahneman, Knetsch, & Thaler, 1990; Knetsch, Tang, & Thaler, 2001).[1] The easy assumption of standard theory that "we shall normally expect the results to be so close together that it would not matter which we choose" (Henderson, 1941, p. 121), is clearly contradicted by these results, and those of many other similar studies (Camerer, 2000).

Although some reports have suggested that the difference between valuations of gains and losses diminishes, or even disappears, with repeated trials, the evidence in most of these demonstrations has come from experiments using a second price Vickrey auction. (In a second price Vickrey auction the highest bidder buys at the second highest bid, and the lowest seller sells at the second lowest offer.) Although substituting a ninth price for a second price in a Vickrey auction should have absolutely no effect on people's valuations, the findings of controlled tests showed that it gave rise to large and rapidly widening differences (Knetsch, Tang, & Thaler, 2001). This finding leaves the conclusions from earlier reports of convergence very much in doubt. Other reports that people in the business of trading are less likely to exhibit endowment effects, at least with respect to buying and selling goods (for example, List, 2003), is not an unexpected result; it says little, however, about the many other instances of an endowment effect on other types of valuations.

Although differences in the evaluation of gains and losses may not be universal, current evidence strongly suggests that it is pervasive among individuals involved in economic activities or weighing the advantages and disadvantages of proposed changes. Field studies of people's real investment and consumption decisions and choices indicate that this is especially likely to be the case for most consumer dealings and for changes that are likely to be the subject of benefit-cost or other forms of policy analyses. These would include, for example, those involving sustainable forest management. Consequently, the common practice of valuing losses of some forest outputs by using the WTP measure is very likely to seriously understate their value – perhaps by one half or less – and thereby distort management choices. This understatement is not that which might result from errors of estimation, but is due entirely to the wrong choice of measure.

### *4.2 Different Measures and Different Values*

The different valuations of gains and losses give rise to four different measures of sacrifice, as indicated by the 2 x 2 array of Figure 5.1.[2]

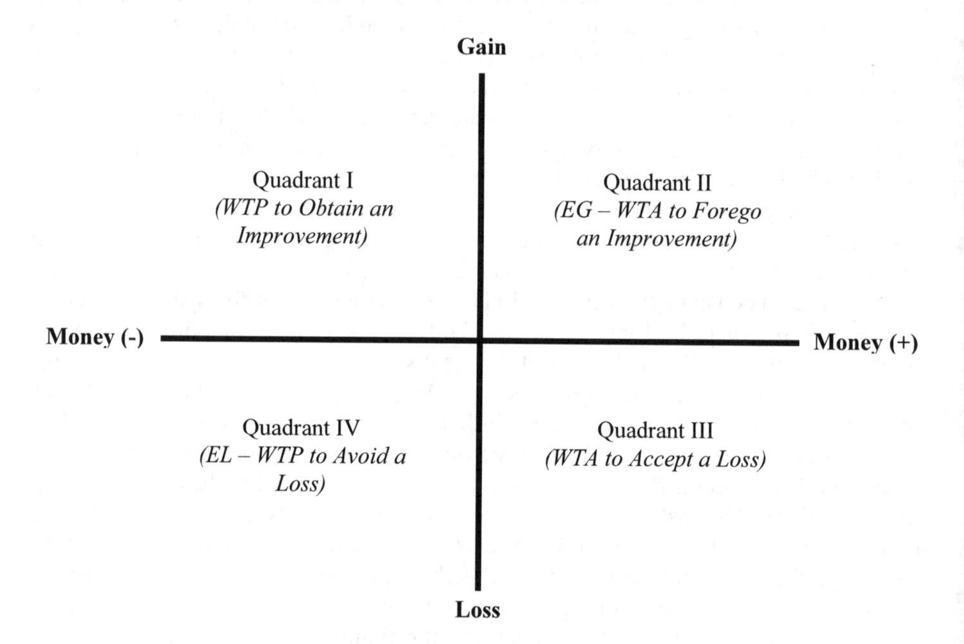


***Figure 5.1.*** *Combinations of Gains and Losses and Differing Valuations.*

The Quadrant I measure is the most an individual would pay to acquire a gain – the WTP measure. The Equivalent Gain (EG) measure of Quadrant II provides a choice between two gains, and values a gain in terms of the sum an individual feels is equivalent to it – the WTA to forego the gain. The Quadrant III measure is the minimum sum a person would demand to accept a loss – the WTA measure of its value. The Equivalent Loss (EL) valuation of Quadrant IV measure assesses the value of a loss in terms of the loss equivalent to it – a measure frequently posed, or framed, as the amount people are willing to pay to avoid a loss.

There are, then, two measures of a gain: (1) the WTP to obtain it, and (2) the EG, or WTA to forego it. Similarly, there are two measures of a loss: (1) the WTA to accept it, and (2) the EL, or WTP to avoid it.

If people's valuations of gains and losses are fully equivalent, as generally prescribed by standard theory and assumed in common practice, then not only would the two gain measures be equal to each other and the two loss measures be equal, but all four measures would yield the same estimates of value[3]. However, rather than being fully equivalent, in many (and likely the vast majority of) cases for which tradeoff rates are estimated or asserted, the tradeoffs and valuations can be expected to vary in a systematic and predictable pattern. As losses of either the entitlement or the numeraire good (usually money) are valued more than gains of the entitlement or money, the WTP measure (Quadrant I) can be expected to yield the smallest valuation (as it involves a *loss* of money to *gain* the entitlement), the WTA measure (Quadrant III) will yield the largest (as it entails *gaining* money and *losing* the entitlement). The Equivalent Gain (Quadrant II) and Equivalent Loss (Quadrant IV) values will be intermediate between the others (Knetsch & Tang, in press).[4]

### *4.3 The Choice of Measure*

A major implication of the valuation disparity evidence is that, given that different valuations will result from the use of different measures, the choice of measure will become an issue of substantial practical importance. Agreements on which measures are appropriate for valuing particular kinds of management or policy outcomes are, however, in short supply – likely due in no small part to the lack of much attention to the disparity issue by economists and the near total absence of interest on the part of public and private agencies and organizations.

The appropriate choice of measure appears to turn on what people regard as the reference state, and the directions of change for both the entitlement and the numeraire good from that position. The choice is akin to the distinction between compensating variation (CV) and equivalent variation (EV) measures of the welfare gains and losses associated with price changes and changes in availability of an entitlement. The CV measures take the initial state, for example before a price change, as the reference state for valuing the change in welfare caused by the change. The EV values the welfare change on the basis of the reference being the position after the change.

The parallel here is that the WTP and WTA measures take the state without the change as the reference positions, and are, therefore, CV measures. The maximum

WTP is the sum that an individual would pay to acquire the entitlement and be left as well off as without the exchange – the amount that would leave the person on the same indifference curve as without the exchange. The minimum WTA is the sum that would fully compensate the individual – the amount that would leave the person on the same indifference curve as if the loss had not been encountered.

In contrast, the EG and EL are EV measures as they are based on a reference that incorporates the change. The EG measure is based on a choice between two gains – two points on an indifference curve indicating an improved level of welfare associated with having gained the entitlement. The reference for the valuation is, therefore, the improved state after the change. The EL measure is given by the choice between two losses – two points on an indifference curve indicating a lower level of welfare brought about by the changed circumstances.

The two measures of the value of gains and the two measures of the value of losses, and the relationship to the reference state, can be summarized in the following array:

| *Welfare Measure* | *Implied Reference State* | *Valuation Measures* | |
|---|---|---|---|
| | | *Gain* | *Loss* |
| Compensating Variation | Present | WTP to Obtain | WTA to Accept |
| Equivalent Variation | After Change | EG to Forego | EL to Avoid |

It is not just the knowledge of a change that is likely to determine the reference state from which valuations are made. It is the expected state, or norm, that is likely to be the determining factor. If people regard their current position, or expectations, as the reference, the WTA measure of losses is needed. The EL measure of a loss – the WTP to Avoid – is correct only if their reference welfare level is that of the changed state. The value of gains is measured by people's WTP if their reference is pre-change, and is aptly assessed with the EG measure – the WTA to Forego – only if their reference is the changed circumstance.

It is often suggested that the alternative measures here being distinguished on the basis of the reference state, are mostly determined as a matter of extant legal entitlements. However, the preferred choice appears instead to depend on what people regard as the appropriate basis for judging the consequences of a change. This seems unlikely to be determined by legal rights, as these are about other issues reflecting not only efficiency, equity, fairness, and other justice goals, but also asymmetries in avoidance costs and costs of enforcement, compliance, and transfer of original entitlements. The choice of measure is about choosing a metric that best reflects actual changes in economic welfare resulting from particular changes in entitlements.

Rather than being determined by legal entitlements, discriminating between the CV and EV measures of gains and losses, and the appropriate choice of measure, may more usefully be determined by what Zerbe (2001) refers to as "psychological ownership" (p. 20). While not an entirely operational definition, the determining reference state may be one reflected in what people regard as the expected or normal state (Kahneman & Miller, 1986), a differentiation similar to the good neighbour test of what is acceptable or unacceptable behavior (Ellickson, 1973), and the harm/benefit test for legal liability. As Kahneman and Miller suggest, an out of the ordinary event commonly prompts a question of what or how, whereas continuation of the norm would not. Changes assessed from the original, expected, reference state appear likely to call for CV measures; changes that bring about a post-change reference state would call for EV measures.

The common practice of, for example, determining the "value of damages to health (both morbidity and mortality) due to air pollution" on measurement of people's "willingness to pay to avoid such effects" (Alberini & Krupnick, 2000, p. 37), would seem to be justified only on a showing that people regard suffering health damages due to human-caused pollution as being the normal or reference state. In this case, and in many others for which this EL (Quadrant IV, WTP to Avoid) measure is used, this seems unlikely to be an easy task.

Determination of the reference state is also needed to discriminate between gains and reductions in losses, and between losses and foregone gains. While it is common to treat all positive changes as gains, and measured as such, mitigation of losses and reductions in the risk of loss are more appropriately assessed by the "individual's willingness to accept compensation to tolerate a loss" (Pearce & Seccombe-Hett, 2000, p. 1420). For example, the widely cited value of life study based on "asking over 3,000 members of the general public" (Cropper, Aydede, & Portney, 1994, p. 244), takes the saving of future lives as a future gain, thereby necessarily assuming that people's reference state for valuing premature deaths is one of "exposure to a pollutant, often a cancer-causing one" (p. 243). However, the reference state for such a change seems far more likely to instead be that of being free of such death causing pollution, suggesting that the change is more realistically framed as reducing a loss, and best valued in terms of the WTA needed to put up with this less-than-normal condition.

Determining the appropriate reference state appears to be largely an empirical matter of which state is likely to best describe people's feeling about changes. Although the reference state will often be the status quo, in important cases it may not be: soiled foreshores may be the reality after a marine oil spill, but most people in the area would no doubt regard unspoiled shores as the norm. This would then be the reference for their subjective reactions and valuations of both the loss caused by the spill and the benefit of cleanup activities. As Kahneman and Miller (1986) suggest, a spill would be considered out of the ordinary and would prompt questions of why and how it happened. Another day without a spill would not be out of the ordinary; it would be considered the norm and people would be in no need of an explanation of how it came about. The reference state in this case is the absence of the spill. The loss of welfare resulting from the change would therefore best be measured by the compensation required to retain the level enjoyed in the reference state level, the CV measure of the WTA.

While what most people regard as a reference state is an empirical matter, most changes that are likely to be subject to any form of weighing or valuation appear to call for the CV measures: the WTP for gains, and WTA for losses. This may be most apparent in cases such as oil spills and sudden discharges of toxic wastes, although it appears likely that people would also weigh the loss of wildlife habitat or water quality or any number of other consequences on a similar basis, as changes from a reference state exclusive of the adverse change. This may not be conclusive, but it does suggest a broad presumption in favour of CV over EV measures.[5]

To the extent that the present reference state is the dominant case, then resource losses and damages will generally need to be assessed in terms of the WTA measure, and not by the amounts people are willing to pay to avoid a loss. While people's being willing or not willing to pay to avoid a loss is a common framing of policy debates, it can also be a very misleading one – posing an issue as an EV test such as "whether it is possible for the losers to bribe the gainers to obtain their consent to forgo the proposed policy change" (Freeman, 2003, p. 62), may not be completely compatible with most people's intuitions about the appropriate reference state and measure.

The distinction between a change being in the domain of losses, for which the WTA is the better measure, or being in the domain of gains (gains and foregone gains), for which the WTP is the better measure, points again to the critical importance of determining the reference state appropriate to the specific valuation at hand. A presumption of most people's reference being one for which most resource changes, or at least those of much concern, appear to fall in the domain of losses, is supported by what seems to be wide agreement with suggestions such as, "The benefits derived from pollution control are the damages prevented" (Tietenberg, 1996, p. 71). To the extent that this is the case, the value is then measured by the compensation people require to be left with no pollution control. Loss of scenic amenities, wildlife habitat and others associated with particular forms of forest harvest, and management generally, would be assessed similarly.

## 5. DISCOUNTING FUTURE GAINS AND LOSSES

The implications of the behavior findings of gain-loss valuation disparities extend to future outcomes as well as present ones. It is generally understood that gains and losses that occur in the future are worth less than commensurate present outcomes – $100 now is worth more than $100 a year from now. Apart from important questions involving intergenerational comparisons, which seems to be far more than a simple discounting issue, it is generally agreed that intertemporal outcomes can be made comparable by discounting to a common time. In practice, a single rate is taken to reflect people's time preferences, or tradeoffs, for evaluating both future gains and future losses.

While the evidence of some particular patterns of time preferences is a good deal weaker than on others, it does seem clear that people do not use a single rate to discount the value of all future outcomes. Specifically, people discount the value of future losses at a lower rate than they use to discount the value of future gains.

This difference in rates appears to be a predictable extension of the more general findings that individuals commonly value losses more than commensurate gains

(Donkers, Gregory, & Knetsch, in process). Just as people are willing to pay less for a gain than they demand to accept a loss, they can be expected to be willing to pay less for a *future* gain than they require to accept a *future* loss. The present value of a future gain is, of course, the sum that an individual is willing to pay now. Similarly, the present value of a future loss is the sum demanded now. The smaller present value of future gains implies that individuals use a higher rate to discount them, and the larger present value of future losses implies they use a lower rate to discount such future outcomes. Clear empirical demonstrations of such differences are not yet plentiful, but the reported evidence that is available appears to be fully consistent with this interpretation (for example, Loewenstein, 1988; Donkers, Gregory, & Knetsch, in process).

The likelihood that different measures would give rise to different discount rates raises again the issue of choosing an appropriate measure. And again, the criteria remain much the same: the choice depends on the reference state people use to value future outcomes. As in the case of present gains and losses, the use of rates based on how much people would pay to reduce the risks of a future harm, for example, or how much they would demand to forego a future gain, would call for a showing that these equivalent variation measures were justified. Casual observation suggests that quite the opposite is more likely to be the case; the compensating variation WTA and WTP measures appear to be the rule rather than the exception. Here again, knowledge that a future loss is likely to occur does not necessarily change the reference state, it is likely to be viewed as a loss from the current state regardless of any forewarning.

Given that many consequences of management, policy, and project options extend over lengthy time periods, the current practice of using a single rate for discounting gains and losses may very well provide quite distorted views of people's preferences. Taking account of the evidence of differing discount rates would point to quite different policy responses. Rates reflecting observed preferences would, for example, give more, and probably much more, weight to future losses, and justify greater present sacrifices to deal with them, than would be the case following normal present practice. The difference in rates would also likely call for more actions that reduce the risks of future losses (as the lower rates would indicate larger present values) relative to ones that provide future gains (as these are discounted at a higher rate).

## 6. CONCLUDING COMMENTS

As a result of extensive empirical studies, it is becoming increasingly clear that most economic analyses of resource issues, including those that guide forest management and policy decisions, could be markedly improved by including the insights from the findings of behavior economics. While this is likely the case for a wide range of topics, it seems particularly true of resource valuations, where present exercises based on the conventional assumptions of standard theory seem likely to provide very flawed guidance.

While the empirical results from behavior studies suggest many opportunities to greatly increase the explanatory power and usefulness of economics, the potentials for improvement remain largely unrealized. There is probably no single explanation for the tenacious grip that standard theory has over how economics is done, but the asymmetric incentives and restraints facing individuals and organizations may be at least a partial explanation. Continued use of the accepted and conventional carries fewer risks to careers and support than departures, and the textbook writer's explanation for ignoring behavior findings was undoubtedly correct: "If I put this in my books, no one would adopt them".

## NOTES

[1] Hanemann (1991) has correctly pointed out that standard theory can, under particular conditions, allow for a large difference in gain and loss values for an identical entitlement. These include a positive income effect and a lack of substitutes for the good at issue. However, large differences have been observed under conditions that violate those required for this standard theory explanation. The endowment effect is, as Hanemann notes, "a different phenomenon" (1991, p. 645n), but it seems to be a far more general explanation for the observed pervasive differences than the narrow possibilities offered by standard theory.

[2] There may well be other differences depending on other valuation contexts, but only those related to the differing valuations of gains and losses are considered here.

[3] Absent an income, or wealth, effect, which for most cases can be safely ignored.

[4] Bateman, et al. (1997) provide an example of the expected pattern of different valuations for present gains and losses: the proportions of people preferring four tins of Cola to £0.80 was 40 percent, 74 percent, 84 percent, and 50 percent for the four quadrants, I through IV, respectively. Another example, for three of the measures, is the report that people were willing to pay $2.00 to buy a mug, $7.00 to give one up, and chose receiving $3.50 as equivalent to gaining a mug (Kahneman, Knetsch, & Thaler, 1990).

[5] Most policy analyses appear to be consistent with this position, "The CV measure is generally the standard for benefit-cost analysis" (Zerbe, 2001, p. 7n).

## REFERENCES


Alberini, A., & Krupnick, A. (2000). Cost-of-illness and willingness-to-pay estimates of the benefits of improved air quality: Evidence from Taiwan. *Land Economics*, 76, 37-53.

Arrow, K., Cropper, M.L., Eads, G.C., Hahn, R.W., Lave, L.B., Noll, R.G., Portney, P.R., Russell, M., Schmalensee, R., Smith, V.K., & Stavins, R.N. (1996). Is there a role for benefit-cost analysis in environmental, health, and safety regulation? *Science,* 274, 221-222 (12 April).

Bateman, I.J., Munro, A., Rhodes, B., Starmer, C., & Sugden, R. (1997). A test of the theory of reference-dependent preferences. *Quarterly Journal of Economics*, 112, 479-505.

Camerer, C.F. (2000). Prospect theory in the wild. In D. Kahneman, A. Tversky (Eds.), *Choices, values, and frames (pp.288-300)*. Cambridge: Cambridge University Press.

Camerer, C., Babcock, L., Loewenstein, G., & Thaler, R.H. (1997). Labor supply of New York cabdrivers: One day at a time. *Quarterly Journal of Economics,* 112, 407-442.

Cohen, D., & Knetsch, J.L. (1992). Judicial choices and disparities between measures of economic values. *Osgoode Hall Law Journal*, 30, 737-770.

Cropper, M.L., Aydede, S.K., & Portney, P.R. (1994). Preferences for life saving programs: How the public discounts time and age. *Journal of Risk and Uncertainty*, 8, 243-265.

Diamond, P. (1996). Testing the internal consistency of contingent valuation surveys. *Journal of Environmental Economics and Management,* 30, 337-347.

Diamond, P.A., Hausman, J.A., Leonard, G.K., & Denning, M.A. (1993). Does contingent valuation measure preferences? Experimental evidence. In J. Hausman (Ed.), *Contingent valuation: A critical assessment (pp.41-85)*. Amsterdam: North-Holland.

Donkers, B., Gregory, R., & Knetsch, J.L. (in process). Discounting future gains and future losses: Predictable differences and appropriate measures.

Ellickson, R.C. (1973). Alternatives to zoning: Covenants, nuisance rules, and fines as land use controls. *University of Chicago Law Review*, 40, 581-781.

Freeman, A.M. (2003). *The measurement of environmental and resource values: Theory and methods* (2nd ed.). Washington, D.C.: Resources For the Future.

Hanemann, W.M. (1991). Willingness to pay and willingness to accept: How much can they differ? *The American Economic Review*, 81, 635-647.

Henderson, A.M. (1941). Consumer's surplus and the compensation variation. *Review of Economic Studies*, 8, 117-121.

Johnson, E.J., Hershey, J., Mesaros, J., & Kunreuther, H. (1993). Framing, probability distortions, and insurance decisions. *Journal of Risk and Uncertainty*, 7, 35-51.

Kachelmeier, S.J., & Shehata, M. (1992)."Examining risk preferences under high monetary incentives: Experimental evidence from the People's Republic of China. *The American Economic Review, 82, 1120-1140.*

Kahneman, D., Knetsch, J.L., & Thaler, R.H. (1986). Fairness as a constraint on profit seeking: Entitlements in the market. *The American Economic Review*, 76, 728-741.

Kahneman, D., Knetsch, J.L., & Thaler, R.H. (1990). Experimental tests of the endowment effect and the Coase theorem. *Journal of Political Economy*, 98, 1325-1348.

Kahneman, D., Knetsch, J.L., & Thaler, R.H. (1991). The endowment effect, loss aversion, and status quo bias. *Journal of Economic Perspectives*, 5, 193-206.

Kahneman, D., & Miller, D.T. (1986). Norm theory: Comparing reality to its alternatives. *Psychological Review*, 93, 136-153.

Kahneman, D., & Tversky, A. (1995). Conflict resolution: A cognitive perspective. In K.J. Arrow, R.H. Mnookin, L. Ross, A. Tversky, & R.B. Wilson (Eds.), *Barriers to conflict resolution (pp. 44-60).* New York: Norton.

Knetsch, J.L., & Tang, F. (in press). The context, or reference, dependence of economics values: Further evidence and predictable patterns. In M. Altman (Ed.), *Foundations and extensions of behavioral economics: A handbook.* New York: Sharpe Publishers.

Knetsch, J.L., Tang, F., & Thaler, R.H. (2001). The endowment effect and repeated market trials: Is the Vickrey auction demand revealing? *Experimental Economics*, 4, 257-269.

List, J.A. (2003). Does market experience eliminate market anomalies? *Quarterly Journal of Economics*, 118, 47-71.

Loewenstein, G.F. (1988). Frames of mind in intertemporal choice. *Management Science*, 34, 200-214.

Michelman, F.I. (1967). Property, utility, and fairness: Comments on the ethical foundations of 'just compensation. *Harvard Law Review*, 80, 1165-1258.

Nautiyal, J.C., & Rezende, J.L. (1985). Forestry and benefit-cost analysis. *Journal of World Forest Resource Management*, 1, 189-198.

Odean, T. (1998). Are investors reluctant to realize their losses? *The Journal of Finance*, 53, 1775-1798.

Pearce, D.W., & Seccome-Hett, T. (2000). Economic valuation and environmental decision-making in Europe. *Environmental Science and Technology*, 34, 1419-1425.

Putler, D.S. (1992). Incorporating reference price effects into a theory of consumer choice. *Marketing Science*, 11, 287-309.

Rabin, M. (1998). Psychology and economics. *Journal of Economic Literature*, 36, 11-46.

Samuelson, W., & Zeckhauser, R. (1988). Status quo bias in decision making. *Journal of Risk and Uncertainty*, 1, 7-59.

Shefrin, H., & Statman, M. (1985). The disposition to sell winners too early and ride losers too long: Theory and evidence. *Journal of Finance*, 40, 777-790.

Thaler, R.H. (1999). Mental accounting matters. *Journal of Behavioral Decision Making*, 12, 183-206.

Thaler, R.H., & Benartzi, S. (2004). Save more tomorrow: Using behavioral economics to increase employee saving. *Journal of Political Economy*, 112, S164-S182.

Tietenberg, T. (1996). *Environmental and natural resource economics* (4th ed.). New York: Harper Collins Publishers.

Zerbe, R.O. (2001). *Economic efficiency in law and economics*. Cheltenham: Edward Elgar.

# CHAPTER 6

# HOW SUSTAINABLE IS DISCOUNTING?

COLIN PRICE

*School of Agricultural and Forest Sciences, University of Wales, Bangor, Gwynedd LL57 2UW, United Kingdom.*
*Email: c.price@bangor.ac.uk*

**Abstract**: Discounting has caused disquiet because it trivializes the very long term. Several arguments have been advanced for reducing the discount rate over time. They include justice to future generations, and conformity with the discount profile that people seem in practice to apply. However, a declining discount rate leads to inconsistent preferences through time. In a variety of circumstances where components of value are aggregated across consumers, products and scenarios, components with a low rate of diminishing marginal utility become increasingly dominant in forming the aggregate discount factor. The reasons given for disaggregation throw light on weaknesses in the justifications of discounting: reinvestment at compound interest to compensate for future damage may not take place; time preference, if interpreted as preference for immediacy, has no implications for discounting of futurity. The declining aggregate discounting protocol has political allure, but may lead to indefinite postponement of worthwhile investment.

## 1. INTRODUCTION

Discounting is a process applied by economists ostensibly to give appropriate weight[1] to benefits and costs at some future time, and, collaterally, to allocate investment funds and resources among projects competing for their use. Classically, and almost invariably in practice, discounting has taken negative exponential form, such that:

$$[\text{present equivalent value}] = \frac{[\text{expected future value}]}{(1 + [\text{discount rate}])^{[\text{future time}]}} \quad (1)$$

The conversion factor, $1 \div (1 + [\text{discount rate}])^{[\text{future time}]}$, is called the discount factor, and will be frequently alluded to in the following sections.

The justifications given for this process and format have been many and varied. For example, the capital rationing justification refers to allocative efficiency: projects whose present equivalent value of costs exceeds the present equivalent



*Kant and Berry (Eds.), Economics, Sustainability, and Natural Resources: Economics of Sustainable Forest Management,* 105-135.

value of revenues have low financial growth potential, and should be rejected. The time preference justification acknowledges the popular desire for early consumption. It refers not only to psychological impatience, but also to the persuasive argument, that less weight should be given to increments of consumption enjoyed by future citizens, since they will already be more affluent – and hence less needful – than present citizens.

But in recent years there has been an unremitting debate about which discount rate is appropriate; and increasingly the entire process is under challenge. The elements of the controversy are reviewed in Price (1993a). Particularly, in the decade following the Rio Earth Summit, concern was widely expressed about the implications of discounting for the importance given to very long-term costs and benefits (see for example the selection of views compiled by Portney & Weyant, 1999).

This widespread disquiet had two roots:

- distaste for the results
- dissatisfaction with the reasoning by which they were derived.

Notably, discounting between generations seemed to conflict with the requirements of sustainability, which are now central to the declared policies of many governments and organizations (e.g. Gummer, 1994).

Stepping aside from the sustainable development debate, which has been confused by the wanton proliferation of definitions, "sustainable" actually means:

a. capable of continuing indefinitely
b. susceptible of being logically defended.

This paper discusses the relationship of discounting to sustainability as interpreted in each of these senses. It examines whether, in the light of concerns centering on sustainability, discounting

a. ought to be continued indefinitely, and especially at a constant rate; and, if not,
b. whether arguments for discounting are logically and consistently defensible, *even within a short time horizon.*

After considering the consequences of discounting according to different protocols, it reviews ethical, democratic and empirical arguments for discounting at a rate which falls through time. It explores in detail how the effects of diminishing marginal utility of consumption may also lead to falling discount rates. It shows how discounting may actually favor future generations – but that a declining discount protocol is not always the best one for this purpose. Then it extends the arguments from which declining rates have been derived, to suggest that in some circumstances no discounting at all may be justified. Finally, it casts a skeptical eye at the political advantages of not sustaining the discount rate at the current high level.

## 2. EXPONENTIAL DISCOUNTING: SOME CONSEQUENCES AND ALTERNATIVES

A dramatic but typical example of the "unacceptable results" of discounting is its impact on the importance ascribed to future climate change. Most of the serious global warming effects are expected to arise at a time between several decades and many centuries into the future. According to some predictions, polar ice-caps will be disintegrating in about 500 years' time, with results that could, for example, compromise the gross domestic product of the UK. Yet at a 6% discount rate (until recent years advocated by Her Majesty's Treasury (1991)) the effect of total loss of gross domestic product would be discounted, as shown in Figure 6.1, to a value equivalent to a small bag of potato crisps (Price, 1996). Even rates much lower than this have a similar effect, though over a longer time period.

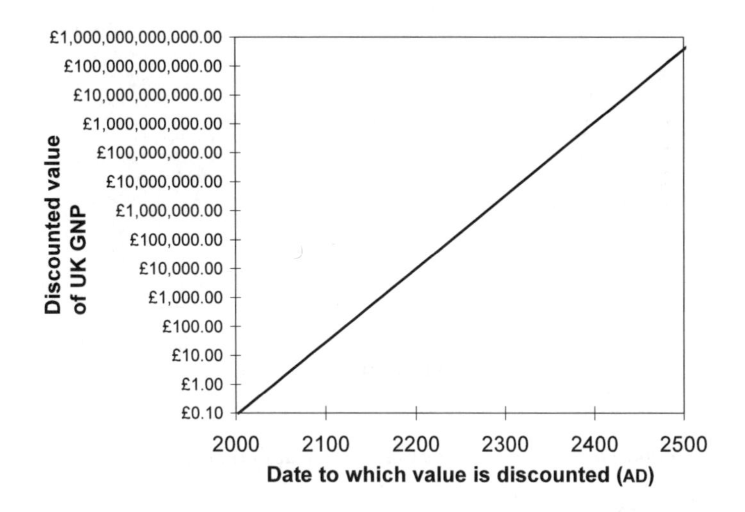


***Figure 6.1.*** *Effect of Discounting on the Value of a Huge Distant-future Sum*

***Source***: Price (1996)

Another result which perturbs non-economists is that an indefinitely prolonged stream of benefit has a finite discounted value – a small multiple of the annual benefit according to the standard capitalization formula[2].

Thus sustainability, often seen as a good not to be sacrificed in trade for any limited amount and period of benefit, appears after all to be commensurable with, and therefore vulnerable to outweighing by, such transient benefit.

Aversion to such results may prompt a more thorough scrutiny of discounting processes, but does not itself constitute an adequate reason for rejecting the processes. It is errors in the justifications given for the processes, if they exist, that should prompt reform of appraisal techniques.

Several recent publications have constructed a case for a declining discount rate, and lately both Her Majesty's Treasury (undated) and the Office of the [UK] Deputy Prime Minister (OXERA, 2002) have shown interest in this mode of discounting. Figure 6.2 shows the discount factors produced by one such stepped reduction of discount rates over a 300-year period. While the arguments against uniform discounting reviewed in these sources appear diverse, most of them arise from a requirement intrinsic to cost–benefit analysis: to aggregate values across time periods, stakeholders, kinds of goods, and possible scenarios, in circumstances where different discount rates might be held to apply to different periods, costs or benefits.

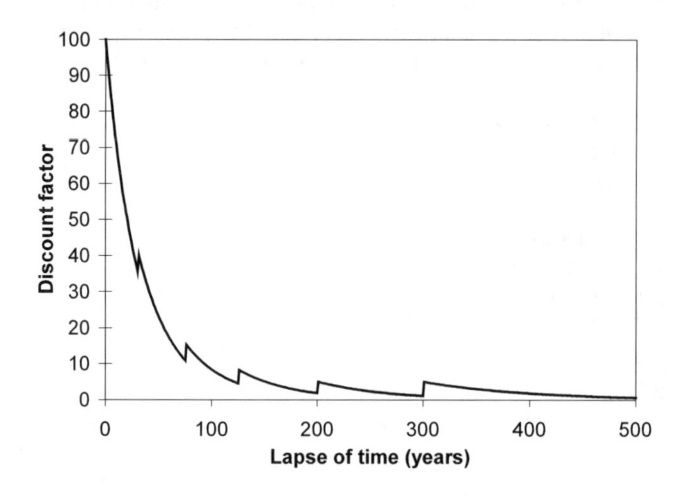


***Figure 6.2**. Discount factors based on rates proposed by OXERA*

***Source***: OXERA (2002)[3]

A regime of declining discount rates is more favorable to forestry than to many other investments. Take for example a 150-year rotation of some slow-growing but valuable timber crop, costing £2000 to establish and predicted to yield £60 000 in timber revenues after 150 years. (Intermediate expenditures and revenues are ignored for clarity.) Compare this with

a. some investment of the £2000 yielding a constant annual revenue over 150 years, sufficient to generate an internal rate of return of 4%

b. some exploitation yielding immediate revenue of £2000, but with constant environmental cost over 150 years, again resulting in an internal rate of return of 4%.

Table 6.1 shows the results of discounted cash flow comparison of the three possible uses of £2000 under conventional discounting at 3.5% and 1%, and under the regime of discount rates proposed by OXERA (2002), declining from 3.5% initially to 1% beyond 300 years. Project values are expressed as net present value (NPV): that is, the discounted value of benefits minus the discounted value of costs.

***Table 6.1****. Discounting Regimes and Project Choice (NPVs in £)*

| | Project | | |
|---|---|---|---|
| Discounting protocol | Slow-growing timber | Annual revenue | Exploitation |
| At a constant 3.5% | –1656 | 331 | –331 |
| At a declining rate | 1077 | 1025 | –1025 |
| At a constant 1% | 11 488 | 6060 | –6060 |

The declining discounting regime makes slow-growing timber the best use, whereas with conventional discounting at 3.5% it is the worst. Note, however, that the constant 1% discount rate favors forestry much more clearly: the *low rate*, rather than the *profile of rates*, is the key factor here.

Typically, environmental costs [benefits] are long delayed, and the revenues [expenditures] associated with these environmental effects are short term. Thus the declining discounting regime is also likely to promote environmental interests, compared with what conventional discounting does. But, once again, simply lowering the discount rate would normally be even more favorable.

Nevertheless, before welcoming these outcomes as evidence that the regime is benign, forestry and environmental economists also owe the declining discount rate protocol a critical appraisal. The arguments in favor of a declining rate, and some critiques of them, are presented in the following sections.

## 3. SHIFTING TIME PREFERENCE

Any evaluation protocol that discounts the future at a heavy and constant rate might be *judged* unfair to future generations: it gives less weight to a particular increment of consumption accruing to future generations than it would give to the same increment of consumption accruing to the present generation. It is more definitely unfair if it gives less weight to a given increment of *utility* or of *happiness* merely because it accrues in future. ("… the time at which a man exists cannot affect the value of his happiness …" (Sidgwick, 1874, p. 414).)

The perception that future generations are thus improperly treated by conventional discounting underlies several protocols in which the high rate of immediate discounting is not sustained as the time horizon is extended.

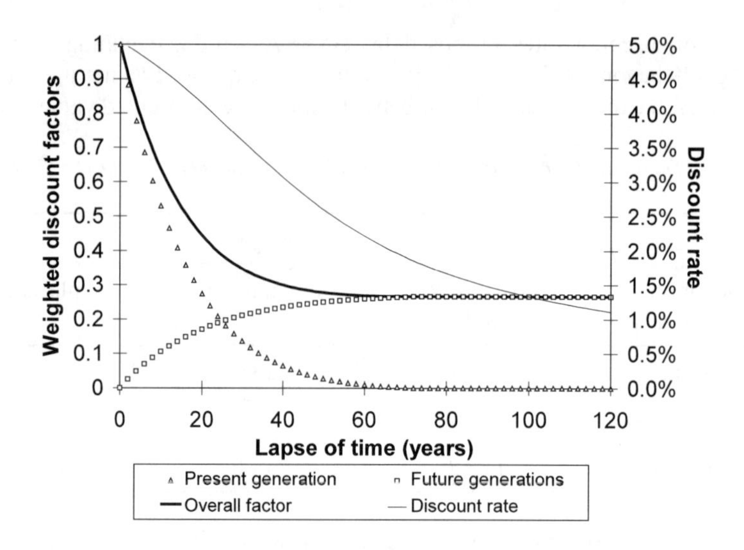


***Figure 6.3**. Discount Factors and Rate According to the Protocol of Kula*

*3.1 Modified Discounting*

The ethical imperative could be seen as to deal more even-handedly with future generations, without overriding the apparent preferences of the present generation for early consumption within its own life-span. Thus discounting protocols which differ in treatment of intra- and inter-generational discounting have been devised by several authors (Kula, 1988; Bellinger, 1991; Bayer, 2003). According to these, the present generation is entitled to discount its future well-being as it pleases, but the consumption of future generations should not be discounted – possibly at all, and particularly not at the whim of the present generation's preference. A practical difficulty is the absence of a clearly defined moment at which "the present generation" cedes place to "future generations": generations overlap, and, because their well-being is discounted differently, intragenerational discounting and intergenerational discounting need to be mixed in changing ratios as time passes. The above-mentioned authors address this problem by calculating discount factors for each future year. For these, different proportions of the present generation survive; also changing are the proportions of successive recruits to the population who survive for various periods up to the reference date. Figure 6.3 shows the contribution to Kula's discount factor made by present and future generations, with discounting, if appropriate, at 5%, and using weights according to proportions of each age cohort in the population. The discount rate graphed is the "whole period discount rate" – the rate which, applied over the whole period up to that point in time, would yield the discount factor shown.

Kula's discount factors decrease over a few decades to a limit – in this case to 0.26619. Beyond this time, the whole period discount rate declines asymptotically to zero as the time period is extended indefinitely.

Even this formulation is in practice unfair to future generations. For people alive at present, the discount factors applied to consumption over their remaining life range downwards from unity to a defined limiting low value; for people born over the next human life-span, the range of discount factors is downwards from a number less than unity to the limiting low value; for people born after that, the discount factor is constant at the limiting low value (Price, 1989).

An intergenerational discount rate lower than the intragenerational rate also leads to what Strotz (1956) terms "dynamic inconsistency" of preference, given that future generations are likely to view *their* future from their own perspective, rather than merely accepting discount factors handed down to them by previous generations (Price, 1989). Consider a plantation of oak (*Quercus spp.*) with alternative rotations of 80 and 120 years. Discount factors are calculated according to Kula (1988), with an intragenerational time preference rate of 5%, and an intergenerational rate of 0%. The discount factors are as in figure 6.3: for 0 years 1.00000; for 40 years, 0.29927; for 80 years, 0.26619; for 120 years, 0.26619 also. Table 6.2 shows discounted cash flows seen from two different time perspectives.

***Table 6.2**. Options for Oak (Quercus spp.) Rotation*

| Event | Cash flow /ha | Discounted value seen from AD2000 | Discounted value seen from AD2080 |
|---|---|---|---|
| Plant | –£2000 | –£2000 | |
| Fell at age 80 | £6000 | £6000 × 0.26619 = £1597 | £6000 |
| Fell at age 120 | £12000 | £12 000 × 0.26619 = £3194 | £12 000 × 0.29927 = £3591 |

***Source***: Modified from Price (1984)

Viewed from the perspective of AD2000, the forest is worth planting for its timber benefits on a 120-year rotation. However, in AD2080, revenue from immediate felling will appear more desirable than the heavily discounted revenue from a rotation prolonged by 40 years. And yet, had it been expected that a future manager would fell at 80 years, the present decision maker would consider that the crop was not worth planting.

Kula (1989) asserts that the initial discount factors are to be respected, so that a future decision maker may not revise the originally decided rotation period. But future generations, not having been party to any agreement on discount factors, are not bound ethically – nor are they by self-interest – to accept the inviolability of their predecessors' protocols.

*3.2 Weighted Mean Rates*

A similar result is derived in several recent studies which distinguish discount rates, not between generations, but between circumstances. For example, Li and Löfgren (2002) ascribe discount rates to two stereotypic individuals with different perspectives on future values. "The conservationist" has a zero discount rate, weighting all future time periods equally: "the [conventionally defined] utilitarian" discounts conventionally at a positive rate, in this case 10%[4]. The social choice rule is to discount according to a (weighted) mean of the discount factors used by each, as shown in figure 6.4.

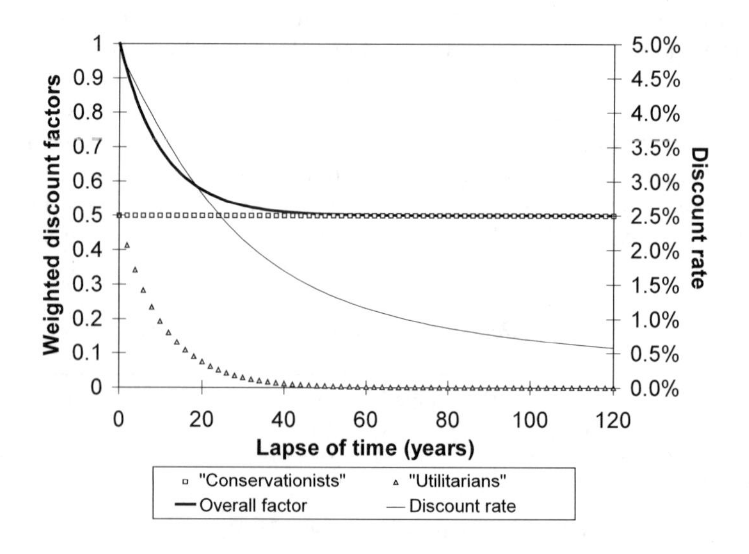


***Figure 6.4**. Discount Factors and Rate According to the Protocol of Li and Löfgren*

Over time, the utilitarian's discount factor is asymptotic to zero, while the conservationist's remains at unity. If equal weight is given to the two perspectives, the overall discount factor is asymptotic to 0.5. Similarly to Kula's protocol, the discount rate yielding this factor approaches zero asymptotically as the time period of discounting is extended indefinitely, although unlike Kula's, their discount factors are only asymptotic, never actually reaching this limit. As with Kula's protocol also, dynamic inconsistency is inherent. For example, a conservationist and a utilitarian in 50 years' time will together compile different relative discount factors for (say) AD2054 (1.0000) and AD2104 (0.5043), compared with factors of 0.5043 and 0.5000 respectively as seen from the present. Indeed dynamic inconsistency exists whenever there is a shift through time in the ratio between factors for any pair of dates.

*3.3 Hyperbolic and Like Discount Functions*

The literature in the border zone between economics, psychology and sociology (e.g. Ainslie, 1991) is rich in results which show that even a single individual discounts at different rates over different time periods (see figure 6.5). The high rates typically found over short time periods become progressively lower as the period of discounting lengthens. This result is the more remarkable, as it contradicts the expectation and evidence from financial investment markets, that sacrifice of long-term liquidity would require a *higher* offered interest rate. Empirically, an hyperbolic functional form has been found to fit the data (Henderson & Bateman, 1995; Cropper & Laibson, 1999), though no underlying reason for this mathematical representation seems to have been advanced.[5]

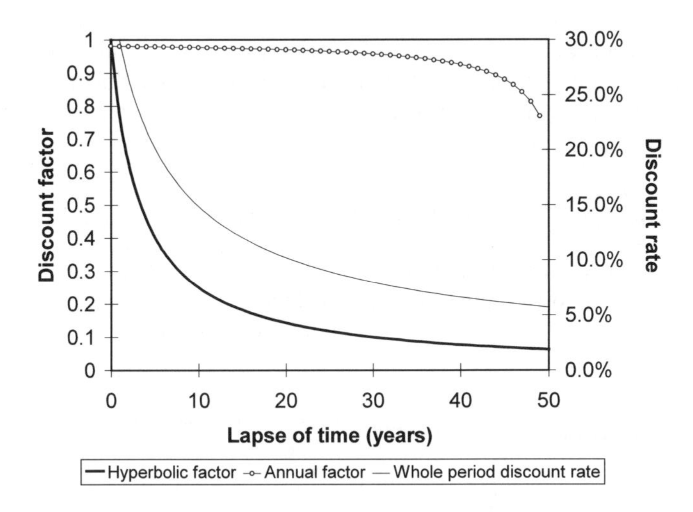


***Figure 6.5**. Hyperbolic Discount Factors and Equivalent Rates*

Unlike negative exponential discounting, hyperbolic discounting does not yield a finite value for an indefinitely prolonged flow of benefit, though the value rises only very slowly, in the logarithm of the time period considered. It is therefore *technically* not subject to the objection that indefinite sustainability may be outweighed by transient benefits.

The ratio between discount factors for any pair of points in time changes as the baseline for discounting moves forwards. The "annual factor" in figure 6.5 is the ratio between factors for years 50 and 49 as it appears after increasing lapse of time. Once again, dynamic inconsistency results.

While there are similarities between all the cases discussed so far – notably, dynamic inconsistency – they produce different profiles of discount factors and discount rates.

The problems encountered in these formulations stem from the ephemeral nature of one element in the discount rate: pure time preference – that is, a weight given to consumption merely on the grounds of the time at which it takes place. Many distinguished writers have queried the validity of pure time preference as an appropriate element in social discounting (Hume, 1739; Ramsey, 1928; Harrod, 1948; Sen, 1957). This theme is taken up later under the heading "How defensible is the argument for discounting?"

There is much wider agreement that diminishing marginal utility provides an intellectually robust justification for discounting the consumption of future generations, who, by conventional assumption, are treated as having higher consumption than the present one (Tullock, 1964). Conveniently, diminishing marginal utility applies equally to the presumed increasing affluence of the present generation, so that there is no necessary distinction between intra- and inter-generational discount rates[6].

## 4. DIMINISHING MARGINAL UTILITY AND PERSONAL AFFLUENCE

Nonetheless, just as pure time preference rates may vary between individuals, so individuals may be differently affected by income change and consequently by diminishing marginal utility of income and of consumption. Let the marginal utility of income be given as

$$[\text{marginal utility of income}] = a \times [\text{income}]^{[\text{elasticity of marginal utility of income}]} \quad (2)$$

Elasticity of marginal utility of income being negative, the relationship is an inverse one. The marginal utility discount rate is the rate of change with time of this function, given by:

$$[\text{discount rate}] = [\text{growth rate of income}] \times \left| [\text{elasticity of marginal utility of income}] \right| \quad (3)$$

The same reasoning may be applied in the international community. Suppose that the costs of global warming are expected to be borne equally by two representative nations, having growth rates of income per head of 0.5% and 2.5%. For purposes of demonstration, elasticity of marginal utility of income is taken as –2. Initially, suppose that the two nations have the same income per head. Figure 6.6 illustrates the change of discounted values through time, discount factors being based on marginal utilities calculated from equation (2).

Evidently a social discount rate based on the mean growth rate of national income would not be appropriate to groups having a slower or faster rate of income growth. Even a discount rate based on the aggregate income growth only of a project's beneficiaries may *still* give inadequate weight to the future, if the poorer beneficiaries' income is growing relatively slowly (Price & Nair, 1985). This is because growth of *mean income* is dominated by the rapid income growth of the rich, whereas change in *utility* is dominated by values to the poor (their poverty means that marginal utility starts at a high level, and their slow income growth means that it remains so). An illustration using data from the UK is given in Price (2003).

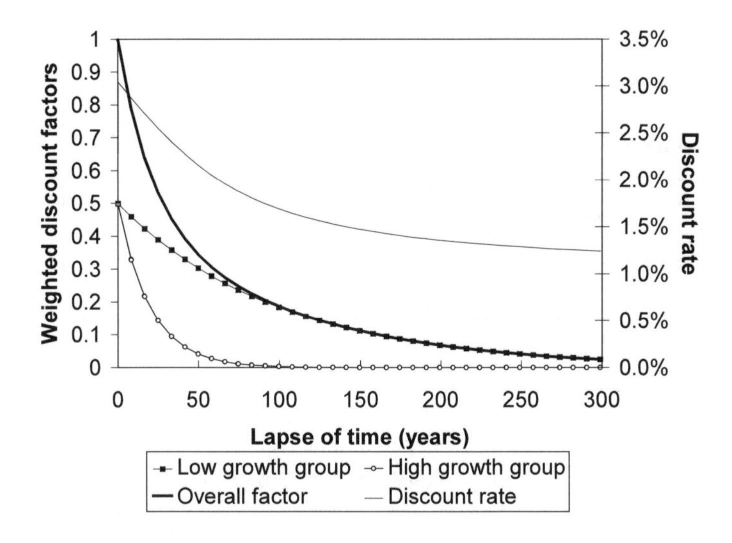


***Figure 6.6**. Discounting the Effect of Climate Change: Two Different Income Growth Rates*

The time path of marginal utility of consumption summed for the two groups (the overall discount factor) gives a profile of whole period discount rates which is similar to OXERA's.

Now suppose that the low-income-growth country has initial income only one-twentieth that of the high-income-growth country. This combination of circumstances may not be unrealistic: recorded growth of income in low-income countries may for a number of reasons overstate the real change of consumption per head (see Price, 2003).

By contrast with cases treated earlier, the discount rates in figure 6.7 change only slightly over 300 years. The upper curve shows the discount rate based on overall income growth: the rate increases only slightly towards the rate appropriate to the rich country, as its *income* comes to dominate overall income somewhat more. That based on weighted consumption declines only slightly towards the rate for the poor group, as its *utility* comes to dominate overall utility somewhat more. The problem with discounting based on aggregate income growth is that the discount rate is inappropriate from the start, leading to serious errors even in the short term. A more-or-less *constant* discount rate through time, based on overall income growth, is no guarantee that the rate is *appropriate*.

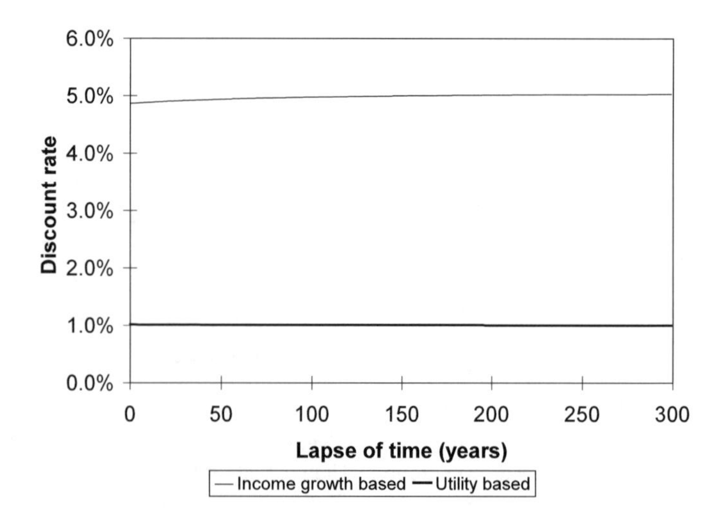


***Figure 6.7.*** *Discounting the Effect of Climate Change: Two Different Initial Income Levels*

## 5. DIMINISHING MARGINAL UTILITY AND THE BASKET OF GOODS

What is true of aggregating individuals is true also of aggregating goods (Price, 1993a, chapter 18). Suppose that two goods with very different elasticities of marginal utility of consumption[7] compose the consumption basket. At low income levels, consumption is dominated by the good having high elasticity of marginal utility of consumption. This high elasticity is often associated with basic needs goods, such as forest-based food and fuel wood. As income rises, at first the (rapid) diminution of marginal utility of consumption is attributable largely to this basic good. But the luxury good, for example access to aesthetic or recreational services of trees, comes increasingly important in marginal purchases.

As figure 6.8 shows, after an initial flat section where the basic good dominates, the rate declines asymptotically to the rate for the luxury good. Note that, while the circumstances structurally resemble Li and Löfgren's formulation for two individuals with different discount rates, the profile of whole period discount rate is quite different.

It would be possible to hypothesize the existence of utility functions having varying elasticity of marginal utility of consumption, as consumption increased, such that a constant discount rate arose. But it is not very evident why such a shift of elasticity would take place.

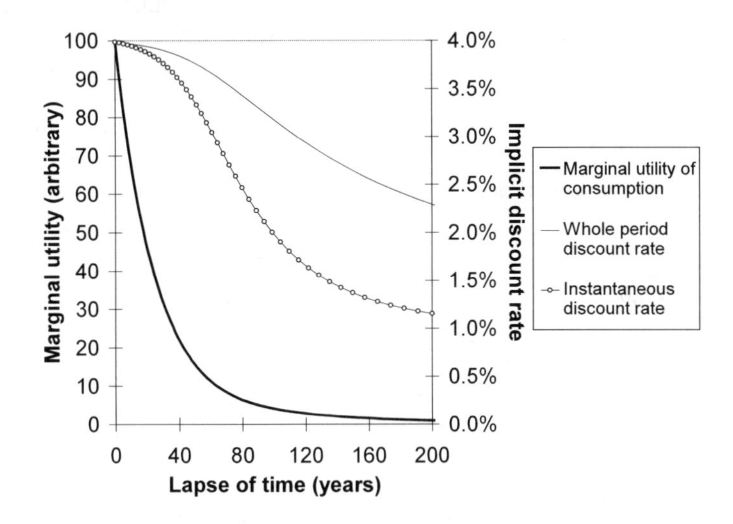


*Figure 6.8. Summative Discount Rates with One Basic and One Luxury Good, and Constant Income Growth Rate*

## 6. DIMINISHING MARGINAL UTILITY AND INELASTIC SUPPLY

The case above assumes that the two goods in the mixed basket are in such elasticity of supply, that relative prices will be the same throughout the period of growing consumption. This may not be so, particularly for environmental qualities such as wilderness (Fisher, Krutilla, & Cicchetti, 1972) and landscape (Price, 2000a), for which substitution between human-made and natural capital may be impossible. The increasing absolute or relative scarcity of such qualities might be reflected in an increase in imputed price. On the other hand, to the extent that a wilderness site has public goods nature (congestion is not a problem), it is *capable* of offering a number of wilderness experiences that increases through time; and if consumption of other goods and services is rising, it would be *expected* that increasing visits would be demanded.

Fisher *et al.* (1972) incorporate these factors by projecting increases of price and visit numbers which partially offset the effect of discounting. Fisher and Krutilla (1975) embrace these changes in a reduced effective discount rate[8].

However, even an adjusted discount rate is incapable of reflecting the shift of the site's values in these circumstances (Price, 1993b). Suppose that the rate of visitation is increasing at 2% per year, and the elasticity of marginal utility of consumption is –2, leading to the *marginal* utility of *additional* visits diminishing at 4% per year. But, since the value of the site embraces the summation of utilities of *all* visits, including the intramarginal ones, the value actually increases through time, though (because of diminishing marginal utility) at a decreasing rate.

Figure 6.9 shows this summed utility, and its value after [illustrative and controversial] discounting for time preference at 2%. As in all previous cases, no single discount rate would track the profile of value change through time; but in this case the rate rises through time.

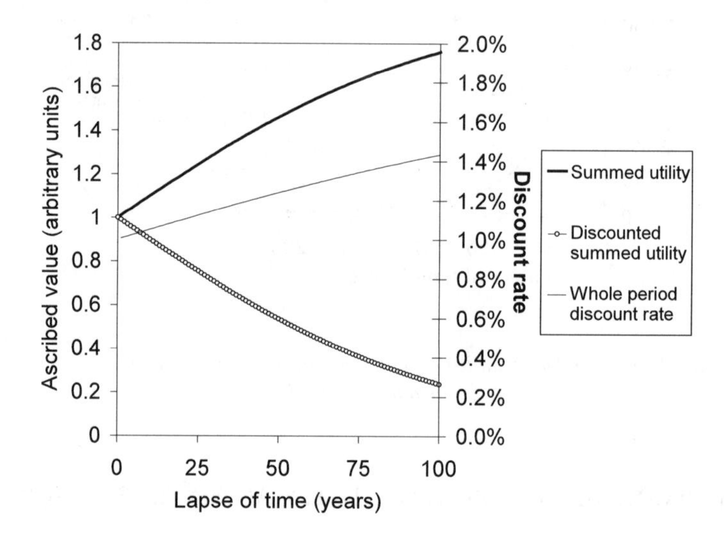


***Figure 6.9.*** *Valuing Recreational Visits by Various Discounting Protocols*

## 7. DIMINISHING MARGINAL UTILITY AND AGGREGATING SCENARIOS

One of the great economic debates of the twentieth century, continuing presently, concerned whether the advance of technology would suffice to outweigh the combined effects of natural resource limitations and population growth. Increasingly, the possibility of global climate change was seen as a key factor. It seems appropriate to treat the situation as one of risk, in which no one scenario is considered certain, but all are aggregated by giving weights reflecting some notion of each one's probability of eventuation.

The following example considers the value of projected future effects of climate change, with two polar economic projections. (In practice a greater number of scenarios would be proposed, but two suffice to make the point.) Under the so-called optimistic scenario, real economic growth is maintained at 5% per year, with zero population growth: under the pessimistic scenario, real economic growth is constrained to 2% per year, with 1.5% per year population growth. The elasticity of marginal utility of consumption is taken as –2. Growth of income per head is economic growth rate minus population growth rate. As is often done (Fankhauser, 1995), the damage resulting from climate change is taken to be proportional to the level of economic activity, so that the rate of economic growth acts against discounting due to diminishing marginal utility.

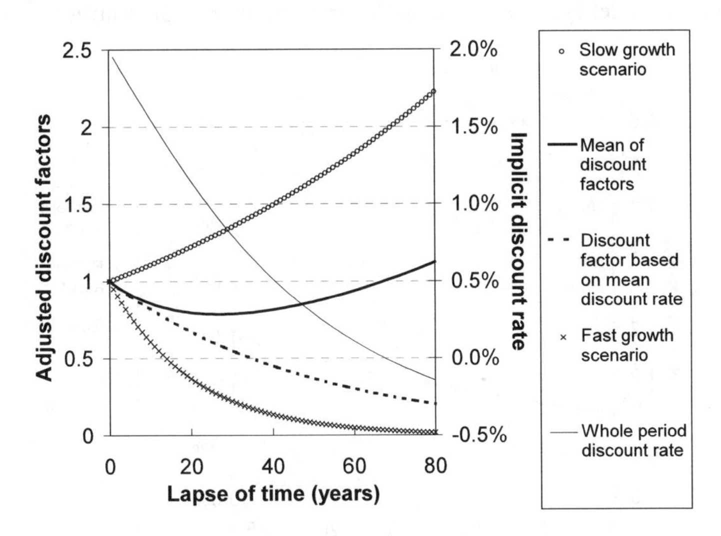


***Figure 6.10****. Averaging Discount Rates and Discount Factors for Climate Change Damage*

Thus under the optimistic scenario the effective discount rate for diminishing marginal utility is

$$(5\% - 0\%) \times |-2| - 5\% = 5\%,$$

under the pessimistic scenario it is

$$(2\% - 1.5\%) \times |-2| - 2\% = -1\%;$$

and, it might be thought, if both scenarios were deemed equally probable, the mean discount rate would be 2%. But by now it should not be expected that this averaging of discount rates will yield the correct result. Figure 6.10 shows that the whole period discount rate changes not only in magnitude through time, but also in sign, being positive initially, but eventually becoming negative, increasingly so.

Many authors (e.g. Lind, 1982; Price, 1993a, chapter 12) have criticized the adjustment of discount rates to include a risk element: the example above shows how unexpected the real impact of an unsure future might be, compared with the customary prescription of adjusting discount rates upwards, uniformly, to "allow" for it.

Figure 6.11 indicates how a risk premium on the discount rate might develop at a smaller spatial and temporal scale. A poor household's income is derived entirely by gathering and selling a portfolio of non-timber forest products, whose price fluctuates randomly by anything up to 80% over one time period, and also has a slow upward tendency. The household's elasticity of marginal utility of income is –2.

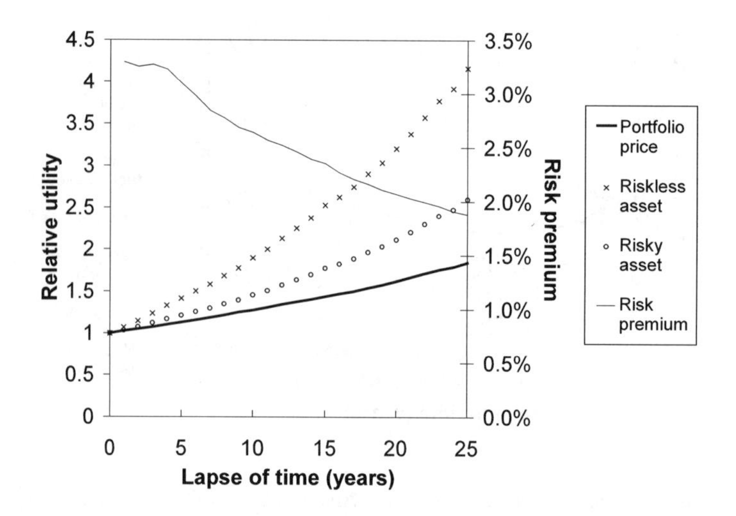


***Figure 6.11**. Relative Values of Investments, with Prices Fluctuating Stochastically by up to 80% per Period, and Tending Upwards at up to 5% per Period*

The household may invest either in a riskless income-yielding asset, or in the means of increasing the efficiency of utilization of the non-timber forest products. Owing to the price fluctuation, the latter option is risky (technically, it has a beta coefficient of +1[9]). At times of low prices, both investments have enhanced value, because of the high marginal utility of income, which has no theoretical upper bound. This more than offsets the low value at times of high prices, which has a lower bound of zero. Thus both assets would grow in value through time, as the limits of possible variation of the portfolio price became wider. However, the riskless asset is more valuable, because in times of low prices the income derived from it remains constant. The declining risk premium represents the difference in rates of value change between risky and riskless investments. Given the large potential scale of fluctuation, the risk premium rate might seem surprisingly small: yet it declines, as in all the other instances where values which change at different rates are aggregated.

The result shown is derived by aggregating 10 000 replicates of a 25-year time period. It might be expected that with so many replicates the curves would all be smooth, but there remain irregularities due to the strong influence of a few extreme sequences.

The effect on asset value of different elasticities of marginal utility of income is explored in Price (1993a, chapter 12). A similar argument for declining discount rates in the face of risk to household income has recently been made by Gollier (2002).

The most extreme risk facing humankind is extinction, and the probability of the event has been proposed as one rational component of a discount rate (Price, 1973; Dasgupta & Heal, 1974). Lately it has been suggested that humanity has a 50/50 chance of surviving the 21st century, equivalent to a 0.7% discount rate. But this projection of risk is itself not certain. Others might believe there are good grounds for thinking that the risk is 2%, or that there is no significant risk. If each of these projections is given equal weighting, the combined discount rate has a similar profile to that of Li and Löfgren (2000) given in figure 6.4: the lowest rate eventually dominates.

## 8. AGGREGATION: CONSPECTUS

Several features are common to all these examples of diminishing marginal utility and discounting. The process of aggregation, which is central to cost–benefit analysis, plays a key role in determining these features.

- Averaging of initial discount rates (across incomes, goods, scenarios) is a crude and inaccurate mode of aggregation, increasingly so with lapse of time.
- It is feasible to aggregate the separate discount factors which result from applying different discount rates to different income groups, goods and scenarios. However, the resulting composite discount factors correspond to a whole period discount rate which changes through time.
- While in most (but not all) cases the whole period discount rate declines through time, the profile may differ according to the underlying reasons for discounting.
- The circumstances which generate the lowest rate of diminishing marginal utility eventually dominate any discount rate derived from aggregate discount factors. This outcome has already been noted in the literature:

> It might be thought that this balanced approach would be to take the average [of 6% and 0% =] 3% discount rate: £1 millions worth of a resource conserved for use in 100 years' time has value discounted at 3% of £52 033. But averaging exponentials is misleading. It would be more appropriate to average the values of the conserved resource: £2947 at 6%, and £1 million at 0%, giving a mean of £501 474 …. (Price, 1993a, p.275)

> The key insight here is that what should be averaged over states of the world is not discount *rates* at various times, but discount *factors*. In the limit, the properly-averaged certainty equivalent discount *factor* corresponds to the *minimum* discount *rate* (Weitzman, 1998, p.206).

In all the cases of aggregation it is possible to calculate, for any particular point in time, a discount rate which, applied in the conventional discounting formula, would give the same result as discounting at separate, circumstance-specific rates (OXERA, 2002; Her Majesty's Treasury, undated). But, since this rate is not sustained through time, the derivation of such a parameter appears pointless, particularly as it is only a derivative from, not a determinant of, discount factors. The point *is*, that rate of diminution of marginal utility is a function of circumstances; and because each situation requiring evaluation includes a different mix of circumstances, discount rates differentiated only according to lapse of time remain too crude in their disaggregation. There is more possibility of intellectual rigor, and less risk of being led into error, if consumption streams differently affected by the lapse of time remain disaggregated throughout the calculations of present values.

The effective discount factors that emerge from following this process are unlike those given by conventional discounting. However, dynamic inconsistency does not necessarily arise: discount factors are calculated, not according to lapse of time, but directly from expected circumstances at a particular point in Earth history. Relative values are thus independent of the point in time from which any decision is viewed. Only a deviation from expected circumstances would lead – as of course it should – to an altered preference ordering.

## 9. DISCOUNTING THAT IS GOOD FOR FUTURE GENERATIONS

### *9.1 The Floodgate Argument*

Not all commentators agree that discounting is bad for future generations. Discounting may make projects with a long lead time unprofitable. The higher the discount rate, the smaller the volume of accepted projects (Scott, 1958). If projects tend to be persistently damaging to environment or resource base, then discounting does in this sense act in the interest of future generations. (The contrary case, where projects accumulate sufficient productive capital to more-than-compensate future generations for loss of environment and resources, is treated later.)

This "floodgate" effect may sometimes be enhanced by a declining rate of discount. Consider Table 6.3, which shows the cash flows projected for the UK's Sizewell B Nuclear Power Station. The figures are the official ones, with the sole addition of a notional £12 million annual cost following decommissioning in year 33: this allows for long-term environmental and health damages, which seem not to have been evaluated in the original, official cost–benefit analysis. Cash flows are discounted at two rates, and also at a rate declining from 3.5% for the short term, to 1% for periods exceeding 300 years, as advocated by OXERA (2002). The declining rate protocol potentially combines lowering the positive value of the period of benefit (electricity generation), with enhancing the late costs relative to benefits. Only this protocol gives a negative net present value for the project.

***Table 6.3.*** *Cash Flows for a Nuclear Power Station*

| Start time | End time | Annual cash flow | Variable discount rate | NPV with variable rate | NPV @ 3.5% | NPV @ 1.0% |
|---|---|---|---|---|---|---|
| 0 | 1 | –100 | 3.5% | –98 | –98 | –100 |
| 1 | 2 | –200 | 3.5% | –190 | 190 | 197 |
| 2 | 4 | –300 | 3.5% | –541 | –541 | –582 |
| 4 | 5 | –500 | 3.5% | –428 | –428 | –478 |
| 5 | 6 | –200 | 3.5% | –166 | –166 | –189 |
| 6 | 7 | –300 | 3.5% | –240 | –240 | –281 |
| 7 | 30 | 160 | 3.5% | 1999 | 1999 | 3068 |
| 30 | 32 | 160 | 3.0% | 128 | 110 | 235 |
| 23 | 30 | –2 | 3.5% | –6 | –6 | –11 |
| 30 | 33 | –2 | 3.0% | –2 | –2 | –4 |
| 33 | 34 | –500 | 3.0% | –186 | –158 | –358 |
| 34 | 75 | –12 | 3.0% | –104 | –82 | –288 |
| 75 | 125 | –12 | 2.5% | –54 | –22 | –224 |
| 125 | 200 | –12 | 2.0% | –39 | –4 | –183 |
| 200 | 300 | –12 | 1.5% | –32 | 0 | –104 |
| 300 | perpetuity | –12 | 1.0% | –61 | 0 | –61 |
| Total | | | | –21 | 172 | 242 |

Once again, however, it is unsatisfactory to adopt a discounting protocol simply because it gives an intuitively plausible or desirable result. In other circumstances there may be significant medium-term project costs (like those arising from the flush of $CO_2$ into the atmosphere following logging and before forest regrowth). These are treated rather dismissively by the protocol, compared with the treatment given by a uniformly low rate.

It is even more unsatisfactory to select, from a portfolio of discounting protocols, the one that gives a politically expedient outcome. The intellectual foundations of a discounting protocol should be robust, irrespective of the circumstances to which it applies. It is the nature of the circumstances, not the protocol, that differentiates how important each aspect of those circumstances should be. If the objective is to give emphasis to long-term environmental costs, then the obvious approach is to discount them at a very low rate: a constant 0.5% rate gives this project a decisively negative net present value.

### *9.2 The Filter (Maximum Endowment) Argument*

On the other hand it has been argued that high discount rates favor future generations, by ensuring that investment funds are channeled into projects which offer, through investment and reinvestment, the highest possible accumulated revenue at a given future date[10].

The generally prevailing rate of return should (it is said) be used as a discount rate for specific projects, to filter out the lower-yielding ones. Because small but persistent long-term benefits are treated rather favorably by the declining discount rate protocol, this is a less effective filter than a high discount rate in perpetuity would be. It could be considered that future generations are best endowed by adopting projects capable of showing a profit even if a sustained high discount rate is applied to its long-term benefits.

### *9.3 The Compensation (Adequate Endowment) Argument*

The discounting of future environmental damage has sometimes been justified as follows. Suppose damage of £1000 is expected after 50 years as a result of a project which yields immediate cash benefits. Suppose cash is invested in a capital fund, at 6% interest, in order to provide compensation for this damage. A sum of £1000 $\div 1.06^{50}$ = £54.29 would be needed. But if compensation were required after 100 years, only £1000 $\div 1.06^{100}$ = £2.95 would be needed. Hence the later the damage, the more heavily the required compensation can (legitimately) be discounted. Obversely, projects yielding early benefit are, all else equal, to be preferred. To pay £1000 in compensation after 100 years requires investment of £2.95 from revenue accruing immediately, but of £54.29 from revenue not accruing until the end of year 50. In these calculations of provision to be made for compensation, the discount rate remains constant, at the expected return on investment, according to conventional theory.[11]

However, Weitzman (1998) and Newell and Pizer (2001) argue that future returns to investment (and particularly future interest rates) are not certain, but follow a random walk or other stochastic process. This being the case, the issue becomes once again the aggregation of scenarios, in which some possible sequences – with generally declining returns to investment – compromise the ability to compensate future generations. Rates of return may rise as easily as fall: the mean expected rate of return in stochastic simulations was equal to the initial rate of return. However, the mean expected investment required to achieve compensation corresponds to a mean expected discount rate lower than the current rate of return, the more so as time elapses and random walks take rates of return to greater possible extremes (see figure 6.12).

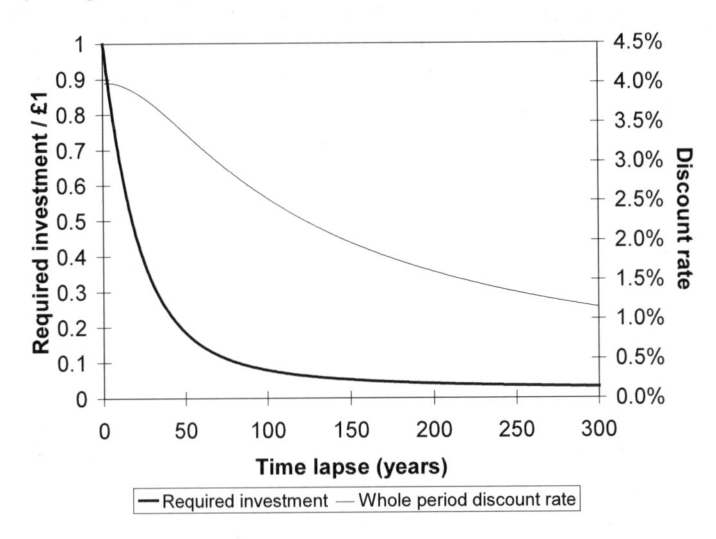


***Figure 6.12.*** *Mean Investment Required for Compensation with Risky Returns, Based on 10 000 Iterations of a Stochastic Model, with 4% Starting Rate of Return and Maximum Variation ±20% per Period*

Simply put, if rates rise continuously, required investment tends to zero, whereas if rates decline continuously towards zero, required investment tends to equality with the required compensation. The mean of these two extreme cases, tending to a discount factor of 0.5, is much greater than the discount factor for a long time period based on a constant rate of return, and the discount rate producing it must therefore be lower than the present rate of return. Figure 6.12 illustrates this.

This profile of rates is closely similar to that derived for OXERA (2002), by a similar process.

*9.4 The Filter Argument Revisited*

In the context of variable returns, the filter argument provides a quite different conclusion from the above. Suppose the objective is to supply a generation (at some arbitrary point in the future) with the maximum endowment from investment. Figure 6.13 shows the mean rate at which the endowment grows.

In the context of forestry decision making, figure 6.13 has the following implications. If a forest rotation lasts 50 years, an alternative investment would yield on average a rate of return of 4.3%. This is the mean opportunity cost to future generations of the forest investment: it is only slightly above the present rate of return. However, if the forest rotation is 100 years, the mean opportunity cost of forest investment has risen markedly to 6.0%. (Naturally, the predicted returns from forestry should also be treated stochastically in making such a comparison.)

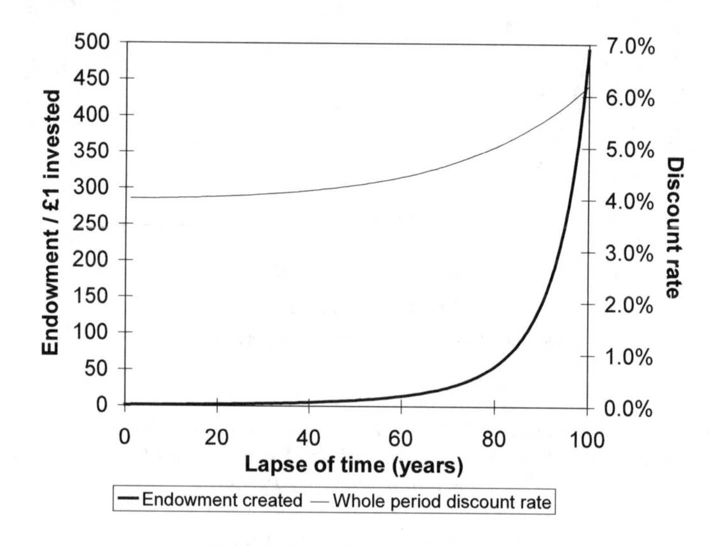


***Figure 6.13**. Mean Endowment Generated with Risky Returns, Based on 10 000 Iterations of a Stochastic Model, with 4% Starting Rate of Return and Maximum Variation ±20% per Period*

In determining the growth of the endowment, the dominant replicates are the rather small number for which the random walk produces a long sequence of high returns, yielding a massive final endowment. By contrast, the worst that can happen at the end of a long sequence of low rates of return is that the endowment fund has not grown at all.

The filter (maximum endowment) and compensation (adequate endowment) arguments, which superficially resemble one another, in fact generate diametrically opposite conclusions for long-term discount rates[12].

*9.5 Diminishing Marginal Utility and Compensation*

A complicating factor is the diminishing marginal utility of monetary endowments or compensations in the context of a growing economy. Suppose, for example, that the growth rate of financial income per head, $g$ is equal to half the rate of return on investment, $i$, and that the elasticity of marginal utility of income is –2. The discount rate on consumption, $r$, is given by

$$r = g \times |-2| = \frac{i}{2} \times |-2| = i \qquad (4)$$

which is equal to the growth rate of the endowment or compensation fund. In such an instance, the growth of funds exactly offsets diminishing marginal utility discounting of the *value* of the fund, whether interpreted as the value of the endowment or the cost of adequate compensation. This is unchanged by stochastic variation in the trajectory of rates of return, since $i$ and $r$ remain equal. The corresponding discount rate on utility would thus seem to be constant at zero. However, as demonstrated elsewhere (Price, 2000b, 2004) when amounts of compensation are, or may become, larger than marginal, the discount rate declines through time and may become negative, indefinitely so. This is the case, whether rates of return are stochastic or constant.

## 10. HOW DEFENSIBLE IS THE ARGUMENT FOR DISCOUNTING?

Three lines of thought that have run through previous sections are these.

i. The results of conventional discounting over the long term are becoming less acceptable, from a variety of perspectives.

ii. Discounting at a declining rate, as the time period is extended, may raise practical perplexes (dynamic inconsistency), and in all cases the arguments for so doing are even stronger arguments for a more complete disaggregation of discounting procedures.

iii. Disaggregation throws up cases where discounting is not appropriate at all: individuals with zero time preference or zero income growth; products whose utility does not diminish; scenarios under which environmental impacts are continuingly serious, or capacity to pay compensation is compromised by uncertain returns to investment.

Thus this section considers whether any form of discounting can be defended logically as a default position.

### *10.1 The Compensation Argument*

As d'Arge, Schulze, and Brookshire (1982, p.255) put it:

> Economists often use the notion of "hypothetical" compensation to justify discounting. In an ethical context such arguments play no role whatsoever. Rather, if no actual compensation occurs, the market rate of return has no relevance for discount rates.

Nor does it have much relevance if the compensation fund exists, but is not entirely invested at compound interest. The declining discount rate protocol could here be based on a belief that, while the present generation will be conscientious in reinvesting the compensation fund, future generations are progressively less likely to do so. After all (it could be argued) intermediate future generations had no part in the decision to gain short-term benefit at the cost of providing some compensation. Uncertainty arises now about the amount of reinvestment, as well as about the rate of return.

But there may be doubts even in the short term. El Serafy (1989) puts the case plainly when he states that "the setting aside of part of the proceeds [of natural resource exploitation] in reinvestment is only a metaphor". Metaphorical reinvestment endows no compensation fund, and no justification for discounting, even from day one.

Newell and Pizer (2001) assume an identity between the financial rate of interest and what they call the "consumption rate of interest", the rate at which consumers discount future acts of consumption. This assumed identity of rates of interest, as seen from production and consumption viewpoints, has been in currency at least since the time of Fisher (1907; 1930). Its message for this context is that it does not matter whether reinvestment takes place or not, because consumers are indifferent between the fruits of marginal investment (a higher level of future consumption) and those of marginal immediate consumption (a higher level of utility attributable to earliness in time).

### *10.2 The Time Preference Argument*

But such a conclusion reactivates the issues with which this paper began its search for a declining discount rate.

- Is the present entitled to discount *utility* accruing to future generations at any rate at all? The usual ethical position has been at least to query such an entitlement.
- What is the rational justification for discounting even benefits and costs accruing to oneself, other than the variously applicable diminishing marginal utility of consumption?

If, as Newell and Pizer (2001) suggest, interest rates vary from the presently prevailing ones, then personal investment decisions will be adjusted so that equality at the margin between investment and consumption rates of interest is maintained. This altered *marginal* discount rate does not mean however that people's relative valuation of the future *in its totality* has shifted commensurately. In this sense, the argument behind declining discount rates represents people's time preference as being more unstable than it really is.

More generally, the customary justification for public discounting is that a democratic society should respect the preferences of citizens – and this applies to preference among time periods as much as to preference among goods. However, there is more than one interpretation of people's choices in favor of early consumption, because "early consumption" may also at the instant of choice be characterized as "immediate consumption". There are plentiful accounts – economic (Bohm-Bawerk, 1884; Thaler, 1981) and literary[13] – of subsequent regret about impulsive choices made in favor of early consumption, with the consequence of heavy later costs. And, just as the view of the far future embedded in discounting minimizes its importance, so we tend to trivialize the significance of events in the dim and distant past (Price, 1997b).

This symmetry of views looking forwards and backwards indicates that it is not *earliness as such* that is favored, but *immediacy in time* – preference for now over any other time, and preference for close proximity to now over lesser proximity to now. Reflective exploration of individuals' motives for choice of "early" has confirmed that "earliness" as such is rarely considered to confer a premium.

Suppose there were reasonable doubts about which of these two views of time preference was correct. Suppose equal weight is given to the two views in deriving a discount factor. Even so, with the lengthening of time period, the "nowness" view, with no premium on earliness, will come to dominate the discount rate derived from the mean discount factor. This result is mathematically equivalent to Li and Löfgren's (2000) result for two individuals with different time preference rates. But the weight of evidence and of reason is such, that the "nowness" view should be judged overwhelmingly the more plausible, as well as the more popularly held. In this perspective, it is not of particular relevance that some notional conservationist discounts at 0% and an equally stereotypic utilitarian at a positive $r$%: if the utilitarian discounts only for *non-immediacy*, future benefits are not to be discounted for their mere *futurity*. What is to be estimated is the value of benefits, as it will be perceived at the instant of consumption, whether this is value to the conservationist, or to the utilitarian, or to some future consumers bearing equivalent or different labels. A range of changed circumstances affects the value reasonably to be attributed. But it is those circumstances *as such* that should be addressed in the judgment of future values, and not futurity in itself.

By this reinterpretation of time preference, the problems emerging from different discounting treatment of future and present generations' consumption disappear, as do those of hyperbolic time preference. If time preference means preference for immediacy, all acts of consumption, when they occur, have that immediacy: all points in time have an equivalent profile of importance as they are anticipated, experienced and remembered. The predicted circumstances of consumption may not eventuate in practice, and better knowledge will lead to revised judgments of value and preferred courses of action. Again, however, it is not the passing of time as such that changes valuations: if this better knowledge had been available at present, then the present evaluation of consumption would have been the same as that made later. Only in that sense is there dynamic inconsistency: unlike the inconsistency that results from a time-varying discount rate, the direction and magnitude of shift of preference are not determinate. Present judgments should consider a best-guess range of outcomes, including, as relevant, changes of economic circumstances; and flexible options that allow adaptive future decision making should be given some premium value.

### *10.3 The Diminishing Marginal Utility Argument*

Of the circumstances that condition future value, diminishing marginal utility is the most evident[14], though its importance varies by product, by consumer group and by scenario. The case for discounting on these grounds is as sustainable (in either sense) as the ability of the Earth and its economic systems to provide increased consumption per capita of the focus product or condition to the focus group of consumers: no more and no less. Discounting both affects future resource availability (by favoring immediate exploitation), and belittles the significance of that availability: there is need to be vigilant against circular, inconsistent and irresponsible arguments. Gordon's dictum (1967, p.267) – "exhaustion [of non-renewable resources] will occur in the distant future – present generations may wisely ignore it" – is symptomatic of a discounting philosophy which seems explicitly and aggressively at odds with sustainability philosophy.

Lack of certainty about resource and environmental futures (and even opponents of measures against global climate change have often based their case on the uncertainty of the science) shifts the balance towards a zero discount rate. This is quite contrary to traditional financial treatment of risk.

## 11. POLITICAL SUSTAINABILITY

Discounting for the circumstances of each product, scenario or income group is a time-intensive and controversial task. Using a tariff of discount rates which varies only by time period represents a relatively manageable alternative for project evaluation. But there are subtler advantages than this.

A democratic government may discount on the grounds that its electorate wishes it to do so. The citizens may be quite mistaken in interpreting their own time preference as a wish for earliness rather than a wish for immediacy. But – until time has passed and they have formulated regrets[15], and while the past remains unalterable – they will consider themselves well served by a government that continues to discount conventionally on their behalf. Of course, governments do not do all the unethical things that an electorate may wish of them. But when the political advantage of acting rightly bears fruit only during the office of some future government, it has limited appeal. A government seeking popular support, while claiming moral high ground, may well be attracted by the declining discount rate philosophy. It gives the impression that the government cares about the distant future, while being considerate of the present generation's impatience.

The protocol does not on the whole make heavy demands on the present: Newell and Pizer (2001) find that it brings *at maximum* an increase from \$5.74 to \$10.44 in the discounted long-term cost of emitting a ton of $CO_2$ to the atmosphere[16]. That justifies only minor present mitigation measures, as anyone familiar with the economics of carbon-fixing forestry will appreciate. In contrast, indefinite increase in cost and in defensible mitigation measures arises when discounting is abandoned. Neither does the protocol make the value of a sustainable benefit flow indefinitely large: the discounted value of £1 every year in perpetuity is a modest £44 as opposed to the £29 given by discounting conventionally at the initial 3.5% rate.

The net result is likely to be business-not-very-different-from-usual.

As with all protocols in which discount rates decline through time, a requirement for rather undemanding present sacrifice to favor the medium term (a generation or two ahead), is combined with desirability of much greater medium-term sacrifice to favor the long term (several generations ahead). And, if long-term good may be done either by modest present sacrifice, or by much greater medium-term sacrifice, this greater medium-term sacrifice may well be preferred. Figure 6.14 shows the discounted cash flows from a major forestry program costing £1000 million immediately, and yielding a modest return (in environmental or material benefit) of 2.5% (dark bars). It has positive net present value (£95M) under the OXERA discount rates – but a much greater net present value (£282M) if postponed for 30 years (light bars).

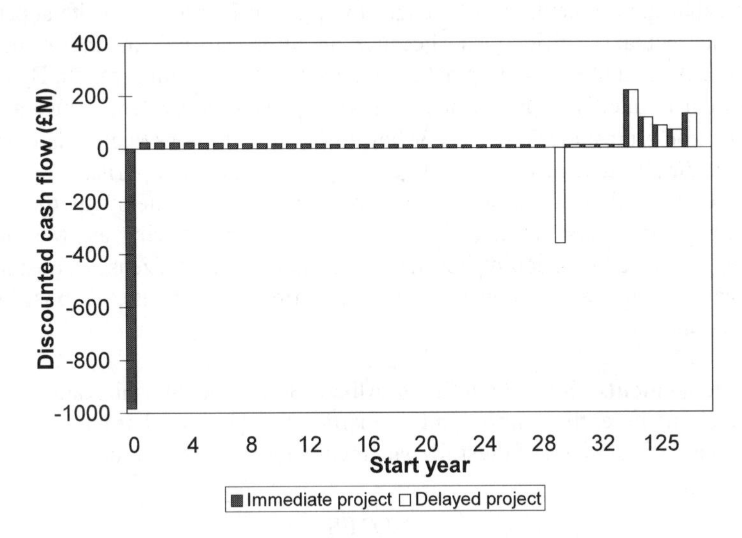


***Figure 6.14**. Discounted Cash Flows of Immediate and Delayed Project*

However, on the same argument the sacrifice or the investment is also likely to be postponed by the decision makers of the next generation, and by those of the next. Thus the long-term good may never arise, though all generations agree that it is worth achieving, by whatever is the least-sacrifice means. Casual introspection will reveal that many of us as individuals also act in this way ("I'll get that paper written tomorrow") under the influence of hyperbolic discounting. This is picoeconomics (Ainslie 1992): and adopting the declining discount protocol as national economic policy is pico-economics writ large.

## 12. CONCLUSION

The problem with adopting a standard protocol for a discount rate declining with *time* is that it obscures the underlying causes of decline: that is, the aggregation of components in which the low – or no – discount components progressively become more dominant than the high discount components. By combining the effects of aggregation into a single function of time, the protocol itself suffers the problem of aggregation – lack of context sensitivity. The mistake is to vary discount rates only across time, rather than across circumstances. On the whole high discount rates should not be sustained through time. But in many circumstances their application cannot be logically sustained even in the short term. Whatever the time scale, discounting as a means of evaluating and selecting projects is hostile to the ethic underlying sustainability.

Despite these fundamental weaknesses of the declining-rate protocol, it is to be expected that governments will be eager to embrace it, because of its superficially sound intellectual justifications, because it offers a nod in the direction of sustainability[17], and because in practice it does not change things much. By contrast, the protocol implicitly approved here – giving present equivalent values according to predicted circumstances, not according to lapse of time as such – is demanding both procedurally and in terms of its consequences. Perhaps purists should not let the perfect (not discounting at all for time lapse) be the enemy of the marginal improvement (not discounting the distant future as heavily as was formerly practiced). But neither should they let governments – or citizens – subside into a complacent belief that, by lowering the long-term discount rate, all possible good has been done.

**Acknowledgements:** I am grateful to Albert Berry and Shashi Kant for helpful comments on an earlier draft, and to participants in the Toronto International Conference on Sustainable Forest Management for their observations.

## NOTES

---

[1] Weighting is a process of adjusting an ascribed value to allow for factors specific to the circumstances of its accrual. In this chapter, these circumstances include time-related conditions, and the income level of the beneficiary or cost-bearer.

[2] E.g. the present value of £1000 received every year in perpetuity, discounted at 6%, is £1000÷0.06=£16 667.

[3] It might reasonably be thought that the discontinuities in this curve could cause problems. For example, the discount factor for 301 years is 4.3 times *greater* than that for 300 years, and the same as that for 201 years. The steps in discount factors may seem rather small on the scale of the axis. However, if values after 300 years have any relevance at all to a decision, they would have to be big, so a 4.3-fold step would

be significant, even after discounting. This casual overlooking of what in fact is a powerful objection to stepped discounting could be taken as symptomatic of how superficially some unconventional discounting protocols have been considered.

[4] Broome (1991) argues that utilitarianism embraces a number of possible interpretations of value, not all of which necessarily imply discounting.

[5] Although all are loosely referred to as "hyperbolic", the various functional forms appearing in the literature under this title are not strictly hyperbole.

[6] There *may*, however, be a distinction, since individuals' affluence relative to mean income changes through their life-cycle. This adds weight to the case made in the following sections for circumstance-specific utility evaluation.

[7] For one individual, this may be identified with the reciprocal of the income-adjusted price elasticity of demand.

[8] Effective discount rate: a discount rate adjusted for specific time trends which might be expected to change future values.

[9] The beta coefficient expresses the relationship between the yield of an individual asset and the yield of the whole portfolio.

[10] Actually, whatever the underlying rationale, to select the most efficient investments does not require a high discount rate. A low rate may lead to low return investments' being accepted, if the question is *whether* to do them or not. But if the question is *which* projects to do, low rates actually increase the margin by which high revenue projects are economically superior to low revenue projects.

[11] The entire compensation argument is, as it happens, based on the unreasonable assumption that a constant trade-off exists between future damage to interests, and cash compensation. In practice, the value placed on particular damage to health or environment is likely to remain more-or-less constant through time, while the marginal utility of given cash compensation declines with increasing income (Price, 2000b). This, too, may result in a discount rate which changes in both magnitude and sign through time.

[12] This opposite trend of the two derived discount rates persists even when the same sequences of stochastic variation in rates is the basis for both compensation and filter versions of the argument.

[13] It might convincingly be argued that failure to take proper account of the future costs resulting from short-term gains is the most common theme of literary tragedies.

[14] Others are considered in Price (1993, chapters 10-14).

[15] For example, as about inadequate provision for state pensions or insufficient investment in the railway network.

[16] These figures are both, as it happens, too low, being based on a misunderstanding of the durability of $CO_2$ in the atmosphere (Price, 1995).

[17] The widely advocated strategy of imposing sustainability constraints to limit the malign effects of discounting results in inflexibility: this may obstruct beneficial changes, while preserving an unsatisfactory status quo (Price 2000b).

## REFERENCES


Ainslie, G. (1991). Derivation of 'rational' economic behavior from hyperbolic discount curves. *American Economic Review Papers and Proceedings*, 81, 334-340.

Ainslie, G. (1992). *Picoeconomics*. Cambridge : Cambridge University Press.

Bayer, S. (2003). Generation-adjusted discounting in long-term decision-making. *International Journal of Sustainable Development*, 6, 133-49.

Bellinger, W.K. (1991). Multigenerational value: Modifying the modified discounting method. *Project Appraisal*, 6, 101-108.

Böhm-Bawerk, E.C. (1884). *Positive theory of capital*, translated by G.D. Huncke and H.F. Sennholz. Libertarian Press, 1959.

Broome, J. (1991). *Weighing goods*. Oxford: Blackwell.

Cropper, M., & Laibson, D. (1999). The implications of hyperbolic discounting for project evaluation. In: Portney and Weyant (q.v.), pp. 163-172.

d'Arge, R.C., Schulze, W.D., & Brookshire, D.S. (1982). Carbon dioxide and intergenerational choice. *American Economic Review Papers and Proceedings,* 72, 251-256.

Dasgupta, P.S., & Heal, G. (1974). The optimal depletion of exhaustible resources. *Review of Economic Studies Symposium*, 3-28.

El Serafy, S. (1989). The proper calculation of income from depletable natural resources. In Y.J. Ahmad, S. El Serafy, E. Lutz. (Eds), *Environmental accounting for sustainable development* (pp. 10-18). Washington: The World Bank.

Fankhauser, S. (1995). *Valuing climate change*. London: Earthscan.

Fisher, A.C., Krutilla, J.V., & Cicchetti, C.J. (1972). The economics of environmental preservation. *American Economic Review*, 62, 605-619.

Fisher, A.C., & Krutilla, J.V. (1975). Resource conservation, environmental preservation and the rate of discount. *Quarterly Journal of Economics,* 89, 358-370.

Fisher, I. (1907). *The rate of interest*. London: Macmillan.

Fisher, I. (1930). *The theory of interest*. London: Macmillan.

Gollier, C. (2002). Discounting an uncertain future. *Journal of Public Economics*, 85, 149-166.

Gordon, R.L. (1967). A reinterpretation of the pure theory of exhaustion. *Journal of Political Economy*, 75, 274-286.

Gummer, J.S. (1994). In Anon., *Sustainable development: The UK strategy*. London: Cmnd 2426, HMSO.

Harrod, R.F. (1948). *Towards a dynamic economics*. London: Macmillan.

Henderson, N., & Bateman, I. (1995). Empirical and public choice evidence for hyperbolic social discount rates and the implications for intergenerational discounting. *Environmental and Resource Economics*, 5, 413-423.

Her Majesty's Treasury (1991). *Economic appraisal in central government: A technical guide for government departments*. London: HMSO.

Her Majesty's Treasury (undated). *The green book: Appraisal and evaluation in central government.* TSO, London. [Downloaded 2003.]

Hume, D. (1739). *A treatise of human nature*. Oxford: Oxford University Press.

Kula, E. (1988). Future generations: The modified discounting method. *Project Appraisal,* 3, 85-88.

Kula, E. (1989). The modified discounting method – a rejoinder. *Project Appraisal*, 4, 110-13.

Li, C-Z., & Löfgren, K-G. (2000). Renewable resources and economic sustainability: A dynamic analysis with heterogeneous time preferences. *Journal of Environmental Economics and Management*, 40, 236-50.

Lind, R.C. (Ed.) (1982). *Discounting for time and risk in energy policy*. Baltimore: Johns Hopkins University Press.

Newell, R., & Pizer, W. (2001). *Discounting the benefits of climate change mitigation: How much do uncertain rates increase valuations?* Arlington: Pew Center on Global Climate Change.

OXERA (2002). *A social time preference rate for use in long-term discounting*. The Office of the Deputy Prime Minister, Department for Transport, and Department of the Environment, Food and Rural Affairs, London.

Portney, P.R., & Weyant, J.P. (Eds) (1999). *Discounting and intergenerational equity*. Washington: Resources for the Future.

Price, C. (1973). To the future: With indifference or concern? – The social discount rate and its implications in land use. *Journal of Agricultural Economics*, 24, 393-398.

Price, C. (1984). The sum of discounted consumption flows method: Equity with efficiency? *Environment and Planning A*, 16, 829-833.

Price, C. (1989). Equity, consistency, efficiency and new rules for discounting. *Project Appraisal,* 4, 58-64.

Price, C. (1993a). *Time, discounting and value.* Oxford: Blackwell. Also freely available electronically from c.price@bangor.ac.uk.

Price (1993b). Discounting diversity. *Scandinavian Forest Economics*, 34, 37-57.

Price, C. (1995). Emissions, concentrations and disappearing $CO_2$. *Resource and Energy Economics,* 17, 87-97.

Price, C. (1996). Discounting the environment – a critique. In P. Hyttonen, A. Nilson (Eds), *Integrating environmental values into forest planning; EFI proceedings* 13 (pp.89-98). Joensuu: European Forest Institute.

Price, C. (1997a). Analysis of time profiles of climate change. In W.N. Adger, D. Pettenella, M. Whitby (Eds.), *Climate change mitigation and European land use policies* (pp. 71-87). Wallingford: CAB International.

Price, C. (1997b). *Dim and distant futures? Sustainability and the abominable practice of discounting.* Inaugural lecture paper, University of Wales, Bangor.

Price, C. (2000a). The landscape of sustainable economics. In J.F. Benson, M. Roe (Eds.), *Landscape and sustainability* (pp. 33-51). London: Spon.

Price, C. (2000b). Discounting compensation for injuries. *Risk Analysis*, 20, 239-249.

Price, C. (2003). Diminishing marginal utility: The respectable case for discounting? *International Journal of Sustainable Development*, 6, 117-32.

Price, C. (2004). The rate of return may not be equal to the social discount rate: So what follows? *Presented at the International Conference on Economics of Sustainable Forest Management, Faculty of Forestry, University of Toronto, May 20-22, 2004.*

Price, C. (in press). An intergenerational perspective on effects of environmental changes: Discounting the future's viewpoint? *The Socio-economic Implications of Environmental Change with particular relevance to Forestry*. Wallingford, UK: CAB International.

Price, C., & Nair, C.T.S. (1985). Social discounting and the distribution of project benefits. *Journal of Development Studies*, 21, 525-32.

Ramsey, F.P. (1928). A mathematical theory of saving. *Economic Journal*, 38, 543-59.

Scott, A. (1958). *Natural resources: The economics of conservation*. Toronto: University of Toronto Press.

Sen, A.K. (1957). A note on Tinbergen on the optimum rate of saving. *Economics Journal*, 67, 745-748.

Sidgwick, H. (1874). *The methods of ethics*. London: Macmillan.

Strotz, R.H. (1956). Myopia and inconsistency in dynamic utility maximization. *Review of Economic Studies*, 23, 165-80.

Thaler, R. (1981). Some empirical evidence on dynamic inconsistency. *Economics Letters*, 8, 201-7.

Tullock, G. (1964). The social rate of discount and the optimal rate of investment: Comment. *Quarterly Journal of Economics,* 78, 331-336.

Weitzman, M.L. (1998). Why the far-distant future should be discounted at the lowest possible rate. *Journal of Environmental Economics and Management*, 36, 201-8.

# CHAPTER 7

# INTERGENERATIONAL EQUITY AND THE FOREST MANAGEMENT PROBLEM

TAPAN MITRA

*Department of Economics, Cornell University,*
*Ithaca, NY 14853, USA.*
*Email: tm19@cornell.edu*

**Abstract.** The paper re-examines the foundations of representation of intertemporal preferences that satisfy intergenerational equity, and provides an axiomatic characterization of those social welfare relations, which are representable by the utilitarian ordering, in ranking consumption sequences which are eventually identical. A maximal point of this ordering is characterized in a standard model of forest management. Maximal paths are shown to converge over time to the forest with the maximum sustained yield, thereby providing a theoretical basis for the tradition in forest management, which has emphasized the goal of maximum sustained yield. Further, it is seen that a maximal point coincides with the optimal point according to the well-known overtaking criterion. This result indicates that the more restrictive overtaking criterion is inessential for a study of forest management under intergenerational equity, and provides a more satisfactory basis for the standard forestry model.

## 1. INTRODUCTION

In forestry management, there has been a tradition which claims that "the goal of good policy is to have sustained forest yield, or even maximum sustained yield somehow defined" (Samuelson, 1976, p.146). In an attempt to understand this tradition, Mitra and Wan (1986) formulated the problem of forest management as one of optimizing the sum of (undiscounted) utilities from harvests of timber according to the well-known overtaking criterion. Their formulation allowed forestry economics to be viewed as a particular case of modern capital theory. Exploiting the optimization methods familiar from this general theory, they were able to show that, when the utility function is strictly concave, starting from any initial forestry configuration, the optimally managed forest converges over time to the forest with the maximum sustained yield, which corresponds to the "golden-rule" of the forestry model. This demonstration provided a theoretical basis for the tradition in forestry management.



*Kant and Berry (Eds.), Economics, Sustainability, and Natural Resources: Economics of Sustainable Forest Management,* 137-173.

Over the last twenty five years, work on the general area of sustainable development has brought about a thorough re-examination of the nature of preferences over time relevant to studies on environmental economics and the management of renewable resources. The Pareto efficiency principle is, of course, an essential ingredient in such studies. And, the equal treatment of all generations figures prominently in this literature, unlike other optimal intertemporal allocation problems, where discounting future utilities is common practice.[1]

These two guiding principles in the study of intertemporal preferences (often jointly referred to as the Suppes-Sen grading principle)[2] do not, however, take us far in terms of being able to develop useful qualitative properties of the evolution of renewable resource stocks over time under appropriate management. In aggregating utilities of all generations, it is therefore common practice to choose a more restrictive social welfare relation (SWR), and typically this SWR has been induced by some form of the overtaking criterion.

In this paper, I will re-examine the foundations of intertemporal preferences which respect intergenerational equity, propose a social welfare relation which is weaker (less restrictive) than the one induced by the overtaking criterion, and provide an axiomatic basis for it. I will then apply this SWR to a standard model of forestry and demonstrate the somewhat surprising result that *all* the qualitative properties of optimally managed forests that one can obtain by applying the more restrictive overtaking criterion can be obtained with the application of the weaker and more acceptable SWR.

In proposing the new SWR, I would like to push the point of view that in comparing infinite consumption streams, it is only the comparisons of consumption streams which are "eventually identical" that are truly non-controversial.[3] Consequently, I will put forward a set of axioms which are consistent with this point of view. These axioms are Weak Pareto, Anonymity, Completeness and Continuity for finite horizon comparisons, and Independence. Of these the first two have already been discussed above, except that we postulate a weak version of Pareto, which restricts comparisons to consumption streams which are eventually identical. The third axiom is natural: since we are, after all, extending the theory from the finite-dimensional to the infinite-dimensional case, we cannot hope to weaken the postulates of the standard finite-dimensional theory.[4] In the intertemporal context, the fourth axiom is common in *utilitarian* representations of preferences.

A noteworthy feature of our SWR is that it is axiomatized without postulating any continuity property on the preference relation in the infinite dimensional space containing the set of consumption streams.[5] In contrast, axiomatic characterizations of the more restrictive SWR induced by the overtaking criterion typically involve some form of a continuity axiom.[6] Axioms on the continuity of preferences in infinite-dimensional spaces have been the most controversial in the literature in this area, since the topology in which such continuity is assumed determines to a large extent the nature of allowable preferences.[7]

We apply our SWR to rank consumption streams generated by the model of forestry used in Mitra and Wan (1986). We call a consumption stream *maximal* if it is a maximal point in the feasible set in terms of the SWR, and study properties of maximal paths. We find that maximal paths converge over time to the forest with the maximum sustained yield, which demonstrates that the notion of maximality is enough to provide a theoretical basis for the tradition in forest management, which has emphasized the goal of maximum sustained yield. That is, in studying the long-run properties of paths which are maximal and those which are *optimal* according to the familiar overtaking criterion, the latter concept does not possess any discernible advantage.

Using duality theory, we show that maximal paths have generalized intertemporal profit maximizing (bounded) shadow prices associated with them, just like optimal paths do. So, one cannot have "errant behavior" on maximal paths, compared to optimal paths, in the short-run either. We combine the above two findings to establish the result that the set of maximal paths coincides exactly with the set of optimal paths. This leads us to conclude that in the context of the forestry model, one can completely dispense with the more restrictive overtaking criterion.

## 2. NOTATION

Let $\mathbb{N}$ denote the set of natural numbers $\{1,2,3,...\}$, and let $\mathbb{R}$ denote the set of real numbers. Let $L$ denote the closed interval $[0,1]\subset\mathbb{R}$, let $S$ denote the set $\mathbb{R}^{\mathbb{N}}$, and let $C$ denote the set $L^{\mathbb{N}}$. $C$ is to be interpreted as the set of consumption sequences. Thus, we write $c\equiv(c_1,c_2,...)\in C$ if and only if $c_t\in[0,1]$ for all $t\in\mathbb{N}$.

Given $x\in S$, and $N\in\mathbb{N}$, let us denote by $x(N)$ the vector consisting of the first $N$ elements of $x$ and by $x[N]$ the sequence from term $(N+1)$ onwards. So, $x(N)=(x_1,x_2,...,x_N)$ and $x[N]=(x_{N+1},x_{N+2},...)$. Clearly, $x[N]\in S$. We will denote the sequence $(x_1,x_2,...,x_N,0,0,...)$ by $(x(N),0[N])$.

For $c,c'\in C$ we write $c'\geq c$ if $c_t'\geq c_t$ for all $t\in\mathbb{N}$ ; and, we write $c'>c$ if $c'\geq c$, and $c'\neq c$.

For each $n\in\mathbb{N}$, the *sum-norm* on $\mathbb{R}^n$ is defined by:

$$\|x\|=\sum_{i=1}^{n}|x_i| \quad for\ x\in\mathbb{R}^n$$

The corresponding *sum metric* is defined by:

$$d(x,y)=\|x-y\| \quad for\ x,y\in\mathbb{R}^n$$

The unit vectors in $\mathbb{R}^n$ are denoted by $e^1,...,e^n$.

Denote by $U$ the set of continuous functions $u:L\to\mathbb{R}$, such that $u$ is increasing on $L$, and $u(0)=0,u(1)=1$.

A *social welfare relation* (SWR) is a binary relation, $\succsim$, on $C$, which is reflexive and transitive (a pre-ordering).[8] We associate with $\succsim$ its symmetric and asymmetric components in the usual way. Thus, we write $c\sim c'$ when $c\succsim c'$ and $c'\succsim c$ both hold; and, we write $c'\succ c$ when $c'\succsim c$ holds, but $c\succsim c'$ does not hold.

A SWR $\succsim_A$ is a *subrelation* to a SWR $\succsim_B$ if (a) $x, y \in X$ and $x \succsim_A y$ implies $x \succsim_B y$ and (b) $x, y \in X$ and $x \succ_A y$ implies $x \succ_B y$.

## 3. ON A UTILITARIAN SOCIAL WELFARE RELATION[9]

In this section we will introduce the notion of a utilitarian social welfare relation, and provide an axiomatic basis for it. The social welfare relation will *only* compare consumption sequences which are "eventually identical", so that the relation will be "more incomplete" than the familiar SWR induced by the overtaking criterion.[10] It will be based consequently on axioms that relate only to consumption sequences which are eventually identical, and therefore will be more widely acceptable.

### *3.1 Axiomatic Characterization of a Utilitarian SWR*

**Definition 1.** *Given a utility function, $u \in U$, a utilitarian SWR corresponding to $u$ is a binary relation $\succsim_u$, defined by:*

$$c \succsim_u c' \textit{ iff } \exists N \in \mathbb{N} \textit{ such that } c[N] = c'[N],$$

$$\textit{and} \sum_{t=1}^{N} u(c_t) \geq \sum_{t=1}^{N} u(c'_t) \tag{U}$$

We find the ranking of consumption sequences according to the utilitarian SWR to be persuasive. Thus, one may consider getting together the members of the finite society $\{1, \dots, N\}$, and asking them to rank $c$ versus $c'$. If they apply utilitarian principles to themselves, they will rank $c$ above $c'$ if:

$$\sum_{t=1}^{N} u(c_t) > \sum_{t=1}^{N} u(c'_t)$$

In this case, it is legitimate for the infinite horizon society to rank $c$ above $c'$ because the infinite number of future generations, who are not included in the finite society $\{1, \dots, N\}$, are indifferent between $c$ and $c'$. In other words, in this situation, all future generations beyond $N$ are willing to go along with the (utilitarian) preferences of the finite society $\{1, \dots, N\}$.

The following example illustrates the nature of the utilitarian SWR:

**Example 1.**

$$\left.\begin{array}{lllllllll} c = & (0.2, & 0, & 0.1, & 0.2 & 0.1, & 0.2, & \dots) \\ c' = & (0, & 0.1, & 0.1 & 0.2, & 0.1 & 0.2, & \dots) \end{array}\right\}$$

Here the finite society $\{1,2\}$ applying utilitarian principles to itself prefers $c$ to $c'$. All future generations beyond time period 2 are indifferent between $c$ and $c'$. So, in this case, we would have $c \succ_u c'$, according to every utility function $u \in U$.

We turn now to the set of axioms which characterize the utilitarian SWR. To this end, let us first consider the following two axioms, which are fairly straightforward.

**Axiom 1** *(Weak Pareto). If $c, c' \in C$, and there exists $N \in \mathbb{N}$, such that $c(N) > c'(N)$ and $c[N] = c'[N]$, then $c \succ c'$.*

**Axiom 2** *(Anonymity). If $c, c'$ are in $C$, and there exist $r, s$ in $\mathbb{N}$, such that $c_r = c'_s$ and $c_s = c'_r$, while $c_t = c'_t$ for all $t \in \mathbb{N}$, such that $t \neq r, s$, then $c' \sim c$.*

The first axiom is weaker than the standard *Pareto Rule*, since it postulates the Pareto pre-order only in comparisons of consumption sequences $c$ and $c'$ which are eventually identical. The second axiom, which is sometimes also called the *Equity Axiom* (see Svensson, 1980), embodies a minimal equity principle in that it ensures equal treatment of the consumption of generations over time in the preference structure.[11]

The next axiom incorporates the notion that when we are dealing with "finite horizon intergenerational societies", we have no difficulty in ranking any two consumption sequences. In fact, the axiom makes preferences over such consumption sequences as well-behaved as postulated in the standard theory of numerical representation of preferences, following Debreu (1954, 1959).

In order to formally write this axiom, let us define, for each $T \in \mathbb{N}$,

$$C(T) = \{c \in C : c_t = 0 \text{ for all } t \in \mathbb{N}, \text{satisfying } t > T\}$$

**Axiom 3.** *For each $T \in \mathbb{N}$, the SWR $\succsim$ is a complete pre-order on $C(T)$, and it is continuous in the sum metric on $C(T)$.*

This axiom combines Axioms 1 and 2 used by Brock (1970b) in his axiomatic characterization of the overtaking criterion. Note that assumptions like completeness and continuity of the pre-order $\succsim$ on $C(T)$ have been non-controversial axioms in this literature. It is only when such assumptions are made on the set $C$ that one needs to discuss whether such a complete ranking is always possible, and whether continuity in some topologies on $C$ are more acceptable than others.

Our final axiom is a strong one, but its use is so prevalent in describing *intertemporal* preferences that we shall have little to say about it, beyond what is already known in the literature. It is the *Independence Axiom*, and it follows the postulate, introduced by Debreu (1960) in the finite-horizon context, and by Koopmans (1960) in an infinite-horizon context, in their studies on the representation of preferences by suitable utility functions.[12]

**Axiom 4** *(Independence). If $c, c' \in C$ and $N \in \mathbb{N}$ satisfy:*

$$(c(N), c'[N]) \succsim (c'(N), c[N]) \tag{1}$$

*then they must also satisfy:*

$$(c(N), c'[N]) \succsim (c'(N), c'[N]) \tag{2}$$

Remark 1

i. *Loosely speaking, the independence axiom says that the preference between two consumption sequences is independent of those parts of the two streams which are identical, whatever be the vector representing the identical part.*

ii. *The above independence axiom is similar to Postulate 3 in Koopmans (1960) and to Postulate P3 in Koopmans, Diamond and Williamson (1964), where it is referred to as "Limited Noncomplementarity".*

If a SWR $\succsim$ satisfies the above four axioms, then it is representable by the utilitarian ordering, in comparisons of consumption sequences which are "eventually identical." To state this result formally, we need to introduce the following notation.

Denote by $D$ the subset of $\mathbb{R}^{\mathbb{N}}$, consisting of sequences with at most a finite number of non-zero entries. If $c, c' \in C$, then $c$ and $c'$ are said to be *eventually identical* if $(c - c') \in D$.

**Theorem 1.** *Suppose a SWR $\succsim$ satisfies Axioms 1-4. Then, there is a utility function, $u \in U$, such that for all $c, c' \in C$, satisfying $(c - c') \in D$,*

$$c \succsim c' \text{ if and only if } \sum_{t=1}^{\infty} (u(c_t) - u(c'_t)) \geq 0 \tag{3}$$

Further, the utility function $u \in U$ satisfying (3) is unique.

**Remark 2.**

i. *Note that the sum in (3) is well-defined since $c'_t = c_t$ for all but a finite number of periods $t \in \mathbb{N}$.*

ii. *In the utilitarian representation of preferences in (3), the fact that the same utility function can be used for all periods follows directly from the Anonymity axiom; in fact, the utility function must be the same for all the periods. The fact that the same utility function can be used regardless of how many non-zero components there are in $(c - c')$ is a deeper result, involving the Independence axiom in an essential way.*

Theorem 1 can be used to provide the following axiomatic characterization of the utilitarian SWR $\succsim_u$.

**Theorem 2.** (i) *Suppose there is a utility function, $u \in U$, for which the corresponding utilitarian SWR $\succsim_u$ is a subrelation to a SWR $\succsim$. Then $\succsim$ satisfies Axioms 1-4.*

(ii) *Suppose a SWR* $\succsim$ *satisfies Axioms 1-4. Then, there is a utility function,* $u \in U$, *such that the corresponding utilitarian SWR* $\succsim_u$ *is a subrelation to* $\succsim$.

### *3.2 Comparison with the SWR defined by the Overtaking Criterion*

The standard method of comparing consumption sequences, while respecting the equal treatment of all generations, is by employing the overtaking criterion. The resulting pre-order is a generalization of the one used by Ramsey (1928), and was proposed independently by Atsumi (1965) and von Weizsacker (1965). It was subsequently generalized by Gale (1967), McKenzie (1968) and Brock (1970a). We can formally define it in the following way.

**Definition 2.** *Given a utility function,* $u \in U$, *a Ramsey-Atsumi-von Weizsacker (RAV) social welfare relation corresponding to* $u$ *is a binary relation* $R_u$, *such that for* $c, c' \in C$,

$$cR_u c' \text{ iff there is } \bar{N} \in \mathbb{N}, \text{such that } \sum_{t=1}^{N} u(c_t) \geq \sum_{t=1}^{N} u(c'_t) \text{ for all } N \geq \bar{N} \qquad \text{(R)}$$

*The symmetric and asymmetric components of* $R_u$ *will be denoted by* $I_u$ *and* $P_u$.

Following Brock (1970b), one can obtain an axiomatic characterization of the RAV social welfare relation in terms of Axioms 1-4 of the previous section, and an additional *consistency* axiom[13], which we now state.

**Axiom 5.** *(Consistency) For* $c, c' \in C$,

$$(a) \text{If there is } N' \in \mathbb{N}, \text{such that } (c(N), 0[N]) \succsim (c'(N), 0[N]) \text{ for all } N \geq N', \text{then } c \succsim c' \qquad \text{(4a)}$$

$$(b) \text{If there is } N' \in \mathbb{N}, \text{such that } (c(N), 0[N]) \succsim (c'(N), 0[N]) \text{ for all } N \geq N', \text{ with } (c(N), 0[N]) \succ (c'(N), 0[N]) \text{ holding for a subsequence of } N \geq N', \text{then } c \succ c' \qquad \text{(4b)}$$

Notice that, unlike Axioms 1-4, Axiom 5 does compare consumption sequences which are not eventually identical. Since such comparisons are based on comparisons of consumption sequences which are eventually identical, one might view Axiom 5 also as a continuity restriction on preferences.[14]

The characterization result[15] can be stated as follows.

**Theorem 3.** (i)*Suppose a SWR* $\succsim$ *satisfies Axioms 1-5. Then, there is a utility function,* $u \in U$, *such that the RAV social welfare relation* $R_u$, *corresponding to* $u$, *is a subrelation to* $\succsim$.

(ii) *Suppose there is a utility function,* $u \in U$, *such that the RAV social welfare relation* $R_u$, *corresponding to* $u$, *is a subrelation to a SWR* $\succsim$. *Then,* $\succsim$ *satisfies Axioms 1-5.*

The crucial difference between the utilitarian social welfare relation and the RAV social welfare relation is that the latter has to satisfy (in addition to Axioms 1-4) the consistency (or continuity) Axiom 5. This difference might best be explained by looking at an example[16] of two consumption sequences where, because of Axiom 5, the RAV SWR can compare the two sequences, but where the utilitarian SWR declares them non-comparable.

**Example 2.**

$$\left.\begin{array}{l} c = \;\; 1/2,\;\; 1/4,\;\; 1/8,\;\; 1/16,\;\; 1/32,\;\; 1/64,\;\; ...) \\ c' = \;\; 3/4\;\; 1/8,\;\; 1/16,\;\; 1/32,\;\; 1/64,\;\; 1/128,\;\; ...) \end{array}\right\}$$

Given the utility function $u \in U$, which satisfies $u(c) = c$, consider the utilitarian SWR $\succsim_u$ and the RAV SWR $R_u$ associated with it. We can verify that for $\bar{N} = 1$,

$$\sum_{t=1}^{T} u(c_t') > \sum_{t=1}^{T} u(c_t) \textit{ for all } T \geq \bar{N}$$

so that $c'$ is preferred to $c$ according to the RAV SWR, corresponding to $u$.

The question arises whether $c'$ *should* be preferred to $c$ by the infinite horizon society. This is not altogether clear. The problems with judging $c'$ to be better than $c$ in such a case can be seen as follows.

If we look at any finite-horizon society, and ask the society to rank $c$ versus $c'$ they will indeed rank $c'$ higher than $c$, if they apply utilitarian principles to themselves. However, no matter how large the finite horizon, there are an *infinite* number of future generations who rank $c'$ below $c$; in fact, *all* generations beyond the first prefer $c$ to $c'$. Thus, it is never possible to have consensus of opinion between any finite horizon society and the infinite number of future generations not included in that finite society. [This should be compared with the "consensus" obtained when the utilitarian SWR can be used to compare consumption sequences].

If one considers the infinite horizon society, one sees that the (infinite) utility sum for both consumption sequences is precisely equal to 1, and so the RAV SWR in fact violates the utilitarian principles on which it is supposed to be based.

It appears to us that Axiom 5 is not as obvious an axiom to accept as Axioms 1-4 in this context. In other words, we find the RAV SWR less persuasive than the utilitarian SWR. Nevertheless, almost all of the theory of optimal intertemporal allocation, in which generations are treated equally in its preference structure, uses the RAV SWR, and therefore accepts Axiom 5 (in addition to Axioms 1-4). The reason for this is that even though Axiom 5 is not an obvious axiom to accept, it gives sufficient structure to intertemporal preferences so that the theory of optimal intertemporal allocation has some predictive power: a path which is optimal according to this pre-ordering in the typical intertemporal model is unique, and the nature of such an optimal path can be described quite accurately, both in terms of short-run characteristics (the *Ramsey-Euler* or *competitive* conditions), and long-run behavior (the *turnpike* property).

The presumption appears to be that if we wanted to proceed with intertemporal preferences satisfying only Axioms 1-4 (that is, *without* imposing something like Axiom 5), one would not have a useful theory of optimal behavior over time. In principle, there could be many maximal points according to the utilitarian SWR, and it might be the case that their behavior (as a group of paths) fails to have some unifying mode worth characterizing. Unfortunately, this issue has not been explored in the literature, and therefore such misgivings about a theory based solely on the utilitarian SWR might be premature.[17]

We will establish the rather surprising result that in the standard forestry model, *a maximal point according to the utilitarian SWR is in fact unique and coincides with the optimal point according to the RAV SWR*. This shows that we do not need the SWR induced by the overtaking criterion, and consequently the comparisons entailed by Axiom 5, to present the theory of forest management, respecting intergenerational equity. This puts the theory on a more robust and satisfactory basis in terms of the postulated intertemporal preference structure.

### *3.3 Intertemporal Inequality Aversion*

In order to obtain the result mentioned above, in the standard forestry model, we need a bit more structure on the utility function derived in Theorem 1, based on Axioms 1-4. Specifically, we will need the utility function to be concave, and strictly mid-concave, on $I$. While this is standard fare for many models of intertemporal allocation, the axiomatic basis of this requirement is often not stressed.[18] We present an exposition of this aspect of the theory by imposing the following additional axiom on preferences.

**Axiom 6.** *(Strong Convexity) For* $c, c' \in C$ *with* $(c - c') \in D$, *and* $c \neq c'$,

$$c \sim c' \text{ implies } (\frac{1}{2})c + (\frac{1}{2})c' \succ c$$

The terminology "strong convexity" of preferences follows Debreu (1959, p.61).[19] The standard way to interpret this axiom is that the underlying preference structure exhibits inequality aversion, where the inequality in question here is with respect to the intertemporal consumption pattern.

A function $u \in U$ will be called *concave* if for all $x, x' \in L$ and $\lambda \in (0,1)$, we have $u(\lambda x + (1-\lambda)x') \geq \lambda u(x) + (1-\lambda)u(x')$. It will be called *strictly mid-concave* if for all $x, x' \in L$ with $x \neq x'$,

$$u((\frac{1}{2})x+(\frac{1}{2})x')>(\frac{1}{2})u(x)+(\frac{1}{2})u(x')$$

We define the class of utility functions $U^c = \{u \in U : u$ is concave and strictly mid-concave on $L\}$.

**Theorem 4.** *Suppose a SWR $\succsim$ satisfies Axioms 1-4 and 6. Then, there is a utility function, $u \in U^c$, such that for all $c, c' \in C$, satisfying $(c-c') \in D$,*

$$c \succsim c' \text{ if and only if } \sum_{t=1}^{\infty} (u(c_t)-u(c'_t)) \geq 0 \qquad (5)$$

*Further, the utility function $u \in U^c$ satisfying (5) is unique.*

## 4. A FORESTRY MODEL

The standard model of forest management under intertemporal equity can be described in terms of the objects $(f, A, b, u)$, where $(f, A, b)$ represent the technological aspects, and $u$ represents the preferences[20]. We describe each in turn in what follows.

We begin with $f$, a production function from $\mathbb{R}_+$ to $\mathbb{R}_+$, which relates the timber content of a tree $f(a)$ to the age of the tree, $a$. The following assumptions on $f$ are maintained:

(A.1) There is $\tilde{a} \geq 1$, such that $f(a)=0$ for $a \in [0, \tilde{a}]$.

(A.2) $f$ is continuous for $a \geq \tilde{a}$, and there is a positive integer $n > a$, such that (i) $f$ is increasing for $a \in [\tilde{a}, n]$, and (ii) $f$ is decreasing for $a > n$.

(A.3) There is a positive integer $m$, satisfying $\tilde{a} < m < n$, such that (i) $[f(a)/a]$ is maximized at $a = m$ among all $a \in \{1,...,n\}$, and (ii) $[f(a)/a] < [f(m)/m]$ for all $a \in \{1,...,n\}$, with $a \neq m$.

We will denote the vector $(f(1),..., f(n))$ by $P$; note that $P \in \mathbb{R}^n_+$. In what follows, we normalize $f(n)=1$ by choice of units in which the timber content is measured.

Next, we describe $(A, b)$, which indicate the transition possibilities of forestry land occupied by trees of different ages. It is useful, in this connection, to indicate the nature of evolution of the forest informally, before introducing the matrix-vector notation.

Given (A.2), for any reasonable objective function of the forest manager, trees will never be allowed to grow beyond age $n$. We therefore take this as a condition of feasibility itself. The total land available for forestry is taken to be one unit (by appropriate choice of units in which land is measured). Imagine that we are starting at the end of time period 1 with a standing forest, which might be described by the land occupied by output of trees of ages 1 to $n$. This will be written as $(y_1^1, \ldots, y_1^n)$. At the end of time period 1, two things are supposed to happen instantaneously, by the nature of our point-input, point-output framework. First trees of different ages are harvested. Second, new seedlings (trees of age 0 ) are planted in the cleared areas. The land released by harvests are described by $(h_1^1, \ldots, h_1^n)$. Since trees are never allowed to grow beyond age $n$, we must have $h_1^n = y_1^n$. The land occupied by input of trees of various ages at the end of time period 1 can then be described by $(x_1^0, \ldots, x_1^{n-1}, x_1^n)$, where $(y_1^1, \ldots, y_1^n) - (h_1^1, \ldots, h_1^n) = (x_1^1, \ldots, x_1^{n-1}, x_1^n) = (x_1^1, \ldots, x_1^{n-1}, 0)$, and $x_1^0 = h_1^1 + \cdots + h_1^n$.

The trees grow during time period 2, with trees of age $a \in \{0, \ldots, n-1\}$ at the end of time period 1, becoming trees of age $a+1$ at the end of time period 2. Thus, the land occupied by output of trees at the end of time period 2 is $(y_2^1, \ldots, y_2^n) = (x_1^0, \ldots, x_1^{n-1})$. The above process is now repeated indefinitely.

This informal description can be formalized as follows. Define the $n \times n$ matrix $A$ and the vector $b \in \mathbb{R}^n$ as:

$$A = \begin{bmatrix} 0 & 1 & 0 & 0 & \ldots & 0 \\ 0 & 0 & 1 & 0 & \ldots & 0 \\ \ldots & \ldots & \ldots & \ldots & \ldots & \ldots \\ 0 & 0 & 0 & 0 & \ldots & 1 \\ 0 & 0 & 0 & 0 & \ldots & 0 \end{bmatrix} = \begin{bmatrix} 0_n & I_{n-1} \\ 0 & 0_n \end{bmatrix}; b = (1, 1, \ldots, 1)$$

Denote the set $\{x \in \mathbb{R}_+^n : bx = 1\}$ by $Q$, and define a *transition possibility set* by:

$$\Omega = \{(y, z) \in Q^2 : Az \le y\}$$

A *path* starting from $y \in Q$ is a sequence $(y_t)_1^\infty = (y_t^1, \ldots, y_t^n)_1^\infty$ satisfying:

$$(y_t, y_{t+1}) \in \Omega \text{ for } t \in N, \text{and } y_0 = y.$$

We can associate with a path $(y_t)_1^\infty$ a sequence $(h_t)_1^\infty = (h_t^1, \ldots, h_t^n)_1^\infty$ defined by:

$$h_t = y_t - Ay_{t+1} \text{ for } t \in \mathbb{N}$$

and a sequence $(x_t)_1^\infty = (x_t^0, x_t^1, \ldots, x_t^n)_1^\infty$ defined by:

$$(x_t^0, x_t^1, \ldots, x_t^n) = (1 - \| Ay_{t+1} \|, Ay_{t+1}) \; for \; t \in \mathbb{N}$$

Note that $h_t \in \mathbb{R}_+^n$, with $h_t^n = y_t^n$ for $t \in \mathbb{N}$, and $x_t \in \mathbb{R}_+^{n+1}$, with $x_t^n = 0$ and $\| x_t \| = 1$ for $t \in \mathbb{N}$.

The timber content obtained by harvest $h_t$ is $Ph_t$. So, we can associate with a path $(y_t)_1^\infty$, the consumption sequence $(c_t)_1^\infty$ defined by:

$$c_t = Ph_t \; for \; t \in \mathbb{N}$$

Notice that (by choice of $f(n) = 1$ ), we must have $c_t \in [0,1]$ for $t \in \mathbb{N}$.

Finally, we turn to preferences defined on consumption sequences. We will suppose that there is a SWR $\succsim$ satisfying Axioms 1-4 and 6, so there is a unique utility function, $u \in U^c$, as described in Theorem 4 of Section 3.

A path $(y_t')_1^\infty$ from $y \in Q$ *dominates* a path $(y_t)_1^\infty$ from $y \in Q$ if the associated consumption sequences $(c_t')_1^\infty$ and $(c_t)_1^\infty$ satisfy the condition that $(c_t')_1^\infty \succ_u (c_t)_1^\infty$ [where the SWR $\succsim_u$, as defined by (U) of Section 3, corresponds to the $u \in U^c$, mentioned above]. That is,

$$\exists N \in \mathbb{N} \; such \; that \; c'[N] = c[N] \; and \sum_{t=1}^{N} u(c_t') > \sum_{t=1}^{N} u(c_t)$$

A path $(y_t)_1^\infty$ is called *maximal* if there is no path $(y_t')_1^\infty$ from $y$ which dominates it.

In words, a maximal path is a path, such that there is no path (from the same initial conditions) which is better in terms of the social welfare relation $\succsim_u$. It is, perhaps, worth emphasizing that this does *not* necessarily mean that given any path (from the same initial conditions), a maximal path is at least as good as the given path; in fact, the two paths need not be comparable. Thus, the definition of maximality (like the definition of Pareto efficiency) allows for multiple maximal points on the feasible set, given the pre-order.[21]

## 5. MAXIMUM SUSTAINED YIELD FOREST

Suppose the forest does not change at all from one period to the next. What is the "best" composition of the forest in that case ? This is what is meant when one speaks of the forest with the maximum sustained yield. In the terminology of intertemporal allocation theory, one is concerned here with the "golden-rule" forest.

We show that there is a unique golden-rule forest, and obtain dual variables (shadow prices) which provide "price support" to such a forest. These shadow prices are in fact prices (expressed in terms of the utility good) of the various subplots of land, according to the age of trees standing on each subplot. The golden rule prices maximize "generalized profit" [the utility plus the valuation of terminal subplots of land minus the valuation of initial plots of land, terminal and initial referring to a unit time period] at the golden-rule forest among all possible activities in the transition possibility set.

We go on to show (in Proposition 1) that those points in the transition possibility set which stay away (uniformly) from the golden-rule must suffer a (uniform) loss of "generalized profit" (compared to the golden-rule forest), when evaluated at the golden-rule prices.

To discuss the golden-rule, let us introduce the welfare function, $w:\Omega \rightarrow L$, defined by:

$$w(y,z)=u(P(y-Az)) \textit{ for all } (y,z)\in \Omega$$

A *golden-rule* forest is a vector $\overline{y}\in Q$ which solves the problem:

$$\left.\begin{array}{ll} Max & u(P(y-Ay)) \\ subject\ to & (y,y)\in \Omega \end{array}\right\} \qquad \text{(GR)}$$

Since $u$ is increasing, the above problem is the same as:

$$\left.\begin{array}{ll} Max & P(y-Ay) \\ subject\ to & (y,y)\in \Omega \end{array}\right\} \qquad \text{(GR')}$$

To characterize the golden-rule, note that for $(y,y)\in \Omega,$ we have $y_i \geq y_{i+1}$ for $i\in \{1,\dots,n-1\}$. So, we get:

$$\begin{aligned} P(y-Ay) &= P_1(y_1-y_2)+\cdots+P_{n-1}(y_{n-1}-y_n)+P_n y_n \\ &= P_1(y_1-y_2)+[P_2/2]2(y_2-y_3)+\cdots \\ &\quad +[P_{n-1}/(n-1)](n-1)(y_{n-1}-y_n)+[P_n/n]ny_n \\ &\leq [P_m/m][(y_1-y_2)+2(y_2-y_3)+\cdots+(n-1)(y_{n-1}-y_n)+ny_n] \\ &= [P_m/m][y_1+y_2+\cdots+y_{n-1}+y_n] \\ &= [P_m/m] \end{aligned}$$

the inequality following from assumption (A.3)(i) on $f$. Defining $\overline{y}=[(1/m),\dots,(1/m),0,\dots,0]$ in $Q$, it follows that $\overline{y}$ is a golden-rule forest.

By (A.3)(ii), a strict inequality is produced in the above calculation, unless $y_i - y_{i+1}=0$ for all $i\in \{1,\dots,n-1\}$, with $i\neq m$ and $y_n=0$. Thus, if $y$ is a golden-rule forest, then $y=(y_m,\dots,y_m,0,\dots,0)$, and since $y\in Q$, we must in fact have

$y=[(1/m),\ldots,(1/m),0,\ldots,0]$. Thus, $\overline{y}=[(1/m),\ldots,(1/m),0,\ldots,0]$ is the *unique* golden-rule forest. Clearly, the consumption associated with a golden-rule forest is stationary, and equal to $[P_m/m]$; we will denote it by $\overline{c}$.

A convenient "price support property" of the golden-rule may now be noted. Define:

$$q=\overline{c}(1,\ldots,n) \textit{ and } p=u_{+}'(\overline{c})q \tag{GRP}$$

where $u_{+}'(\overline{c})$ is the right-hand derivative of $u$ at $\overline{c}$. Then, we have:

$$u(P(y-Az))+pz-py\leq u(\overline{c}) \textit{ for all } (y,z)\in\Omega \tag{6}$$

To see this, note that $P\leq q$ by (A.3). Thus, for $(y,z)\in\Omega$, we have:

$$P(y-Az)\leq q(y-Az)=qy-qz+q(z-Az) \tag{7}$$

*Now* $(q-qA)=\overline{c}(1,2,\ldots,n)-\overline{c}(0,1,\ldots,n-1)=\overline{c}(1,\ldots,1,1)$, *so* $q(z-Az)=(q-qA)z=\overline{c}\,\|z\|=\overline{c}$. Thus, (7) yields:

$$P(y-Az)\leq qy-qz+\overline{c} \textit{ for all } (y,z)\in\Omega \tag{8}$$

Using the concavity of $u$, we have:

$$u(P(y-Az))-u(\overline{c})\leq u_{+}'(\overline{c})[P(y-Az)-\overline{c}]\leq u_{+}'(\overline{c})[qy-qz] \tag{9}$$

the last inequality in (9) following from (8). Using $p=u_{+}'(\overline{c})q$ and rearranging terms in (9) yields (6). We will refer to the $p$ defined in (GRP) as the *golden-rule price.*

Following McKenzie (1968), let us define the *value loss* of operating at $(y,z)\in\Omega$ as:

$$\delta(y,z)=u(\overline{c})-[u(P(y-Az))+pz-py] \tag{10}$$

Clearly, $\delta(y,z)\geq 0$ for all $(y,z)\in\Omega$, and $\delta(\overline{y},\overline{y})=0$.

The points which have zero value-loss have a very special structure, as we note in the following result[22]. For this purpose, we denote the vector $(1/m)e^m$ [where $e^m$ is the $m-th$ unit vector in $\mathbb{R}^n$ ] by $g$.

**Proposition 1.** (i)*If* $(y,z)\in\Omega$ *and* $\delta(y,z)=0$, *then:*

$$(y - Az) = g \tag{11}$$

(ii) *Given* $\varepsilon > 0$, *there is* $\delta > 0$, *such that if* $(y,z) \in \Omega$, *and* $d((y - Az), g) \geq \varepsilon$, *then* $\delta(y,z) \geq \delta$.

**Proof.** (i) Note that the inequality in (6) depends on the inequalities in (7) and (9). Using (7), we can see that in order to have $\delta(y,z) = 0$, we must have:

$$y_i - z_{i+1} = 0 \text{ for all } i \in \{1,...,n-1\}, i \neq m, \text{and } y_n = 0$$

Thus, $(y - Az)$ must be of the form $(0,0,...,y_m - z_{m+1}, 0,...,0)$. Thus, we must have $P(y - Az) = P_m(y_m - z_{m+1})$. Denoting $P_m(y_m - z_{m+1})$ by $c$, we see from the left-hand inequality in (9),

$$u(c) - u(\overline{c}) \leq u'_+(\overline{c})[c - \overline{c}] \tag{12}$$

In order to have $\delta(y,z) = 0$, we must have equality in (12).

If $c \neq \overline{c}$, then defining $c' = (\frac{1}{2})c + (\frac{1}{2})\overline{c}$, we have by concavity of $u$, and strict mid-concavity of $u$,

$$(\frac{1}{2})[u(c) - u(\overline{c})] = (\frac{1}{2})u(c) + (\frac{1}{2})u(\overline{c}) - u(\overline{c}) < u(c') - u(\overline{c})$$
$$\leq u'_+(\overline{c})[c' - \overline{c}] = (\frac{1}{2})u'_+(\overline{c})[c - \overline{c}]$$

and we cannot have equality in (12). Thus, for equality in (12), we must have $c = \overline{c}$. Thus, $(y_m - z_{m+1}) = (1/m)$, so that $(y - Az) = g$.

(ii) Suppose, on the contrary, there is a sequence $(y^s, z^s) \in \Omega$ with $s \in \mathbb{N}$, such that $d((y^s - Az^s), g) \geq \varepsilon$, but $\delta(y^s, z^s) \to 0$ as $s \to \infty$. Since $\Omega$ is compact, there is a subsequence $(y^{s'}, z^{s'})$ of $(y^s, z^s)$, converging to $(\hat{y}, \hat{z}) \in \Omega$. Since $d((y^{s'} - Az^{s'}), g) \geq \varepsilon$, we must have $d((\hat{y} - A\hat{z}), g) \geq \varepsilon$. By (11), we must have $\delta(\hat{y}, \hat{z}) > 0$.

Since $(y^{s'}, z^{s'})$ converges to $(\hat{y}, \hat{z})$, we must have $\delta(y^{s'}, z^{s'})$ converging to $\delta(\hat{y}, \hat{z})$, by (10) and the continuity of $u$. Since $\delta(y^{s'}, z^{s'}) \to 0$, we have $\delta(\hat{y}, \hat{z}) = 0$, a contradiction to the result obtained in the previous paragraph.

## 6. LONG-RUN PROPERTIES OF MAXIMAL PATHS

In this section, we establish the following long-run property of maximal paths: there is asymptotic convergence to the maximum sustained yield forest, starting from an arbitrary initial forest. This provides a theoretical justification for focusing on the maximum sustained yield forest as a goal of forestry policy.

We show that from every initial forest, one can reach the steady-state of the golden-rule forest in a finite number of periods (Lemma 1). Thus, there are always paths which suffers finite value losses (in terms of the sum of generalized profits evaluated at the golden-rule prices) compared to the golden-rule forest, over the infinite horizon. Such paths are called "good paths", and a maximal path is shown (in Proposition 2) to be a good path. Because activities which stay away uniformly from the golden rule must suffer uniform per period value loss (Proposition 1 in the previous section), paths which stay away uniformly from the golden-rule for an infinite number of periods must suffer infinite value losses. Thus good paths (which, by definition, suffer only finite value losses) must exhibit asymptotic convergence to the golden-rule. Consequently, maximal paths must also exhibit this property, since they are good paths (Theorem 5).[23]

The principal technical result needed for our analysis is the ability to move from any initial forest to any other forest in a finite number of periods, independent of the initial and terminal compositions. In fact, we show that this transition can always be made in $(n+1)$ periods.

**Lemma 1.** *Given any* $(y, \hat{y}) \in Q^2$, *there exist* $(z_1, ..., z_{n+2})$ *such that:*

$$(i)\ (z_s, z_{s+1}) \in \Omega \ \text{for}\ s = 1, ..., n+1; (ii)\ z_1 = y, z_{n+2} = \hat{y} \tag{13}$$

**Proof.** The proof consists of simply defining $(z_1, ..., z_{n+2})$ appropriately and checking that it satisfies (13). Let us define $(z_1, ..., z_{n+2})$ as follows:

$$\left.\begin{array}{lcc} z_1 & = & y \\ z_2 & = & (1, 0, ..., 0) \\ z_3 & = & (\hat{y}^n, 1-\hat{y}^n, 0, ..., 0) \\ z_4 & = & (\hat{y}^{n-1}, \hat{y}^n, 1-\hat{y}^n-\hat{y}^{n-1}, 0, ..., 0) \\ \cdots & \cdots & \cdots \\ z_{n+1} & = & (\hat{y}^2, \hat{y}^3, ..., \hat{y}^n, \hat{y}^1) \\ z_{n+2} & = & \hat{y} \end{array}\right\}$$

It is straightforward to verify from this definition that (13) holds.

Following Gale (1967), let us call a path $(y_t)_1^\infty$ *good* if there is a real number $\overline{G}$ such that:

$$\sum_{t=1}^{T}[u(P(y_t - Ay_{t+1})) - u(\overline{c})] \geq \overline{G} \textit{ for all } T \in N$$

Informally, a good path is one which is at most finitely worse than the golden-rule path in terms of partial sums of its utilities.

We now show, using Lemma 1, that a maximal path is necessarily good, so that it inherits all the well-known long-run properties of good paths.

**Proposition 2.** *Let* $(y_t)_1^\infty$ *be a maximal path from* $y \in Q$. *Then* $(y_t)_1^\infty$ *is good.*

**Proof.** Define $G = -2(n+1)u(\overline{c})$. We claim that for all $T > 2(n+2)$, we have:

$$\sum_{t=1}^{T}[u(P(y_t - Ay_{t+1})) - u(\overline{c})] \geq G \tag{14}$$

Suppose, on the contrary, there is some $T \in \mathbb{N}$, such that $T > 2(n+2)$, and :

$$\sum_{t=1}^{T}[u(P(y_t - Ay_{t+1})) - u(\overline{c})] < G \tag{15}$$

Since $(y, \overline{y}) \in Q^2$, we can use Lemma 1 to obtain $(z_1, ..., z_{n+2})$ such that:

$$(i)\ (z_s, z_{s+1}) \in \Omega \textit{ for } s = 1, ..., n+1; (ii)\ z_1 = y, z_{n+2} = \overline{y} \tag{16}$$

Since $(\overline{y}, y_{T+1}) \in Q^2$, we can use Lemma 1 to obtain $(z'_1, ..., z'_{n+2})$ such that:

$$(i)\ (z'_s, z'_{s+1}) \in \Omega \textit{ for } s = 1, ..., n+1; (ii)\ z'_1 = \overline{y}, z'_{n+2} = y_{T+1} \tag{17}$$

Define a sequence $(y'_t)_1^\infty$ as follows:

$$\left.\begin{array}{lllll} y'_t & = & z_t & \textit{for} & t = 1, ..., n+2 \\ y'_t & = & \overline{y} & \textit{for} & t = n+3, ..., T-(n+1) \\ y'_t & = & z'_{t-(T-n-1)} & \textit{for} & t = T-n, ..., T+1 \\ y'_t & = & y_t & \textit{for} & t > T+1 \end{array}\right\} \tag{18}$$

Using (16), (17) and (18) it is easy to check that $(y'_t)_1^\infty$ is a path from $y \in Q$. Since $y'_t = y_t$ for $t \geq T+1$, we have $c'_t = c_t$ for $t \geq T+1$, and $c'_t = \overline{c}$ for $t = n+3, ..., T-(n+1)$. Using (18), and $u \in U$, it follows that:

$$\sum_{t=1}^{T} [u(P(y'_t - Ay'_{t+1})) - u(\overline{c})] \geq -2(n+1)u(\overline{c}) = G \tag{19}$$

Thus, using (15) and (19), we have:

$$\sum_{t=1}^{T} u(P(y'_t - Ay'_{t+1})) > \sum_{t=1}^{T} u(P(y_t - Ay_{t+1}))$$

But, this implies that $(y_t)_1^\infty$ is not maximal, a contradiction. Thus, our claim (14) must hold, and it follows that $(y_t)_1^\infty$ is good.

Using the above proposition, we can now summarize the long-run properties of maximal paths in the following theorem[24]. To this end, given a path $(y_t)_1^\infty$ from $y \in Q$, we associate with it a *value-loss sequence* $(\delta_t)_1^\infty$ defined by:

$$\delta_t = \delta(y_t, y_{t+1}) \quad for\ t \in N$$

**Theorem 5.** *Let* $(y_t)_1^\infty$ *be a maximal path from* $y \in Q$. *Then, we have:*

$$\sum_{t=1}^{\infty} \delta_t < \infty \tag{20}$$

$$\delta_t \to 0 \ as\ t \to \infty \tag{21}$$

$$d((y_t - Ay_{t+1}), g) \to 0 \ as\ t \to \infty \tag{22}$$

$$d(y_t, \overline{y}) \to 0 \ as\ t \to \infty \tag{23}$$

**Proof.** Given any $T \in \mathbb{N}$, we can use (6) to write:

$$\begin{aligned} \sum_{t=1}^{T} [u(P(y_t - Ay_{t+1})) - u(\overline{c})] &= \sum_{t=1}^{T} [py_t - py_{t+1}] - \sum_{t=1}^{T} \delta_t \\ &= [py - py_{T+1}] - \sum_{t=1}^{T} \delta_t \end{aligned} \tag{24}$$

Since $(y_t)_1^\infty$ is maximal, it is good by Proposition 2, so there is $G \in \mathbb{R}$ such that:

$$\sum_{t=1}^{T} [u(P(y_t - Ay_{t+1})) - u(\overline{c})] \geq G \quad for\ all\ T \in \mathbb{N} \tag{25}$$

Using (24) and (25), we obtain:

$$\sum_{t=1}^{T} \delta_t \leq py - G \leq n - G \quad \textit{for all } T \in \mathbb{N} \tag{26}$$

Since $\delta_t \geq 0$ by (6), (26) establishes (20). Clearly, (21) follows immediately from (20). Using Proposition 1 and (21), we obtain (22).

We now verify (23) as follows. Using (22), given any $\varepsilon > 0$, we can choose $T \in \mathbb{N}$, such that:

$$d((y_t - Ay_{t+1}), g) < (\varepsilon/n^4) \quad \textit{for } t \geq T \tag{27}$$

We will show that:

$$d(y_{s+m}, \overline{y}) \leq \varepsilon \quad \textit{for all } s > T \tag{28}$$

Using (27), we note that for each $t \geq T$, we have:

$$\left.\begin{aligned} (a)\ & y_t^i - y_{t+1}^{i+1} < (\varepsilon/n^4) \quad \textit{for } i = 1,\dots,m-1, m+1,\dots,n \\ (b)\ & y_t^m - y_{t+1}^{m+1} > (1/m) - (\varepsilon/n^4) \end{aligned}\right\} \tag{29}$$

Then, for $t > T$, we must have:

$$y_{t+1}^1 = \sum_{i=1}^{n} h_t^i = \sum_{i=1}^{n} (y_t^i - y_{t+1}^{i+1}) > (1/m) - (\varepsilon/n^3) \tag{30}$$

And, for $j = 1,\dots,m-1$ and $t > T$,

$$\begin{aligned} y_{t+j+1}^{j+1} &= y_{t+1}^1 + \sum_{i=1}^{j} (y_{t+i+1}^{i+1} - y_{t+i}^i) \\ &> (1/m) - (\varepsilon/n^3) - r(\varepsilon/n^4) \\ &\geq (1/m) - (\varepsilon/n^2) \end{aligned} \tag{31}$$

Using (30) and (31), we have for $t > T$,

$$y_{t+j}^j > (1/m) - (\varepsilon/n^2) \quad \textit{for } j = 1,\dots,m \tag{32}$$

Now pick any $s > T$. Then, using (32), we have:

$$\sum_{i=m+1}^{n} y^i_{s+m} < 1 - m[(1/m) - (\varepsilon/n^2)] = m\varepsilon/n^2 \tag{33}$$

Also, using (32), we claim that:

$$y^i_{s+m} < (1/m) + (\varepsilon/n) \quad for\ i = 1, \dots, m \tag{34}$$

For if $y^i_{s+m} \geq (1/m) + (\varepsilon/n)$ for some $i \in \{1, \dots, m\}$, then:

$$\sum_{i=1}^{n} y^i_{s+m} \geq \sum_{i=1}^{m} y^i_{s+m} > (1/m) + (\varepsilon/n) + (m-1)[(1/m) - (\varepsilon/n^2)] \geq 1$$

which contradicts the fact that $y_{s+m} \in Q$. This establishes our claim (34). Now, using (32) and (34), we have:

$$| y^i_{s+m} - (1/m) | < (\varepsilon/n) \quad for\ i = 1, \dots, m \tag{35}$$

Using (33) and (35), and $m \leq (n-1)$, we have for $s > T$,

$$\begin{aligned} d(y_{s+m}, \bar{y}) &= \sum_{i=1}^{m} | y^i_{s+m} - (1/m) | + \sum_{i=m+1}^{n} y^i_{s+m} \\ &< m(\varepsilon/n) + m(\varepsilon/n^2) \\ &\leq (n-1)\varepsilon[(1/n) + (1/n^2)] \\ &= \varepsilon(n^2 - 1)/n^2 < \varepsilon \end{aligned}$$

This establishes (28) and therefore (23).

## 7. SHADOW PRICES FOR MAXIMAL PATHS

Our investigation of the forestry model so far has shown that its central results under intergenerational equity can be obtained without any reference to the overtaking criterion. Our notion of maximality, based completely on ranking of paths which are eventually identical, suffices for this purpose.

However, in order to tie up our present study with the earlier study by Mitra and Wan (1986), we need to investigate the relation of maximal paths to paths which are "optimal" according to the overtaking criterion, a precise definition of which is given below.

We continue to suppose that there is a SWR $\succsim$ satisfying Axioms 1-4 and 6, so there is a unique utility function, $u \in U^c$, as described in Theorem 4 of Section 3.

A path $(y_t')_1^\infty$ from $y \in Q$ *weakly overtakes* a path $(y_t)_1^\infty$ from $y \in Q$ if the associated consumption sequences $(c_t')_1^\infty$ and $(c_t)_1^\infty$ satisfy the condition that $(c_t')_1^\infty P_u (c_t)_1^\infty$ [where the SWR $R_u$ as defined by (R) in Section 3, corresponds to the $u \in U^c$ mentioned above]. This will be the case if and only if there is $\bar{N} \in \mathbb{N}$, such that:

$$\sum_{t=1}^{N} u(c_t') \geq \sum_{t=1}^{N} u(c_t) \text{ for all } N \geq \bar{N}$$

and:

$$\sum_{t=1}^{N'} u(c_t') > \sum_{t=1}^{N'} u(c_t) \text{ for a subsequence } N' \text{ of } N \geq \bar{N}$$

A path $(y_t)_1^\infty$ from $y \in Q$ is called *optimal* if there is no path $(y_t')_1^\infty$ from $y$ which weakly overtakes it.

In words, an optimal path is a path, such that there is no path (from the same initial conditions) which is better in terms of the social welfare relation $R_u$. It is worth noting, by comparing the definitions of maximality and optimality, that an optimal path is necessarily maximal.

This brings us to the part of our paper which distinguishes the present analysis of the forest management problem from the one offered in Mitra and Wan (1986). This distinction parallels the difference in approaches used by Brock (1970a) on the one hand, and by Gale (1967) and McKenzie (1968) on the other, in developing the general theory of optimal intertemporal allocation under the overtaking criterion.

Brock (1970a) showed that the theory of optimal intertemporal allocation under the overtaking criterion can be developed fully without finding shadow prices to support infinite-horizon paths. One only needs the price-support property of the golden-rule, a technically simple problem. In the theory presented by Gale (1967) and McKenzie(1968, 1986), however, shadow prices supporting infinite-horizon paths is an integral aspect.

Unlike the golden-rule prices, the prices supporting a (non-stationary) maximal or optimal path are typically time varying. However, their role is similar. In any time period, at the prices associated with a maximal (or optimal) path, the activity chosen along the maximal (or optimal) path maximizes generalized profit [the utility plus the valuation of terminal subplots of land minus the valuation of initial plots of land, terminal and initial referring to the time period in question] among all possible activities in the transition possibility set.

Roughly speaking, the analysis of forest management offered in Mitra and Wan (1986) follows Brock (1970a) quite closely. When it comes to studying long-run behavior of maximal paths, this theory again turns out to be the most elegant one to apply, as we have indicated in the previous section.

But, now, we need to concern ourselves with short-run behavior as well. If "errant" behavior can occur in the short-run along maximal paths, which would not occur on optimal paths (according to the overtaking criterion), this would destroy to some extent the importance of the concept of maximal paths. The fact that such errant behavior *cannot* occur in the short-run along maximal paths is at the heart of establishing an equivalence between a maximal path and an optimal path. In order to demonstrate this, we need to find shadow prices to support a maximal path. Thus, the theory relating to shadow prices, proposed by Gale (1967) and McKenzie (1968, 1986), becomes very useful in the present context.

Unfortunately, a part of this theory cannot be applied directly, because the model of forestry has characteristics which violate some of the standard assumptions under which the general theory of optimal intertemporal allocation has been developed. We use a version of the Kuhn-Tucker theorem due to Arrow, Hurwicz and Uzawa (1961), and then follow the Gale-McKenzie approach to obtain the appropriate shadow prices. The appropriate result is stated in Proposition 3; its proof is quite involved, although it uses methods familiar in the literature on the general theory of intertemporal allocation that we have repeatedly referred to.

To proceed with our analysis, we will make the following assumption[25] on the utility function, $u$.

(A.4) $u$ is continuously differentiable on $\mathbb{R}_+$, with $B \equiv u'(0) < \infty$.

We define a function $w: \mathbb{R} \to \mathbb{R}$ by:

$$w(c) = \begin{cases} u(c) & for \quad c \in I \\ u'(0)c & for \quad c < 0 \\ u'(1)(c-1) + u(1) & for \quad c > 1 \end{cases}$$

Then $w$ is concave and continuously differentiable on $\mathbb{R}$, with $w(c) = u(c)$ for $c \in L$. Define $W: \mathbb{R}^{2n} \to \mathbb{R}$ by:

$$W(y, z) = w(P(y - Az))$$

Then $W$ is continuously differentiable on its domain.

**Proposition 3.** Suppose $(\tilde{y}_t)_1^\infty$ is a maximal path from $\tilde{y} \in Q$. Then, there is $N \in \mathbb{N}$, $N > 1$, and a sequence of shadow prices $(\tilde{p}_t)_{N-1}^\infty$ such that:

(i) $\tilde{p}_t \in \mathbb{R}_+^n$ *and* $\tilde{p}_t \leq (2n+1)Bb$ *for all* $t \in \mathbb{N}$, *with* $t \geq N-1$;

(ii) If $(y_t)_1^\infty$ is a path from $\tilde{y} \in Q$, then:

$$\sum_{t=1}^{N-1} u(P(y_t - Ay_{t+1})) + \tilde{p}_N y_N \leq \sum_{t=1}^{N-1} u(P(\tilde{y}_t - A\tilde{y}_{t+1})) + \tilde{p}_N \tilde{y}_N \tag{36}$$

(iii) *For* $t \geq N-1$, *and for all* $(y,z) \in \Omega$,

$$u(P(y-Az)) + \tilde{p}_{t+1}z - \tilde{p}_t y \leq u(P(\tilde{y}_t - A\tilde{y}_{t+1})) + \tilde{p}_{t+1}\tilde{y}_{t+1} - \tilde{p}_t\tilde{y}_t \tag{37}$$

**Proof.** Since $(\tilde{y}_t)_1^\infty$ is maximal, given any $T \in \mathbb{N}$, $(\tilde{y}_1,\ldots,\tilde{y}_{T+1})$ must solve the following constrained maximization problem:

$$\left.\begin{array}{lll} \textit{Maximize} & \sum_{t=1}^{T} W(y_t, y_{t+1}) & \\ \textit{subject to} & (y_t, y_{t+1}) \in \Omega & \textit{for } t = 1,\ldots,T \\ & y_1 = \overline{y}_1 & \\ & y_{T+1} = \overline{y}_{T+1} & \end{array}\right\} \tag{F}$$

Suppose on the contrary that there is $(y_1,\ldots,y_{T+1})$, satisfying the constraints of problem (F), for which the following holds:

$$\sum_{t=1}^{T} W(y_t, y_{t+1}) > \sum_{t=1}^{T} W(\tilde{y}_t, \tilde{y}_{t+1}) \tag{38}$$

Then, we can define $(y_t')_1^\infty$ as: $y_t' = y_t$ for $t = 1,\ldots,T+1$, and $y_t' = \tilde{y}_t$ for $t > T+1$. Then, it is easy to check that $(y_t')_1^\infty$ is a path from $y$, for which $c_t' = \tilde{c}_t$ for $t \geq T+1$. Then, using (38), we would have:

$$\sum_{t=1}^{T} u(P(y_t' - Ay_{t+1}')) > \sum_{t=1}^{T} u(P(\tilde{y}_t - A\tilde{y}_{t+1}))$$

a contradiction to the fact that $(\tilde{y}_t)_1^\infty$ is maximal.

We now proceed to examine the constrained maximum problem. In order to apply the Kuhn-Tucker theorem to it, we write it first in the "standard form". Define $X = \mathbb{R}^n$, and $Y = X^{T+1}$. Then, $(\tilde{y}_1,\ldots,\tilde{y}_{T+1})$ must solve the following problem:

$$\left.\begin{array}{lll} \textit{Maximize} & \sum_{t=1}^{T} W(y_t, y_{t+1}) & \\ \textit{subject to} & y_t - Ay_{t+1} \geq 0 & \textit{for } t = 1,\dots,T \\ & y_t \geq 0 & \textit{for } t = 1,\dots,T+1 \\ & y_1 - \tilde{y}_1 \geq 0 & \\ & \tilde{y}_1 - y_1 \geq 0 & \\ & y_{T+1} - \tilde{y}_{T+1} \geq 0 & \\ & \tilde{y}_{T+1} - y_{T+1} \geq 0 & \\ & 1 - \sum_{i=1}^{n} y_t^i \geq 0 & \textit{for } t = 1,\dots,T+1 \\ & \sum_{i=1}^{n} y_t^i - 1 \geq 0 & \textit{for } t = 1,\dots,T+1 \end{array}\right\} \tag{F'}$$

among all $(y_1,\dots,y_{T+1}) \in Y$. Notice that the set $Y$ is open. The constraint functions are all linear, and hence continuously differentiable on $Y$. The objective function is continuously differentiable on $Y$, by the way $W$ was defined and assumption (A.4). The Arrow-Hurwicz-Uzawa constraint qualification is met since the constraint functions are all linear. Thus, we can apply their version of the Kuhn-Tucker theorem to obtain $(r_1,\dots,r_T) \geq 0, (\nu_1,\dots,\nu_{T+1}) \geq 0,$
$\alpha \geq 0, \alpha' \geq 0, \omega \geq 0, \omega' \geq 0, (\mu_1,\dots,\mu_{T+1}) \geq 0, (\mu_1',\dots,\mu_{T+1}') \geq 0,$ such that:

$$-u'(\tilde{c}_t)PA + u'(\tilde{c}_{t+1})P + [r_{t+1} - r_t A] + \nu_{t+1} - \mu_{t+1}b + \mu_{t+1}'b = 0 \quad \textit{for } t = 1,\dots,T-1 \tag{39}$$

$$u'(\tilde{c}_1)P + r_1 + \nu_1 + \alpha - \alpha' - \mu_1 b + \mu_1' b = 0 \tag{40}$$

$$-u'(\tilde{c}_T)PA - r_T A + \nu_{T+1} + \omega - \omega' - \mu_{T+1}b + \mu_{T+1}'b = 0 \tag{41}$$

$$r(t)(\tilde{y}_t - A\tilde{y}_{t+1}) = 0 \quad \textit{for } t = 1,\dots,T-1 \tag{42}$$

$$\nu(t)\tilde{y}(t) = 0 \ \textit{for } t = 1,\dots,T+1 \tag{43}$$

$$\alpha(y_1 - \tilde{y}_1) + \alpha'(\tilde{y}_1 - y_1) = 0 \tag{44}$$

$$\omega(y_{T+1} - \tilde{y}_{T+1}) + \omega'(\tilde{y}_{T+1} - y_{T+1}) = 0 \tag{45}$$

$$\mu_t(1 - \sum_{i=1}^{n} \tilde{y}_t^i) + \mu_t'(\sum_{i=1}^{n} \tilde{y}_t^i - 1) = 0 \tag{46}$$

Let us define $(q_1,...,q_T)$ by $q_t = u'(\tilde{c}_t)P$ for $t = 1,...,T$, and $(p_1,...,p_T)$ by $p_t = q_t + r_t$ for $t = 1,...,T$. Now, let us consider any $(y,z) \in \Omega$, and $t \in \{1,...,T-1\}$. Then, using (39), we have:

$$\begin{aligned} q_t Az &= q_{t+1}z + [r_{t+1} - r_t A]z + v_{t+1}z - \mu_{t+1}bz + \mu'_{t+1}bz \\ &= q_{t+1}z + [r_{t+1} - r_t A]z + v_{t+1}z - \mu_{t+1} + \mu'_{t+1} \end{aligned} \tag{47}$$

Using (47), we obtain:

$$q_t y - q_t Az = q_t y - q_{t+1}z - [r_{t+1} - r_t A]z - v_{t+1}z + \mu_{t+1} - \mu'_{t+1} \tag{48}$$

Using (A.4), we have:

$$\begin{aligned} u(P(y-Az)) - u(P(\tilde{y}_t - A\tilde{y}_{t+1})) &\le u'(\tilde{c}_t)[P(y-Az) - P(\tilde{y}_t - A\tilde{y}_{t+1})] \\ &= q_t y - q_t Az - [q_t \tilde{y}_t - q_t A\tilde{y}_{t+1}] \end{aligned} \tag{49}$$

Using (42), (43) and (48) in (49), we obtain:

$$\begin{aligned} u(P(y-Az)) - u(P(\tilde{y}_t - A\tilde{y}_{t+1})) &\le q_t y - q_{t+1}z - [r_{t+1} - r_t A]z - v_{t+1}z \\ &\quad -[q_t \tilde{y}_t - q_{t+1}\tilde{y}_{t+1} - [r_{t+1} - r_t A]\tilde{y}_{t+1}] \\ &\le q_t y - q_{t+1}z - r_{t+1}z + r_t y \\ &\quad -[q_t \tilde{y}_t - q_{t+1}\tilde{y}_{t+1} - r_{t+1}\tilde{y}_{t+1} + r_t \tilde{y}_t] \\ &= p_t y - p_{t+1}z - [p_t \tilde{y}_t - p_{t+1}\tilde{y}_{t+1}] \end{aligned}$$

Transposing terms, we see that for all $(y,z) \in \Omega$, and $t \in \{1,...,T-1\}$,

$$u(P(y-Az)) + p_{t+1}z - p_t y \le u(P(\tilde{y}_t - A\tilde{y}_{t+1})) + p_{t+1}\tilde{y}_{t+1} - p_t \tilde{y}_t \tag{50}$$

To summarize this first part of the proof, given $T \in \mathbb{N}$, we have obtained $(p_1,...,p_T) \ge 0$, such that for all $(y,z) \in \Omega$, and $t \in \{1,...,T-1\}$, (50) holds.

The vector $(p_1,...,p_T)$ obtained above depends, of course, on the $T \in \mathbb{N}$ that is given to begin with, and one would write $(p_1\{T\},...,p_T\{T\})$ to record this explicitly. In order to obtain an infinite sequence $(\overline{p}_t)_1^\infty$ from these finite vectors, one would use a Cantor diagonal process argument; a pre-requisite to applying this argument is to show that for each $t \in \mathbb{N}$, there is a real number $B_t$, such that $\| p_t\{T\} \| \le B_t$ for all $T \ge t$.

It turns out that showing that such bounds exist for all $t \in \mathbb{N}$ is somewhat problematic in the forestry model. Specifically, showing that there are such bounds on all $r_t\{T\}$ creates difficulties. However, it is possible to show that there is $N \in \mathbb{N}$, such that bounds exist for all $t \geq N$. This is why the statement of our theorem is split up into two parts, instead of the standard Gale-McKenzie formulation, where (38) would be established for all $t \in \mathbb{N}$.

We now proceed with the details of the second part of the proof. Since $(\tilde{y}_t)_1^\infty$ is maximal, it is good. So, we can find $N' \in \mathbb{N}$, such that for all $t \geq N'$,

$$\tilde{y}_t^i \geq (1/2m) \ for \ i = 1, \ldots, m; \tilde{h}_t^m \geq (1/2m)$$

We confine our attention henceforth to those $T \in \mathbb{N}$ which satisfy $T > N' + n$. Given any such $T$, we have non-negative vectors $(r_1\{T\}, \ldots, r_T\{T\}), (q_1\{T\}, \ldots, q_T\{T\})$ and $(p_1\{T\}, \ldots, p_T\{T\})$ as defined above, satisfying for all $(y, z) \in \Omega$, and $t \in \{1, \ldots, T-1\}$,

$$u(P(y - Az)) + p_{t+1}\{T\}z - p_t\{T\}y \leq u(P(\tilde{y}_t - A\tilde{y}_{t+1})) + p_{t+1}\{T\}\tilde{y}_{t+1} - p_t\{T\}\tilde{y}_t \tag{51}$$

Clearly, using the above definition of $q_t\{T\}$ and (A.4), we have:

$$q_t\{T\} \leq BP \leq Bb \ for \ t \in \{1, \ldots, T\} \tag{52}$$

It remains to find appropriate bounds for $r_t\{T\}$.

Define $\eta_{t+1}\{T\} = \mu_{t+1}\{T\} - \mu'_{t+1}\{T\}$ for $t = 1, \ldots, T+1$. Using (39), we get:

$$\eta_{t+1}\{T\}b = -q_t\{T\}A + q_{t+1}\{T\} - r_t\{T\}A + r_{t+1}\{T\} + \nu_{t+1}\{T\} \geq -q_t\{T\}A - r_t\{T\}A \tag{53}$$

Since the first component of the vector on the right hand side of (53) is 0, and $b = (1, \ldots, 1)$, we must have $\eta_{t+1}\{T\} \geq 0$.

Using (39) again and noting that:

$$\begin{aligned} \eta_{t+1}\{T\}b &= -q_t\{T\}A + q_{t+1}\{T\} - r_t\{T\}A + r_{t+1}\{T\} + \nu_{t+1}\{T\} \\ &\leq q_{t+1}\{T\} + r_{t+1}\{T\} + \nu_{t+1}\{T\} \end{aligned}$$

we have:

$$\eta_{t+1}\{T\} \leq q_{t+1}^m\{T\} + r_{t+1}^m\{T\} + \nu_{t+1}^m\{T\} \tag{54}$$

For $t \geq N'$, we have $\tilde{h}_t^m \geq (1/2m)$, and $\tilde{y}_t^m \geq (1/2m)$. So by (42), we have $r_t^m\{T\} = 0$, and by (43), we have $\nu_t^m\{T\} = 0$. Using this information in (54), together with (52) yields:

$$\eta_{t+1}\{T\} \le B \ for\ T \ge t \ge N' - 1 \tag{55}$$

Using (52) and (55) in (39), we have for $T \ge t \ge N' - 1$,

$$\begin{aligned} r_{t+1}\{T\} - r_t\{T\}A &= q_t\{T\}A - q_{t+1}\{T\} - v_{t+1}\{T\} + \eta_{t+1}\{T\}b \\ &\le q_t\{T\}A + \eta_{t+1}\{T\}b \\ &\le 2Bb \end{aligned} \tag{56}$$

We can now use (56) to obtain:

$$\left.\begin{array}{llll} r^1_{t+1}\{T\} & \le 2B & for & t \ge N' - 1 \\ r^2_{t+1}\{T\} & \le 4B & for & t \ge N' \\ r^3_{t+1}\{T\} & \le 6B & for & t \ge N' + 1 \\ \dots & \dots & & \dots \\ r^n_{t+1}\{T\} & \le 2nB & for & t \ge N' + n - 2 \end{array}\right\}$$

This implies that:

$$r_{t+1}\{T\} \le 2nBb \ for\ T \ge t \ge N' + n - 2$$

Thus, denoting $(N' + n)$ by $N$, we have:

$$p_{t+1}\{T\} \le (2n+1)Bb \ for\ T \ge t \ge N - 2 \tag{57}$$

These are the bounds on the shadow prices that we need. This completes the second part of the proof.

Let $(y_t)_1^\infty$ be a path from $y \in Q$. Using (51), we obtain:

$$\sum_{t=1}^{N-1} [u(P(y_t - Ay_{t+1})) - u(P(\tilde{y}_t - A\tilde{y}_{t+1}))] \le p_N\{T\}(\tilde{y}_N - y_N) \tag{58}$$

We have now established that for each $T > N$, there is a vector $(p_{N-1}\{T\}, ..., p_T\{T\})$, such that:

$$0 \le p_{t+1}\{T\} \le (2n+1)Bb \ for\ N - 2 \le t \le T - 1 \tag{59}$$

holds, (58) holds, and for all $(y, z) \in \Omega$, and for all $N - 1 \le t \le T - 1$,

$$u(P(y - Az)) + p_{t+1}\{T\}z - p_t\{T\}y \leq u(P(\tilde{y}_t - A\tilde{y}_{t+1})) + p_{t+1}\{T\}\tilde{y}_{t+1} - p_t\{T\}\tilde{y}_t \quad (60)$$

holds.

Given (59), we can now use the Cantor diagonal process to obtain a subsequence $T'$ of $T(> N)$, and a sequence $(\tilde{p}_t)_{N-1}^{\infty}$ such that, for each $t \geq N-1$

$$p_t(T') \to \tilde{p}_t \ as \ T' \to \infty \quad (61)$$

Using (61) in (58) yields (36). Using (61) in (60) yields (37). Using (61) in (59) yields:

$$0 \leq \tilde{p}_t \leq (2n+1)Bb \ for \ t \geq N-1 \quad (62)$$

This completes the third and last part of the proof.

## 8. MAXIMAL PATHS ARE OPTIMAL

In this section, we use the results of the previous two sections to show that the notions of maximality and optimality coincide for our forestry model.

To this end, recall from the definitions of optimal and maximal paths that an optimal path must be maximal. It is the reverse implication that is non-trivial and of significant interest.

To make the discussion non-void, we show (in Theorem 6) that there exists an optimal path from every initial forest, by following the method of Brock (1970a). Then, using the price support property of maximal paths (established in Proposition 3), we show (in Theorem 7) that a maximal path is necessarily optimal.

We first present our result on the existence of an optimal path. We provide a fairly self-contained treatment of this familiar topic in intertemporal allocation theory, because there are several differences between our framework and the one typically used in the standard version of this theory.

**Theorem 6.** *There exists an optimal path $(\tilde{y}_t)_1^{\infty}$ from every $\tilde{y} \in Q$.*

**Proof.** We first note that there is a good path from $\tilde{y}$. Since $(\tilde{y}, \overline{y}) \in Q^2$ [where $\overline{y}$ is the golden-rule] we can use Lemma 1 to obtain $(z_1, \ldots, z_{n+2})$ such that:

$$(i)\ (z_s, z_{s+1}) \in \Omega \ for \ s = 1, \ldots, n+1; (ii)\ z_1 = \tilde{y}, z_{n+2} = \overline{y}.$$

Define a sequence $(y'_t)_1^{\infty}$ as follows:

$$\left.\begin{array}{lll} y'_t = z_t & for & t = 1, \ldots, n+2 \\ y'_t = \overline{y} & for & t > n+2 \end{array}\right\}$$

It is easy to check that $(y'_t)_1^\infty$ is a path from $y \in Q$. Since $y'_t = \bar{y}$ for $t > n+2$, we have $c'_t = \bar{c}$ for $t > n+2$. It follows that for all $T > n+2$,

$$\sum_{t=1}^{T} [u(P(y'_t - Ay'_{t+1})) - u(\bar{c})] \geq -2(n+1)u(\bar{c})$$

Thus, $(y'_t)_1^\infty$ is good.

Next, we claim that for any good path $(y_t)_1^\infty$ from $\tilde{y}$, we must have:

$$\sum_{t=1}^{\infty} \delta_t < \infty$$

To see this, observe that for any $T \in \mathbb{N}$, we can use (6) to write:

$$\sum_{t=1}^{T} [u(P(y_t - Ay_{t+1})) - u(\bar{c})] = \sum_{t=1}^{T} [py_t - py_{t+1}] - \sum_{t=1}^{T} \delta_t$$
$$= [py - py_{T+1}] - \sum_{t=1}^{T} \delta_t$$

Since $(y_t)_1^\infty$ is good, there is $G \in \mathbb{R}$ such that:

$$\sum_{t=1}^{T} [u(P(y_t - Ay_{t+1})) - u(\bar{c})] \geq G \quad \textit{for all } T \in \mathbb{N}$$

Thus, we obtain:

$$\sum_{t=1}^{T} \delta_t \leq py - G \leq n - G \quad \textit{for all } T \in \mathbb{N}$$

Since $\delta_t \geq 0$ by (6), this establishes our claim.

We now define:

$$\delta(\tilde{y}) = \inf\{\sum_{t=0}^{\infty} \delta_t : (y_t)_1^\infty \textit{ is a good path from } x\}$$

Since there is a good path from $x$, $\delta(\tilde{y})$ is well-defined and $\delta(\tilde{y}) < \infty$.

Given the definition of $\delta(\tilde{y})$, we can choose a sequence of paths $(y_t\{N\})_1^\infty$ from $\tilde{y}$ (for $N = 1, 2, 3, \ldots$ ) such that:

$$\sum_{t=0}^{\infty} \delta_t\{N\} \le \delta(\tilde{y}) + (1/N) \text{ for } N = 1,2,3,...$$

Since for each $N = 1,2,3,...,$ we have $y_t\{N\} \in Q$ for $t \ge 1$, we can use the Cantor diagonal process to find a subsequence $N'$ of $N$ , and a sequence $(\tilde{y}_t)_1^{\infty}$ such that, for each $t \in \mathbb{N}$, we have

$$y_t\{N'\} \to \tilde{y}_t \text{ as } N' \to \infty$$

It can be checked that $(\tilde{y}_t)_1^{\infty}$ is a path from $\tilde{y}$, and for each $t \in \mathbb{N}$, we have:

$$\delta_t\{N'\} \to \tilde{\delta}_t \text{ as } N' \to \infty$$

We claim that the sequence $(\tilde{\delta}_t)_1^{\infty}$ satisfies:

$$\delta(\tilde{y}) = \sum_{t=1}^{\infty} \tilde{\delta}_t$$

If the claim is not true, then we can find $T \in \mathbb{N}$, such that:

$$\delta(\tilde{y}) < \sum_{t=1}^{T} \tilde{\delta}_t$$

Defining $\gamma = \sum_{t=1}^{T} \tilde{\delta}_t$, pick a number $\gamma'$ such that $\gamma > \gamma' > \delta(\tilde{y})$. Then, we can find $\tilde{N}$, such that for all $N' \ge \tilde{N}$,

$$\gamma' < \sum_{t=1}^{T} \delta_t\{N'\}$$

Thus, for all $N' \ge \tilde{N}$, we have:

$$\gamma' < \sum_{t=1}^{T} \delta_t\{N'\} \le \sum_{t=1}^{\infty} \delta_t\{N'\} \le \delta(\tilde{y}) + (1/N')$$

Letting $N' \to \infty$, we get $\gamma' \le \delta(\tilde{y})$, which contradicts the definition of $\gamma'$. This establishes our claim.

Consider the path $(\tilde{y}_t)_1^\infty$ from $\tilde{y}$, defined above. It is easy to check that $(\tilde{y}_t)_1^\infty$ is a good path. We will show that it is an optimal path from $\tilde{y}$. Suppose, on the contrary, there is a path $(y_t')_1^\infty$ from $\tilde{y}$, such that the associated consumption sequences $(c_t')_1^\infty$ and $(\tilde{c}_t)_1^\infty$ satisfy the condition that $(c_t')_1^\infty R_u (\tilde{c}_t)_1^\infty$ [where the pre-order $R_u$ is defined by (R)]. This will be the case if and only if there is $\overline{N} \in \mathbb{N}$, such that:

$$\sum_{t=1}^{N} u(c_t') \geq \sum_{t=1}^{N} u(\tilde{c}_t) \text{ for all } N \geq \overline{N} \tag{63}$$

and :

$$\sum_{t=1}^{N'} u(c_t') > \sum_{t=1}^{N'} u(\tilde{c}_t) \text{ for a subsequence } N' \text{of } N \geq \overline{N} \tag{64}$$

Define a sequence $(y_t'')_1^\infty$ by: $y_t'' = (\frac{1}{2})y_t' + (\frac{1}{2})\tilde{y}_t$ for $t \in \mathbb{N}$. It is easy to check that $(y_t'')_1^\infty$ is a path from $\tilde{y}$, and its associated consumption sequence $(c_t'')_1^\infty$ satisfies: $c_t'' = (\frac{1}{2})c_t' + (\frac{1}{2})\tilde{c}_t$ for $t \in \mathbb{N}$. Using (64), we know that $c_t' \neq \tilde{c}_t$ for some $t = \tau \in \mathbb{N}$. Then, by concavity of $u$, and strict mid-concavity of $u$, we have:

$$\begin{aligned} &(a)\; u(c_t'') \geq (\frac{1}{2})u(c_t') + (\frac{1}{2})u(\tilde{c}_t) \quad \text{for all } t \in N \\ &(b)\; \xi \equiv u(c_t'') - [(\frac{1}{2})u(c_t') + (\frac{1}{2})u(\tilde{c}_t)] > 0 \quad \text{for } t = \tau \end{aligned} \tag{65}$$

Defining $\tilde{N} = \max[\overline{N}, \tau]$, and using (65) in (63), we get:

$$\xi \leq [\sum_{t=1}^{N} u(c_t'') - \sum_{t=1}^{N} u(\tilde{c}_t)] \text{ for all } N \geq \tilde{N} \tag{66}$$

This shows that $(y_t'')_1^\infty$ is also a good path from $\tilde{y}$.

We can now use (6) to write for all $N \in \mathbb{N}$,

$$[\sum_{t=1}^{N} u(c_t'') - \sum_{t=1}^{N} u(\tilde{c}_t)] = \sum_{t=1}^{N} [u(P(y_t'' - Ay_{t+1}'')) - u(P(\tilde{y}_t - A\tilde{y}_{t+1}))]$$

$$=[p\tilde{y}_{N+1}-py''_{N+1}]-\sum_{t=1}^{N}\delta''_t+\sum_{t=1}^{N}\tilde{\delta}_t \tag{67}$$

Since $(y''_t)_1^\infty$ and $(\tilde{y}_t)_1^\infty$ are both good paths from $\tilde{y}$, we have $y''_{N+1}\to\bar{y}$ and $\tilde{y}_{N+1}\to\bar{y}$ as $N\to\infty$. Thus, we can choose $N_1$ such that for $N\ge N_1$, we have $[p\tilde{y}_{N+1}-py''_{N+1}]<(\frac{1}{2})\xi$. Given the definition of $\delta(\tilde{y})$ and the path $(\tilde{y}_t)_1^\infty$, we can choose $N_2$ such that for $N\ge N_2$, we have $[\sum_{t=1}^{N}\tilde{\delta}_t-\sum_{t=1}^{N}\delta''_t]<(\frac{1}{2})\xi$. Thus, for $N>\max[N_1,N_2,\tilde{N}]$, we have from (67) that:

$$[\sum_{t=1}^{N}u(c''_t)-\sum_{t=1}^{N}u(\tilde{c}_t)]<\xi$$

But this contradicts (66) and establishes that $(\tilde{y}_t)_1^\infty$ is optimal from $\tilde{y}$.

Next, we show that a maximal path is necessarily optimal. This depends on the methods used by Gale (1967) and McKenzie (1968), and relies on the shadow prices associated with maximal paths, derived in the previous section.

**Theorem 7.** *A path $(\tilde{y}_t)_1^\infty$ from $\tilde{y}\in Q$ is maximal if and only if it is optimal.*

**Proof.** If a path $(\tilde{y}_t)_1^\infty$ from $\tilde{y}\in Q$ is optimal, then by definitions of optimality and maximality, $(\tilde{y}_t)_1^\infty$ is a maximal path from $\tilde{y}\in Q$.

To establish the converse result, let $(\tilde{y}_t)_1^\infty$ be a maximal path from $\tilde{y}\in Q$. We will show that it is optimal. Suppose this is not the case. Then there is a path $(y'_t)_1^\infty$ from $\tilde{y}$, such that the associated consumption sequences $(c'_t)_1^\infty$ and $(\tilde{c}_t)_1^\infty$ satisfy the condition that $(c'_t)_1^\infty P_u(\tilde{c}_t)_1^\infty$ [where the pre-order $R_u$ is defined by (R)]. This will be the case if and only if there is $\bar{N}\in\mathbb{N}$, such that:

$$\sum_{t=1}^{T}u(c'_t)\ge\sum_{t=1}^{T}u(\tilde{c}_t)\ \textit{for all}\ T\ge\bar{N} \tag{68}$$

and :

$$\sum_{t=1}^{T'}u(c'_t)>\sum_{t=1}^{T'}u(\tilde{c}_t)\ \textit{for a subsequence}\ T'\textit{of}\ T\ge\bar{N} \tag{69}$$

Define a sequence $(y_t)_1^\infty$ by: $y_t=(\frac{1}{2})y'_t+(\frac{1}{2})\tilde{y}_t$ for $t\in\mathbb{N}$. It is easy to check that $(y_t)_1^\infty$ is a path from $\tilde{y}$, and its associated consumption sequence $(c_t)_1^\infty$ satisfies: $c_t=(\frac{1}{2})c'_t+(\frac{1}{2})\tilde{c}_t$ for $t\in\mathbb{N}$. Using (69), we know that $c'_t\ne\tilde{c}_t$ for some $t=\tau\in\mathbb{N}$. Then, by concavity of $u$, and strict mid-concavity of $u$, we have:

$$\begin{aligned}&(a)\ u(c_t)\geq(\frac{1}{2})u(c_t')+(\frac{1}{2})u(\tilde{c}_t)\quad for\ all\ t\in\mathbb{N}\\&(b)\ \xi\equiv u(c_t)-[(\frac{1}{2})u(c_t')+(\frac{1}{2})u(\tilde{c}_t)]>0\quad for\ t=\tau\end{aligned}\tag{70}$$

Defining $\tilde{N}=\max[\bar{N},\tau]$, and using (70) in (68), we get:

$$\xi\leq[\sum_{t=1}^{T}u(c_t)-\sum_{t=1}^{T}u(\tilde{c}_t)]\ for\ all\ T\geq\tilde{N}\tag{71}$$

Since $(\tilde{y}_t)_1^\infty$ is a maximal path from $\tilde{y}\in Q$, it is a good path from $\tilde{y}$, by Proposition 2. The inequality (71) shows that $(y_t)_1^\infty$ is also a good path from $\tilde{y}$.

Since $(\tilde{y}_t)_1^\infty$ is a maximal path from $\tilde{y}\in Q$, we can use Proposition 3 to obtain $N\in\mathbb{N}$, $N>1$, and a sequence of shadow prices $(\tilde{p}_t)_{N-1}^\infty$ such that:

$$0\leq\tilde{p}_t\leq(2n+1)Bb\quad for\ all\ t\in\mathbb{N},\ with\ t\geq N-1\tag{72}$$

and (36) and (37) hold. Defining $\hat{N}=\max[N,\tilde{N}]$, and using (36) and (37), we have for $T>\hat{N}$,

$$\begin{aligned}[\sum_{t=1}^{T}u(c_t)-\sum_{t=1}^{T}u(\tilde{c}_t)]&=\sum_{t=1}^{T}[u(P(y_t-Ay_{t+1}))-u(P(\tilde{y}_t-A\tilde{y}_{t+1}))]\\&\leq\tilde{p}_{T+1}(\tilde{x}_{T+1}-x_{T+1})\end{aligned}\tag{73}$$

Since $(y_t)_1^\infty$ and $(\tilde{y}_t)_1^\infty$ are both good paths from $\tilde{y}$, we have $y_{T+1}\to\bar{y}$ and $\tilde{y}_{T+1}\to\bar{y}$ as $T\to\infty$. Thus, using (72), we must have $\tilde{p}_{T+1}(\tilde{x}_{T+1}-x_{T+1})\to 0$ as $T\to\infty$. Using this in (73) contradicts (71) and establishes the result.

## 9. CONCLUDING REMARKS

Our analysis of forest management has tried to provide a more satisfactory basis for focusing on maximum sustained yield of timber as a key concept, given the objective of intergenerational equity. Since forest management involves much more than timber production, it would be interesting to generalize the analysis to a setting of sustainable forest management, where all forest products are taken into account. Since, in principle, such an exercise would still fit into the general theory of intertemporal allocation, we feel that such a generalization would be theoretically feasible. The inter-relationships that would have to prevail between the various forest products at the maximum sustained forest yield solution would reflect the actual biological features governing the (joint production) of forest products, and should be of interest to foresters.

We note that, even as a model of timber management, our framework is greatly oversimplified, and we comment on possible generalizations. (i) Harvesting and replanting costs can be incorporated into the model fairly easily. (ii) All trees coming to maturity at the same time might have a lower timber yield per tree compared to a situation in which trees come to maturity on different subplots in different years. This aspect is less straightforward to incorporate into our model, because of the nature of the biological interaction among trees that is being captured. In such a scenario, one would expect to see (at least in transition) the phenomenon of "forest thinning" on maximal paths. We feel, however, that such considerations would tend to reinforce the concept of the golden-rule forest in the long-run, since the adverse effect being captured would be fairly minimal for the golden-rule solution. (iii) The assumption of strong convexity (Axiom 6) has been used extensively in our analysis. Without it, the uniqueness of a maximal path is unlikely to obtain. But, then, the uniqueness of an optimal path will not hold in general either.[26] Whether maximal paths nevertheless continue to be optimal in the more general context, where strong convexity is replaced by (weak) convexity, remains an open question.

Our setting for forest management is best applied to governments managing forest resources of a country. Roughly, two-thirds of the world's forests are currently managed by governments, so we feel the model has fairly wide potential use. Of course, our analysis focuses on the managed forest, under intergenerational equity. Thus, the *social* rate of discount is zero, while the market rate of interest is typically positive. In the same vein, the shadow prices supporting maximal paths provide the correct *social* evaluation of forestry land, according to the vintage of trees standing on it, and are likely to differ from market prices of forestry land. This disparity arises because market prices and interest rates are determined by (aggregate) decisions of individuals whose preferences reflect fairly short-run objectives. In the interest of intergenerational equity (of generations yet unborn), governments can play a vital role in managing forests on a sustainable basis.

Our theory of shadow prices supporting maximal paths shows that judicious forest management involves the methods of *social* project evaluation, where the notion of "generalized profit maximization" is still central (as in private project evaluation), but the profits are evaluated at the appropriate shadow prices, and using the appropriate social rate of discount, rather than the corresponding market based magnitudes. In this respect, the considerations which lead a country to preserve its national parks are also the considerations which lead to forest management which respects intergenerational equity.

**Acknowledgements:** This paper owes much to my ongoing collaborative work with Kaushik Basu on the representation of intertemporal preferences, with Swapan Dasgupta on characterizing optimal policies in the forestry model, and with M. Ali Khan on the choice of technique in a dynamic economy. An earlier version of the paper was presented at the International Conference on the Economics of

Sustainable Forest Management at the University of Toronto, May 20-22, 2004. The present version has benefited from detailed comments by Shashi Kant.

## NOTES

---

[1] Ramsey (1928) had maintained that discounting one generation's utility or income vis-a-vis another's to be "ethically indefensible", and something that "arises merely from the weakness of the imagination".
[2] The Grading Principle is due to Suppes (1966). For a comprehensive analysis of it, see Sen (1971).
[3] Such comparisons figure prominently in the theory of efficient capital accumulation of Malinvaud (1953), and in the theory of representation of intertemporal preferences of Koopmans (1972).
[4] For the standard finite dimensional theory, see Debreu (1959), which in turn is based on Debreu (1954).
[5] In this regard, the current work can be seen as a continuation of the study in Basu and Mitra (2003a).
[6] The study by Brock (1970b) uses a "consistency axiom" which is actually a continuity restriction on the underlying preferences. A more recent study by Asheim and Tungodden (2004) also uses a similar continuity axiom.
[7] This is in contrast to postulating continuity properties on preferences in the finite-dimensional theory.
[8] In the economics literature, a pre-ordering is often refereed to as a "partial ordering". However, in the mathematics literature, the term "partial ordering" refers to a binary relation which is transitive and *antisymmetric*. To avoid confusion, we use the mathematical terminology, since the term "pre-order" is never used in any other sense in either discipline. Incidentally, our usage coincides with the terminology introduced in Debreu (1959).
[9] The results in this section are stated without proofs. They can be established along the lines indicated in Mitra (2003).
[10] Our SWR will later be compared in detail with the one induced by the overtaking criterion.
[11] Many authors have felt that a stronger notion than the Anonymity Axiom is needed to reflect intergenerational equity in intertemporal preferences. However, there appears to be general agreement that any notion of intergenerational equity in intertemporal preferences must include the Anonymity Axiom. Since we are attempting an axiomatic characterization of a utilitarian SWR, we feel justified in imposing this weak equity requirement on the SWR.
[12] The independence postulate, developed by Debreu (1960) and Koopmans (1960), follows the pioneering work in this area by Leontief (1947a, 1947b) and Samuelson (1947).
[13] According to Brock (1970a, 1970b), this axiom "captures the notion that decisions on infinite programs are consistent with decisions on finite programs of length $n$ if $n$ is large enough."
[14] Asheim and Tungodden (2004) use an axiom which is similar, and which they view as a continuity restriction.
[15] We present Brock's characterization result in a form so that it may be readily compared with our results reported in the previous section. Actually, Brock does not impose the Anonymity axiom, and therefore gets a sequence of utility functions (rather than a single one) in his characterization result. He does note however that under an Anonymity axiom, these utility functions would be the same. Brock also does not impose the Pareto axiom, although, to apply the representation theorem of Debreu (1960), he does assume that there are at least three factors of $I^N$, which are *essential* in the sense of Debreu (1960).
[16] This example is based on the discussion in Basu and Mitra (2003b). The present form of this example owes much to comments on the Basu-Mitra (2003b) paper by Wolfgang Buchholz.
[17] In this connection, one might note that in a recent paper, Asheim and Buchholz (2005) work with a social preference relation, which has even less structure than ours; the SWR in their paper is restricted only to satisfy the Pareto and Anonymity axioms.
[18] See Yaari (1977) and Debreu, and Koopmas (1982) for the general result relating convexity of preference relations to concavity of utility functions, when the preference relation has an additively separable utility representation.
[19] Actually, Debreu asserts the strict preference for all non-trivial convex-combinations of $c$ and $c'$. We will need the axiom only in its "strictly mid-convex" form.
[20] In terms of notation (but not in terms of content) the present version differs slightly from that used in

Mitra and Wan (1986). References to the earlier literature, on which the model is based, can be found in Mitra and Wan (1986).

[21] It will turn out, as noted later in Section 8, that there is actually a unique maximal path (from given initial conditions) in the model we are analyzing. But, this is a result of the notion of maximality applied to the standard forestry model; it is not definitional.

[22] The second part of this Proposition is a result along the lines of the well-known "value-loss lemma", due to Radner (1961), Atsumi (1965), and McKenzie (1968).

[23] For an excellent non-technical discussion of the methods used in Mitra and Wan (1986), which parallel the methods used in this section (as well as the previous one) see the recent paper by Khan (2005).

[24] The analysis follows Mitra and Wan (1986) closely.

[25] While this assumption is crucial to our method of investigation (duality theory), it is not clear whether it is indispensable for the results of the next section on the relation between maximal and optimal paths. It would seem that a "primal route" to those results should be possible. See Mitra (2003) for such an approach in the context of aggregative models of economic growth and renewable resources. In any case, the duality theory developed in the current section (using (A.4)) might be of interest, independent of its application to obtain the results of the next section.

[26] In the context of a somewhat different dynamic optimization model, this has been demonstrated in a concrete example by Khan and Mitra (2002).

# REFERENCES

Arrow, K.J., Hurwicz, L., & Uzawa, H. (1961). Constraint qualifications in maximization problems. *Naval Research Logistics Quarterly*, 8, 175-191

Asheim, G.B., & Tungodden, B. (2004). Resolving distributional conflicts between generations. *Economic Theory*, 24, 221-230.

Asheim, G.B., & Buchholz, W. (2005). Can stock-specific constraints be justified ? Chapter 8 in this Book.

Atsumi, H. (1965). Neoclassical growth and the efficient program of capital accumulation. *Review of Economic Studies*, 32, 127-136.

Basu, K., & Mitra, T. (2003a). Aggregating infinite utility streams with intergenerational equity: The impossibility of being Paretian. *Econometrica* 71, 1557-1563.

Basu, K., & Mitra, T. (2003b).Utilitarianism for infinite utility streams: A new welfare criterion and its axiomatic characterization. CAE Working Paper 03-05, Cornell University.

Brock, W.A. (1970a). On existence of weakly maximal programmes in a multi-sector economy. *Review of Economic Studies*, 37, 275-280.

Brock, W.A. (1970b). An axiomatic basis for the Ramsey-Weizsacker overtaking criterion. *Econometrica,* 38, 927-929.

Debreu, G. (1954). Representation of a preference ordering by a numerical function. In R.M. Thrall, C.H. Coombs, R.L. Davis, (Eds.), *Decision processes* (pp.159-165). New York: John Wiley.

Debreu, G. (1959). *Theory of value*. New York: John Wiley.

Debreu, G. (1960). Topological methods in cardinal utility theory. In K.J. Arrow, S. Karlin, P. Suppes, (Eds.), *Mathematical methods in the social sciences*. Stanford: Stanford University Press,

Debreu, G., & Koopmans, T.C. (1982). Additively decomposed quasiconvex functions. *Mathematical Programming*, 24, 1-38.

Gale, D. (1967). On optimal development in a multi-sector economy. *Review of Economic Studies*, 34, 1-18.

Khan, M.A. (2005). Intertemporal ethics, modern capital theory and the economics of forestry. Chapter 3 in this book.

Khan, M.A., &. Mitra, T. (2002). Optimal growth in the Robinson-Solow-Srinivasan model: The two-sector setting without discounting. Mimeo, Department of Economics, Cornell University, Ithaca.

Koopmans, T.C. (1960). Stationary ordinal utility and impatience. *Econometrica,* 28, 287-309.

Koopmans, T.C. (1972). Representation of preference orderings over time. In C.B. McGuire, R. Radner, (Eds.), *Decision and organization* (pp.79-100). New York: North-Holland.

Koopmans, T.C., Diamond, P.A., & Williamson, R.E. (1964). Stationary utility and time perspective. *Econometrica,* 32, 82-100.

Leontief, W. (1947a). A note on the interrelation of subsets of independent variables of a continuous function with continuous first derivatives. *Bulletin of the American Mathematical Society*, 53, 343-350.

Leontief, W. (1947b). Introduction to a theory of the internal structure of functional relationships. *Econometrica*, 15, 361-373.

Malinvaud, E. (1953). Capital accumulation and efficient allocation of resources. *Econometrica,* 21, 233-268.

McKenzie, L.W. (1968). Accumulation programs of maximum utility and the von Neumann facet. In J.N. Wolfe, (Ed.), *Value, capital and growth* (pp.353-383). Edinburgh: Edinburgh University Press.

McKenzie, L.W. (1986). Optimal economic growth, turnpike theorems and comparative dynamics. In K.J. Arrow, M. Intrilligator, (Eds.), *Handbook of mathematical economics*, Vol. 3, (pp.1281-1355). New York: North-Holland Publishing Company.

Mitra, T. (2003). Representation of equitable preferences with applications to aggregative models of economic growth and renewable resources. Mimeo, Department of Economis, Cornell University, Ithaca.

Mitra, T., & Wan, Jr. H.Y. (1986). On the Faustmann solution to the forest management problem. *Journal of Economic Theory*, 40, 229-249.

Radner, R. (1961). Paths of economic growth that are optimal with regard only to final states: A turnpike theorem. *Review of Economic Studies*, 28, 98-104.

Ramsey, F. (1928). A mathematical theory of savings. *Economic Journal*, 38, 543-559.

Samuelson, P.A. (1947). *Foundations of economic analysis*. Cambridge: Harvard University Press.

Samuelson, P.A. (1976). Economics of forestry in an evolving society. *Economic Enquiry,* 14, 466-492.

Sen, A.K. (1971). *Collective choice and social welfare*. Edinburgh: Oliver&Boyd.

Suppes, P. (1966). Some formal models of grading principles. *Synthese,* 6, 284-306.

Svensson, L.-G. (1980). Equity among generations. *Econometrica,* 48, 1251-1256.

von Weizsäcker, C.C. (1965). Existence of optimal programs of accumulation for an infinite time horizon. *Review of Economic Studies*, 32, 85-104.

Yaari, M.E. (1977). A note on separability and quasiconcavity. *Econometrica*, 45, 1183-1186.

# CHAPTER 8

# CAN STOCK-SPECIFIC SUSTAINABILITY CONSTRAINTS BE JUSTIFIED?

GEIR B. ASHEIM

*Department of Economics, University of Oslo*
*Blindern, NO-0317 Oslo, Norway*
*Email: g.b.asheim@econ.uio.no*

WOLFGANG BUCHHOLZ

*Department of Economics, University of Regensburg*
*D-93040 Regensburg, Germany*
*Email: wolfgang.buchholz@wiwi.uni-regensburg.de*

**Abstract.** We show that the Suppes-Sen grading principle leads to stock-specific sustainability constraints in a class of resource models, provided that the resource is renewable or utility is derived directly from the resource stock. Decreasing the resource stock is not compatible with Suppes-Sen maximality, unless a smaller stock leads to higher natural growth.

## 1. INTRODUCTION

During the last decade sustainability has become one of the main issues in environmental economics and policy. Even though there exists a multitude of definitions of sustainability, they all boil down to the idea that living conditions on earth should not become worse in the course of time. Consequently, sustainability has in many instances been interpreted as a postulate to keep stocks of natural resources—as part of the whole vector of capital stocks—intact. This does not only conform to common sense but might also be justified from an economic perspective when the natural resource cannot be substituted by man-made capital. With respect to forests, this holds true from an instrumental as well as from a moral perspective. On the one hand, forests are indispensable as a source of biodiversity, and as sinks for carbon emissions—being important for climate protection—as well as a supplier of amenity value and raw material for the pulp and paper industry. On the other hand



*Kant and Berry (Eds.), Economics, Sustainability, and Natural Resources: Economics of Sustainable Forest Management,* 175-189

people—feeling responsible for the preservation of the nature in itself—will also attribute to forests some kind of "existence value".

In economic theory, however, applying the usual discounting criteria to standard renewable resource models will in many cases have adverse effects on sustainability in a stock specific sense: Paths that are optimal w.r.t. to discounted utilitarianism will often lead to a deterioration of resource stocks—at least when the discount rate is sufficiently high.

Using different specific models Heal (1998, 2001) has analysed how such undesirable consequences can be avoided. In particular he has shown that two features are favorable for having non-decreasing resource stocks along optimal paths:

- Utility is not only derived from the flow of resource extraction, but also directly from the resource stock itself.
- Intertemporal paths are evaluated by means of social preferences—like undiscounted utilitarianism (in the form of overtaking) and Rawlsian maximin—that entail equal treatment of all generations. Such equal treatment corresponds to what is referred to as the Weak Anonymity condition in the social choice literature.

In this chapter we extend Heal's (2001) analysis by showing how stock-specific sustainability constraints can be obtained from rather weak ethical axioms. By combining Weak Anonymity with the uncontroversial Strong Pareto condition, the so-called Suppes-Sen grading principle is obtained (Sen, 1970; Suppes, 1966). We show that the Suppes-Sen grading principle leads to stock-specific sustainability constraints, provided that the resource is renewable or utility is derived directly from the resource stock. Under this provision, decreasing the resource stock contradicts Suppes-Sen maximality, unless a smaller stock leads to higher natural growth. Hence, there is an important class of models where extraction leading to a stock smaller than the one corresponding to *maximal sustainable yield* is incompatible with the Suppes-Sen grading principle. By starting from basic ethical axioms for intergenerational social preferences, the analysis of this chapter yields a new justification for stock-specific sustainability constraints in general, and—when applied to forests—for limits on deforestation in particular.

Within the framework of social choice theory, we have in Asheim, Buchholz, and Tungodden (2001) justified sustainability by means of the Suppes-Sen grading principle. In models that satisfy a certain productivity condition, which we refer to as "Immediate Productivity", the set of Suppes-Sen maximal utility paths is shown to equal the set of non-decreasing and efficient paths. This result cannot, however, be applied in the present setting, since none of these models considered in this chapter satisfies "Immediate Productivity". Nevertheless, it turns out that the Suppes-Sen grading principle leads to stock-specific sustainability constraints.

By deriving normative conclusions concerning resource management from *incomplete* social preferences like the Suppes-Sen grading principle, the motivation for this chapter is similar to the previous chapter by Mitra (2005). While Mitra

(2004) weakens a criterion satisfying both Strong Pareto and Weak Anonymity—namely undiscounted utilitarianism in the form of overtaking—by assuming that only paths coinciding beyond some finite point in time are comparable, we go a step further by analyzing social preferences satisfying nothing but Strong Pareto and Weak Anonymity.

We recapitulate in Section 2 the analysis of Asheim et al. (2001), introduce in Section 3 the class of models considered, and show in Section 4 under what conditions the Suppes-Sen grading principle leads to stock-specific sustainability constraints in these models. Since the models abstract from important features of real-world economies, the significance of these results is discussed in the concluding Section 5. Proofs are contained in Section 6.

## 2. THE SUPPES-SEN GRADING PRINCIPLE AND SUSTAINABILITY

There is an infinite number of generations $t = 1, 2, \ldots$. The utility level of generation $t$ is given by $u_t$, which should be interpreted as the utility level of a representative member of this generation. Assume that the utilities need not be more than ordinally measurable and level comparable.

A binary relation $R$ over paths ${}_1\mathbf{u} = (u_1, u_2, \ldots)$ starting in period 1 expresses social preferences over different intergenerational utility paths. Any such binary relation $R$ is throughout assumed to be reflexive and transitive on the infinite Cartesian product $\Re^\infty$ of the set of real numbers $\Re$, where $\infty = |\aleph|$ and $\aleph$ is the set of natural numbers. The social preferences $R$ may be complete or incomplete, with $I$ denoting the symmetric part, i.e. indifference, and $P$ denoting the asymmetric part, i.e. (strict) preference.

In order to define sets of feasible paths, it suffices for the analysis of the present chapter to assume that the initial endowment of generation $t \geq 1$ is given by a stock $x_t$. A generation $t$ acts by choosing a utility level $u_t$ and a capital stock $x_{t+1}$ which is bequeathed to the next generation $t+1$. For every $t$, the function $F_t$ gives the maximum utility attainable for generation $t$ if $x_t$ is inherited and $x_{t+1}$ is bequeathed; i.e., $u_t \leq F_t(x_t, x_{t+1})$ has to hold for any feasible *utility-bequest pair* $(u_t, x_{t+1})$ of generation $t$. Furthermore, it is assumed that the utility level of each generation cannot fall below $0$. If $F_t(x_t, x_{t+1}) < 0$, then the bequest $x_{t+1}$ is infeasible given the inheritance $x_t$. Hence, generation $t$'s utility-bequest pair $(u_t, x_{t+1})$ is said to be *feasible* at $t$ given $x_t$ if $0 \leq u_t \leq F_t(x_t, x_{t+1})$. The sequence ${}_1\mathbf{F} = (F_1, F_2, \ldots)$ characterizes the technology of the economy under consideration. Given the technology ${}_1\mathbf{F}$, a utility path ${}_t\mathbf{u} = (u_t, u_{t+1}, \ldots)$ is feasible at $t$ given $x_t$ if there exists a path ${}_{t+1}\mathbf{x} = (x_{t+1}, x_{t+2}, \ldots)$ such that, for all $s \geq t$, generation $s$'s utility bequest pair $(u_s, x_{s+1})$ is feasible at $s$ given $x_s$.

A utility path ${}_1\mathbf{v}$ weakly Pareto-dominates another utility path ${}_1\mathbf{u}$ if every generation is weakly better of in ${}_1\mathbf{v}$ than in ${}_1\mathbf{u}$ and some generation is strictly better off. A feasible path ${}_1\mathbf{v}$ is said to be efficient if there is no other feasible path that weakly Pareto-dominates this path. A feasible path ${}_1\mathbf{v}$ is said to be $R$*-maximal*, if there exists no feasible path ${}_1\mathbf{u}$ such that ${}_1\mathbf{u}\,P\,{}_1\mathbf{v}$.

Within this framework, the justification for sustainability in Asheim et al. (2001) rests on one technological assumption and two conditions on the social preferences.

First, we in Asheim et al. (2001) impose the following domain restriction on the technological framework.

**Assumption 1** (*Immediate Productivity of* ${}_1\mathbf{F}$). If ${}_t\mathbf{u} = (u_t, u_{t+1}, \ldots)$ is feasible at $t$ given $x_t$ with $u_t > u_{t+1}$, then $(u_{t+1}, u_t, u_{t+2}, \ldots)$ is feasible and inefficient at $t$ given $x_t$.

This assumption means that if a generation has higher utility than the next, then its excess utility can be transferred at negative cost to its successor. It thus generalizes positive net capital productivity to a setting where utilities need not be more than ordinally measurable and level comparable.

Second, we in Asheim et al. (2001) impose the following two conditions on the social preferences $R$ (with $I$ and $P$ as symmetric and asymmetric parts).

**Condition 1** (*Strong Pareto*). *For any* ${}_1\mathbf{u}$ *and* ${}_1\mathbf{v}$, *if* $v_t \geq u_t$ *for all* $t$ *and* $v_s > u_s$ *for some* $s$, *then* ${}_1\mathbf{v}\,P\,{}_1\mathbf{u}$.

**Condition 2** (*Weak Anonymity*). *For any* ${}_1\mathbf{u}$ *and* ${}_1\mathbf{v}$, *if for some finite permutation* $\pi$, $v_{\pi(t)} = u_t$ *for all* $t$, *then* ${}_1\mathbf{v}\,I\,{}_1\mathbf{u}$.

The term 'permutation', as used in Condition 2, signifies a bijective mapping of $\{1, 2, \ \ldots\}$ onto itself, is finite whenever there is a $T$ such that $\pi(t) = t$ for any $t > T$. While Strong Pareto (sometimes referred to as 'Efficiency') ensures that the social preferences are sensitive to utility increases of any one generation, Weak Anonymity (also called 'Equity') can be considered a basic fairness norm as it ensures that everyone counts the same in social evaluation. In the intergenerational context the Weak Anonymity condition implies that it is not justifiable to discriminate against some generation only because it appears at a later stage on the time axis. It thereby rules out discounted utilitarianism.

Define sustainability in the following standard way (cf. the discussion in Pezzey and Toman, 2002, Section 3.1).

**Definition 1** (*Sustainability*). Generation $t$ with inheritance $x_t$ is said to behave in a sustainable manner if it chooses a feasible utility-bequest pair $(u_t, x_{t+1})$ so that the constant utility path $(u_t, u_t, \ldots)$ is feasible at $t+1$ given $x_{t+1}$. The utility path ${}_1\mathbf{u} = (u_1, u_2, \ldots)$ is called sustainable given $x_1$ if there exists ${}_2\mathbf{x} = (x_2, x_3, \ldots)$ such that every generation behaves in a sustainable manner along $({}_1\mathbf{x}, {}_1\mathbf{u}) = (x_1, (u_1, x_2), (u_2, x_3), \ldots)$.

Hence, a generation behaves in sustainable manner if its utility level can also potentially be shared by all future generations. While any feasible non-decreasing path is sustainable, it is not in conflict with sustainability that some generation makes a large sacrifice to the benefit of future generations, leading to its own utility being lower than that of its predecessor.

Our justification for sustainability can now be stated.

**Proposition 1** (Asheim et al. 2001). *If the social preferences $R$ satisfy Strong Pareto and Weak Anonymity, and the technology satisfies Immediate Productivity, then only sustainable utility paths are $R$-maximal.*

As noted in the introduction, this result is not applicable to the models that we consider in this chapter since the assumption of Immediate Productivity will not be satisfied. Instead, we will directly consider the conditions of Strong Pareto and Weak Anonymity, which jointly generate the *Suppes-Sen grading principle*.

**Definition 2** (*Suppes-Sen grading principle*). The Suppes-Sen grading principle $R^S$ deems two paths to be indifferent if one is obtained from the other through a finite permutation, and one utility path to be preferred to another if a finite permutation of the former weakly Pareto-dominates the other.

Strong Pareto and Weak Anonymity generate the Suppes-Sen grading principle $R^S$ in the following sense: It holds that

- ${}_1\mathbf{v}\, I^s\, {}_1\mathbf{u}$ implies ${}_1\mathbf{v}\, I\, {}_1\mathbf{u}$ and
- ${}_1\mathbf{v}\, P^s\, {}_1\mathbf{u}$ implies ${}_1\mathbf{v}\, P\, {}_1\mathbf{u}$,

if and only if the social preferences $R$ satisfy Strong Pareto and Weak Anonymity.

## 3. A CLASS OF MODELS

Consider a class of models, where consumption is derived from resource extraction, where the resource may be renewable, and where, following Krautkraemer (1985), utility may be derived directly from the resource stock. In the framework of Section 2, we have that $F_t$ is independent of time $t$ and given by:

$$F(x_t, x_{t+1}) \begin{cases} = u(x_t + g(x_t) - x_{t+1}, x_t) \text{ if } x_t \geq 0 \text{ and } x_t + g(x_t) \geq x_{t+1} \geq 0, \\ < 0 \text{ otherwise,} \end{cases}$$

indicating that feasibility at time $t$ requires that $x_t \geq 0$ and $x_t + g(x_t) \geq x_{t+1} \geq 0$, that $c_t = x_t + g(x_t) - x_{t+1}$ is the consumption at time $t$, and that $x_t$ is the resource stock at time $t$.

Assume throughout that $u : \Re_+^2 \to \Re$ is a continuously differentiable and quasi-concave utility function that assigns utility $u(c, x)$ to any non-negative consumption-amenity pair and satisfies:

$$u(0,0) = 0, \quad u_c > 0 \text{ if } c > 0, \quad \text{and} \quad u_x \geq 0 \text{ if } x > 0$$

Moreover, assume throughout that $g : [0, \bar{x}] \rightarrow \Re_+$ is a continuously differentiable natural growth function that assigns non-negative natural growth to any stock in $[0, \bar{x}]$ and satisfies:

$$g(0) = 0 \quad \text{and} \quad g(\bar{x}) = 0.$$

Four different models are obtained by considering combinations of the following four assumptions.

**Assumption 2** (*No resource amenities*). $\forall c \geq 0, \forall x > 0, u_x = 0$.

**Assumption 3** (*Positive resource amenities*). $\forall c \geq 0, \forall x > 0, u_x > 0$.

**Assumption 4** (*No natural growth*). $\forall x \in [0, \bar{x}], g(x) = 0$.

**Assumption 5** (*Positive natural growth*). The natural growth function is continuously differentiable and strictly concave and satisfies $\forall x \in (0, \bar{x}), 0 < g(x) \leq \bar{x} - x$.

The restriction of Assumption 5, namely that $g(x) \leq \bar{x} - x$, means that the stock cannot grow beyond its natural biological equilibrium and is satisfied if $g'$ is bounded below and the period length is small enough. We follow Heal (2001) by representing the renewable resource by means of a biomass model, realizing that such modelling is only in special cases adequate for forest management.

Since Assumptions 2 and 3 are mutually exclusive, and so are Assumptions 4 and 5, the following four models are obtained.

**Model 1 (Cake-eating)** satisfies Assumptions 2 and 4.

**Model 2 (Renewable resource)** satisfies Assumptions 2 and 5.

**Model 3 (Non-renewable resource yielding amenities)** satisfies Assumptions 3 and 4.

**Model 4 (Renewable resource yielding amenities)** satisfies Assumptions 3 and 5.

These are the models that Heal (2001) investigates. In additional to considering the applicability of the Chichilnisky (1996) criterion, he applies discounted utilitarianism, undiscounted utilitarianism (in the form of overtaking), and Rawlsian maximin as social preferences over different intergenerational utility paths. Undiscounted utilitarianism and lexicographic versions of Rawlsian maximin satisfy both Conditions 1 (Strong Pareto) and 2 (Weak Anonymity), while discounted utilitarianism satisfies Strong Pareto, but not Weak Anonymity.

It does not come as a surprise that in Model 1 there is no way to have sustainability as a optimal solution, independently of the social preferences used. In Models 2–4, however, all social preferences considered by Heal (1998, 2001) may lead to optimal solutions in which stock specific sustainability constraints are obtained; i.e., in which part of the resource stock is forever kept intact. In the case of discounted utilitarianism, this result holds at least when the discount rate is sufficiently low (and marginal utility of consumption is bounded away from infinity). Therefore, Heal considers that there is no inherent conflict between 'optimality' and 'sustainability'.

Instead of applying specific forms of intergenerational social preferences as Heal does, we here investigate the implications in these four models of imposing the Suppes-Sen grading principle (i.e., the conditions of Strong Pareto and Weak Anonymity), leading to consequences that are shared by undiscounted utilitarianism and Rawlsian maximin, but not necessarily by discounted utilitarianism.

## 4. APPLYING THE SUPPES-SEN GRADING PRINCIPLE

Proposition 1 entails that the Suppes-Sen grading principle leads to sustainable paths in technologies satisfying the assumption of Immediate Productivity. This result cannot be applied to Models 1–4 since they do not satisfy this technological assumption.

**Proposition 2.** *Assumption 1 (Immediate Productivity) is not satisfied by Models 1–4.*

The proofs of this and the other results of this section are contained in Section 6.

Moreover, the direct application of the Suppes-Sen grading principle does not yield any restriction on the depletion policy in Model 1, except that the resource stock must be exhausted as time goes to infinity, so that the path is efficient. Hence, the following result is obtained.

**Proposition 3.** *Consider Model 1 and social preference given by the Suppes-Sen grading principle* $R^S$. *A utility path is* $R^S$ *-maximal if and only if it is efficient.*

Hence, in Model 1 and for any social preferences $R$ satisfying Conditions 1 (Strong Pareto) and 2 (Weak Anonymity), a utility path is $R$ -maximal only if it is efficient.

However, the direct application of the Suppes-Sen grading principle yields a restriction on the depletion policy in Models 2-4, leading to the following stock-specific sustainability constraint.

**Proposition 4.** *Consider Models 2–4 and social preferences given by the Suppes-Sen grading principle* $R^S$. *If the initial stock* $x_1$ *satisfies* $g'(x_1 + g(x_1)) \geq 0$, *then a utility path is* $R^S$ *-maximal only if* $c_1 \leq g(x_1)$ *(so that* $x_1 \leq x_2$*) and* $c_1 \leq c_2$ *(so that* $u_1 \leq u_2$*).*

In the case of Models 2 and 4, and for a "small" period length (so that the maximal per period growth $g(x_1)$ is "small" compared to $x_1$), the condition that $g'(x_1 + g(x_1)) \geq 0$ can be identified with the condition that $x_1$ does not exceed the stock size corresponding to the maximal sustainable yield (MSY); i.e., the stock size maximizing $g(x)$ over all $x \in [0, \bar{x}]$. Hence, Proposition 4 states, unless the stock exceeds the MSY size so that a smaller stock leads to higher natural growth, further depletion of the stock is incompatible with any social preferences $R$ satisfying Conditions 1 (Strong Pareto) and 2 (Weak Anonymity).

In order to show that the results of Propositions 3 and 4 are not empty, we must establish that there exist $R^S$-maximal utility paths in the case of Models 2–4. By the following result, such existence poses no problem.

**Proposition 5.** *Consider Models 2–4 and social preferences given by the Suppes-Sen grading principle* $R^S$. *For any initial stock* $x_1$, *there exists a* $R^S$*-maximal utility path.*

Hence, imposing that the social preferences $R$ satisfy Strong Pareto and Weak Anonymity does not rule out the existence of $R$-maximal utility paths.

While we through Proposition 4 provide conditions that are necessary for $R^S$-maximal paths in Models 2–4, and through the proof of Proposition 5 give a condition that is sufficient for $R^S$-maximality in these models, we do not have available conditions that are both sufficient and necessary and thus characterize the set of $R^S$-maximal paths in these settings.

## 5. THE SIGNIFICANCE OF THE RESULTS

Although the results of the previous section indicate that the seemingly weak and uncontroversial axioms of Strong Pareto and Weak Anonymity entail that a resource stock should not be further reduced if smaller than the size corresponding to MSY, one must keep in mind that the models abstract from factors that are important in the real world.

- The models of Section 3 do not have any production activities other than resource extraction. If production also depends on reproducible capital and the produced output can be split between consumption and accumulation of reproducible capital, then along any Pareto-efficient path there can be no profitable arbitrage possibilities between the two kinds of capital goods, i.e., in any period holding a stock of the natural resource must be as profitable as holding a stock of the reproducible capital. As along a Suppes-Sen maximal utility path the rate of productivity of reproducible capital may well be positive, it therefore follows that for, e.g., a renewable resource that does not yield amenities (cf. Model 2), the marginal rate of growth of the resource stock has to be positive, too. This will reduce the resource stock strictly below its msy size.

- In the real world, natural capital consists of many different types of resources. Since the simple models of Section 3 include only one resource, the results obtained in these models say nothing about how sustainability constraints should be imposed if there are multiple resources. Even though it is quite possible that models with multiple resources would imply sustainability constraints for some or all of these resources, this will naturally depend on how such models are formulated.
- Finally, real world resource stocks are geographically distributed. Since the simple models of Section 3 has no geographical dimension, the results obtained in these models say nothing about how sustainability constraints should be applied to a setting where resource stocks are geographically distributed. Even though it is quite possible that models where resources are geographically distributed would imply sustainability constraints in some or all of the regions, this will also depend on how such models are formulated.

Still, the models suggest that calls for resource conservation and sustainability based on ethical intuition may be provided with a more solid normative underpinning through basic axioms like Strong Pareto and Weak Anonymity.

## 6. PROOFS

*Proof of Proposition 2.* We must show that Assumption 1 is not satisfied in Models 1–4.

*Model 1*: Assume ${}_t\mathbf{u} = (u_t, u_{t+1}, \ldots)$ is feasible at $t$ given $x_t$ with $u_t > u_{t+1}$. Then $(u_{t+1}, u_t, u_{t+2}, \ldots)$ is feasible at $t$ given $x_t$, but is not inefficient, unless ${}_t\mathbf{u}$ inefficient.

*Model 2*: Assume ${}_t\mathbf{u} = (u_t, u_{t+1}, \ldots)$ is feasible at $t$ given $x_t$ with $u_t > u_{t+1}$ and $g'(x_t) < 0$ and $g'(x_{t+1}) < 0$. Then $(u_{t+1}, u_t, u_{t+2}, \ldots)$ is not even feasible at $t$ given $x_t$, unless ${}_t\mathbf{u}$ inefficient.

*Model 3*: Consider the following explicit counterexample. The utility function

$$u(c,x) = \begin{cases} 2c + x & \text{if} \quad c \leq 8 \\ \frac{63}{4} + (c - \frac{127}{16})^{\frac{1}{2}} + x & \text{if} \quad c > 8 \end{cases}$$

is continuously differentiable and satisfies Assumption 3. Let $x_1 = 20$, $x_2 = 14$, and $x_3 = 6$, and let ${}_3\mathbf{u} = (u_3, u_4, \ldots)$ be efficient at time 3 given $x_3 = 6$. We have that $c_1 = x_1 - x_2 = 6$ and $c_2 = x_2 - x_3 = 8$, so that $u_1 = 2*6 + 20 = 32$ and $u_2 = 2*8 + 14 = 30$. Decreasing utility at time 1 to $u_2$ entails decreasing consumption at time 1 to $\tilde{c}_1 = 5$ so that $v_1 = 2*5 + 20 = 30 = u_2$ and $\tilde{x}_2 = 15$. Since ${}_3\mathbf{u} = (u_3, u_4, \ldots)$ is efficient at time 3 given $x_3 = 6$, we can only increase consumption at time 2 to $\tilde{c}_1 = \tilde{x}_2 - x_3 = 15 - 6 = 9$ to keep the remaining utility path unchanged. However,

$$v_2 = \tfrac{63}{4} + (9 - \tfrac{127}{16})^{\frac{1}{2}} + 15 = \tfrac{123}{4} + (\tfrac{17}{16})^{\frac{1}{2}} < 32 = u_1 .$$

Hence, the utility path $(u_2, u_1, u_3 ...)$ is not feasible at time 1 given $x_1$.

*Model 4*: The result follows by combining the features of the proofs in the case of Models 2 and 3. □

*Proof of Proposition 3. Only if.* Assume that ${}_1\mathbf{u} = (u_1, u_2, ...)$ is not efficient. Then it follows, since $R^S$ satisfies Strong Pareto, that there exists ${}_1\mathbf{v} = (v_1, v_2, ...)$ such that ${}_1\mathbf{v}\, P^S\, {}_1\mathbf{u}$.

*If.* Write $u(c)$ since, by Assumption 2, $u$ does not depend on $x$. Assume that ${}_1\mathbf{u} = (u_1, u_2, ...) = (u(c_1), u(c_2), ...)$ is efficient, i.e.,

$$\sum_{t=1}^{\infty} c_t = s_1 .$$

Then any finite permutation of ${}_1\mathbf{u}$ also satisfies

$$\sum_{t=1}^{\infty} c_{\pi(t)} = s_1$$

and is thus efficient. Hence, there is no ${}_1\mathbf{v} = (v_1, v_2, ...)$ such that ${}_1\mathbf{v}\, P^S\, {}_1\mathbf{u}$. □

The following result is helpful for the proof of Proposition 4.

**Lemma 1.** *Consider Models 2–4. Let the feasible consumption path* ${}_1\mathbf{c} = (c_1, c_2, ...)$ *be given with* ${}_1\mathbf{x} = (x_1, x_2, ...)$ *as the accompanying path of resource stocks. If there exists some time* $t$ *such that* $g'(x_t + g(x_t)) \geq 0$ *and* $c_t > c_{t+1}$, *then there exists a feasible consumption path* ${}_1\tilde{\mathbf{c}} = (\tilde{c}_1, \tilde{c}_2, ...)$ *satisfying*

$$\tilde{c}_s \begin{cases} = c_s & \text{for} \quad s = 1, ..., t-1, t+2, ... \\ = c_{s+1} & \text{for} \quad s = t \\ \geq c_{s-1} & \text{for} \quad s = t+1, \end{cases}$$

*where the latter inequality can be made strict if Assumption 5 is satisfied. The accompanying path of resource stocks* ${}_1\tilde{\mathbf{x}} = (\tilde{x}_1, \tilde{x}_2, ...)$ *satisfies*

$$\tilde{x}_s \begin{cases} = x_s & \text{for} \quad s = 1, ..., t, t+2, ... \\ > x_s & \text{for} \quad s = t+1. \end{cases}$$

*Proof.* The path ${}_1\tilde{\mathbf{x}} = (\tilde{x}_1, \tilde{x}_2, ...)$ coincides with ${}_1\mathbf{x}$ up to and including time $t$. For time $t+1$, it follows that $x_{t+1} = x_t + g(x_t) - c_t$, while

$$\tilde{x}_{t+1} = x_t + g(x_t) - \tilde{c}_t = x_t + g(x_t) - c_{t+1} > x_t + g(x_t) - c_t = x_{t+1}$$

since $\tilde{c}_t = c_{t+1} < c_t$. As $g'(x_t + g(x_t)) \geq 0$,

i. $g(x_{t+1}) < g(\tilde{x}_{t+1})$ if Assumption 5 is satisfied, since $g' > 0$ between $x_{t+1} = x_t + g(x_t) - c_t$ and $\tilde{x}_{t+1} = x_t + g(x_t) - c_{t+1}$ by the strict concavity of $g$.

ii. $g(x_{t+1}) = g(\tilde{x}_{t+1})$ if Assumption 4 is satisfied, since $g' = 0$ between $x_{t+1} = x_t + g(x_t) - c_t$ and $\tilde{x}_{t+1} = x_t + g(x_t) - c_{t+1}$.

Let $\tilde{c}_{t+1} = c_t + g(\tilde{x}_{t+1}) - g(x_{t+1})$, so that $\tilde{c}_{t+1} \geq c_t$, with strict inequality if Assumption 5 is satisfied. It follows that the resource stock at time $t+2$, $\tilde{x}_{t+2}$, in the alternative path equals the resource stock at time $t+2$, $x_{t+2}$, in the original path:

$$\begin{aligned}\tilde{x}_{t+2} &= x_t + g(x_t) - \tilde{c}_t + g(\tilde{x}_{t+1}) - \tilde{c}_{t+1} \\ &= x_t + g(x_t) - c_{t+1} + g(\tilde{x}_{t+1}) - (c_t + g(\tilde{x}_{t+1}) - g(x_{t+1})) \\ &= x_t + g(x_t) - c_t + g(x_{t+1}) - c_{t+1} = x_{t+2}.\end{aligned}$$

Hence, it is feasible to keep consumption unchanged from time $t+1$ on. □

*Proof of Proposition 4.* Assume that the initial stock $x_1$ satisfies $g'(x_1 + g(x_1)) \geq 0$, but $c_1 > g(x_1)$ or $c_1 > c_2$. We must show that $u(c_1, x_1)$ cannot constitute the initial period of a $R^S$-maximal utility path.

*Model 2*: If $c_1 > c_2$, then clearly there exists $t \geq 1$ so that $g'(x_t + g(x_t)) \geq 0$ and $c_t > c_{t+1}$. If $c_1 > g(x_1)$ so that $x_1 > x_2$, then $(x_1, x_2, ...)$ would be decreasing at an increasing pace as long as $(c_1, c_2, ...)$ is non-decreasing. Hence, there exists $t \geq 1$ so that $g'(x_t + g(x_t)) \geq 0$ and $c_t > c_{t+1}$ also in this case. Since Model 2 satisfies Assumption 5, it follows from Lemma 1 that there exists a utility path ${}_1\mathbf{v} = (v_1, v_2, ...) = (u(\tilde{c}_1), u(\tilde{c}_2), ...)$ that Pareto-dominates and thus, by Strong Pareto, is preferred to

$$(u(c_1), ..., u(c_{t-1}), u(c_{t+1}), u(c_t), u(c_{t+2}), ...),$$

which, by Weak Anonymity, is equally good as

$$(u(c_1), ..., u(c_{t-1}), u(c_t), u(c_{t+1}), u(c_{t+2}), ...),$$

(where we write $u(c)$ since, by Assumption 2, $u$ does not depend on $x$). By transitivity, the latter utility path is not $R^S$-maximal given $x_1$.

*Models 3–4*: The proof by contradiction consists of two cases.

CASE 1: *There exists* $t \geq 1$ *so that* $g'(x_t + g(x_t)) \geq 0$, $c_t > c_{t+1}$, *and* $g(x_t) \geq c_{t+1}$. By Lemma 1, there exists a feasible consumption path ${}_1\tilde{\mathbf{c}} = (\tilde{c}_1, \tilde{c}_2, \ldots)$ derived from ${}_1\mathbf{c} = (c_1, c_2, \ldots)$ by permuting $c_t$ and $c_{t+1}$, with an accompanying path of resource stocks ${}_1\tilde{\mathbf{x}} = (\tilde{x}_1, \tilde{x}_2, \ldots)$ that coincides with ${}_1\mathbf{x} = (x_1, x_2, \ldots)$, except that $\tilde{x}_{t+1} > x_{t+1}$. The utility path ${}_1\mathbf{v} = (v_1, v_2, \ldots) = (u(\tilde{c}_1, \tilde{x}_1), u(\tilde{c}_2, \tilde{x}_2), \ldots)$ satisfies

$$v_s = \begin{cases} u(c_s, x_s) = u_s & \text{for } s = 1, \ldots, t-1, t+2, \ldots \\ u(c_{s+1}, x_s) > u(c_{s+1}, x_{s+1}) = u_{s+1} & \text{for } s = t \text{ since } x_t > x_{t+1}, \\ u(c_{s-1}, \tilde{x}_s) \geq u(c_{s-1}, x_{s-1}) = u_{s-1} & \text{for } s = t+1, \end{cases}$$

where $\tilde{x}_{t+1} \geq x_t$ follows from $\tilde{c}_t = c_{t+1} \leq g(x_t)$. Hence, there exists a utility path ${}_1\mathbf{v} = (v_1, v_2, \ldots)$ that Pareto-dominates and thus, by Strong Pareto, is preferred to

$$(u_1, \ldots, u_{t-1}, u_{t+1}, u_t, u_{t+2}, \ldots),$$

which, by Weak Anonymity, is equally good as

$$(u_1, \ldots, u_{t-1}, u_t, u_{t+1}, u_{t+2}, \ldots).$$

By transitivity, the latter utility path is not $\mathrm{R}^S$-maximal given $x_1$.

CASE 2: There does not exist $t \geq 1$ so that $g'(x_t + g(x_t)) \geq 0$, $c_t > c_{t+1}$, and $g(x_t) \geq c_{t+1}$. This case clearly rules out $g(x_t) \geq c_1 > c_2$; hence, $c_1 > g(x_1)$. Suppose there exists $t \geq 1$ such that $c_{t+1} \leq g(x_{t+1})$, and let without loss of generality $t$ be the first time at which $c_{t+1} \leq g(x_{t+1})$, so that $x_{t+1} < x_t \leq x_1$. Then, $g'(x_t + g(x_t)) \geq 0$ and $c_t > g(x_t) > g(x_{t+1}) \geq c_{t+1}$, by the assumption that $g'(x_1 + g(x_1)) \geq 0$ and the concavity of $g$ ($g$ is linear under Assumption 4 and strictly concave under Assumption 5). This is also ruled out.

Hence, in this case there does not exist $t \geq 1$ such that $c_{t+1} \leq g(x_{t+1})$. Then, since the resource stock is strictly decreasing, but bounded by a non-negativity constraint, there is some $x^* \geq 0$ such that $\lim_{t \to \infty} x_t = x^*$ and $\lim_{t \to \infty} c_t = g(x^*)$. Hence, since utility is increasing in $c$ and $x$, and $g' \geq 0$ if $x$ does not exceed $x_1 + g(x_1)$, there is some $t > 1$ such that $u(c_s, x_s) < u(g(x_1), x_1)$ for all $s \geq t$.

Consider the alternative feasible consumption path ${}_1\tilde{\mathbf{c}} = (\tilde{c}_1, \tilde{c}_2, \ldots)$ satisfying

$$\tilde{c}_s = \begin{cases} g(x_1) & \text{for } s = 1, \ldots, t-1 \\ c_{s-t+1} & \text{for } s = t, \ldots, 2t-2 \\ x_{s-t+1} + g(x_{s-t+1}) - x_{s+1} & \text{for } s = 2t-1 \\ c_s & \text{for } s = 2t, 2t+1 \ldots, \end{cases}$$

with the accompanying path of resource stocks ${}_1\tilde{\mathbf{x}} = (\tilde{x}_1, \tilde{x}_2, \ldots)$ satisfying

$$\tilde{x}_s = \begin{cases} x_1 & \text{for} \quad s = 1, \dots, t-1 \\ x_{s-t+1} & \text{for} \quad s = t, \dots, 2t-1 \\ x_s & \text{for} \quad s = 2t, 2t+1 \end{cases}$$

To confirm the feasibility of this path, we only have to consider time $2t-1$, where $\tilde{c}_{2t-1} = x_t + g(x_t) - x_{2t} > x_{2t-1} + g(x_{2t-1}) - x_{2t} = c_{2t-1} > 0$ since $x_t > x_{2t-1}$ and $g(x_t) \geq g(x_{2t-1})$. The utility path ${}_1\mathbf{v} = (v_1, v_2, \dots) = (u(\tilde{c}_1, \tilde{x}_1), u(\tilde{c}_2, \tilde{x}_2), \dots)$ satisfies

$$v_s \begin{cases} > u_{s+t-1} & \text{for} \quad s = 1, \dots, t-1 \\ = u_{s-t+1} & \text{for} \quad s = t, \dots, 2t-2 \\ > u_s & \text{for} \quad s = 2t-1 \\ = u_s & \text{for} \quad s = 2t, 2t+1, \end{cases}$$

Hence, there exists a utility path ${}_1\mathbf{v} = (v_1, v_2, \dots)$ that Pareto-dominates and thus, by Strong Pareto, is preferred to

$$(u_t, \dots, u_{2t-2}, u_1, \dots, u_{t-1}, u_{2t-1}, u_{2t}, \dots),$$

which, by Weak Anonymity, is equally good as

$$(u_1, \dots, u_{t-1}, u_t, \dots, u_{2t-2}, u_{2t-1}, u_{2t}, \dots).$$

By transitivity, the latter utility path is not $R^S$-maximal given $x_1$. □

*Proof of Proposition 5.* For given initial stock $x_1$, define $x^*$ as follows:

$$x^* := \arg\max_{x \in [0, x_1]} u(g(x), x).$$

It follows by the properties of $u$ and $g$ under the assumptions of Models 2–4 that $x^*$ is unique. Since, by definition of $x^*$, $x_1 \geq x^*$, we have two cases to consider: $x_1 = x^*$ and $x_1 > x^*$.

CASE 1: $x_1 = x^*$. Consider the path

$${}_1\mathbf{u} = (u_1, u_2, \dots) = (u(g(x_1), x_1), u(g(x_1), x_1) \dots).$$

Since ${}_1\mathbf{u}$ has constant utility, it is Suppes-Sen maximal if and only if it is efficient. Therefore, suppose there exists a path ${}_1\mathbf{v} = (v_1, v_2, \dots)$ (with ${}_1\tilde{\mathbf{x}}$ and ${}_1\tilde{\mathbf{c}}$ as accompanying paths of consumption and resource stocks) that Pareto-dominates ${}_1\mathbf{u}$. Without loss of generality we can assume that $v_1 > u_1$. This entails $\tilde{c}_1 > g(x_1)$ so that

$\tilde{x}_2 = x_1 + g(x_1) - \tilde{c}_1 < x_1 = x^*$. By the definition of $x^*$ and the properties of $u$ and $g$ under the assumptions of Models 2–4, it now follows that if $v_t \geq u_t$ for $t = 2,3,\dots$, then the per time period depletion of the resource is positive and bounded away from zero (with $\tilde{c}_1 - g(x_1) > 0$ as a lower bound). Thus, since the resource stock must be non-negative, such a path is infeasible. Hence, ${}_1\mathbf{u}$ is $R^S$-maximal given $x_1$.

CASE 2: $x_1 > x^*$. This case can only occur in Models 2 and 4, for which $g$ satisfies Assumption 5. Consider the path ${}_1\mathbf{u} = (u_1, u_2, \dots)$, where

$$u_t = \begin{cases} u(x_1 + g(x_1) - x^*, x_1) & \text{for} \quad t = 1 \\ u(g(x^*), x^*) & \text{for} \quad t > 1 \end{cases}$$

By definition, ${}_1\mathbf{u}$ is Suppes-Sen maximal if and only if it is efficient and there does not exist an alternative path Pareto-dominating a finite permutation of ${}_1\mathbf{u}$.

To show that ${}_1\mathbf{u}$ is efficient, suppose there exists a path ${}_1\mathbf{v} = (v_1, v_2, \dots)$ (with ${}_1\tilde{\mathbf{x}}$ and ${}_1\tilde{\mathbf{c}}$ as accompanying paths of consumption and resource stocks) that Pareto-dominates ${}_1\mathbf{u}$. However, if $v_s = u_s$ for $s = 1, \dots, t-1$ and $v_t > u_t$, then $\tilde{x}_{t+1} < x^*$. In line with the proof of Case 1, it now follows that it is infeasible to keep $v_s \geq u_s$ for all $s > t$.

To show that there does not exist an alternative path Pareto-dominating a finite permutation of ${}_1\mathbf{u}$, it is sufficient to show that there exists no finite permutation of ${}_1\mathbf{u}$ (since then no path Pareto-dominating such a permutation is feasible either). By Assumption 5, $g$ is strictly concave and satisfies $\forall x \in (0, \bar{x}), 0 < g(x) \leq \bar{x} - x$, and it follows that $x + g(x)$ is a strictly increasing function of $x$. Since $x_1 > x^*$, we therefore have that $u_1 = u(x_1 + g(x_1) - x^*, x_1) > u(g(x^*), x^*) = u_2 = u_3 = \cdots$. Consequently, any finite permutation amounts to a path ${}_1\mathbf{v} = (v_1, v_2, \dots)$ (with ${}_1\tilde{\mathbf{x}}$ and ${}_1\tilde{\mathbf{c}}$ as accompanying paths of consumption and resource stocks) where

$$v_s = \begin{cases} u_1 = u(x_1 + g(x_1) - x^*, x_1) & \text{for} \quad s = t \\ u_t = u(g(x^*), x^*) & \text{for} \quad s \neq t \end{cases}$$

and $t$ is some period after period 1. Since $x^*$ is the unique maximizer of $u(g(x), x)$, so that in particular, $v_1 = u(g(x^*), x^*)$ only if $\tilde{c}_1 > g(x_1)$, it follows that $\tilde{x}_t < x_1$. By the properties of $u$ under the assumptions of Models 2 and 4, this implies that $\tilde{c}_t \geq c_1$. Since $x + g(x)$ is a strictly increasing function of $x$, we have that $\tilde{x}_t + g(\tilde{x}_t) < x_1 + g(x_1)$. Hence, $\tilde{x}_{t+1} = \tilde{x}_t + g(\tilde{x}_t) - \tilde{c}_t < x_1 + g(x_1) - c_1 = x^*$. As we have argued above, it now follows that it is infeasible to keep $v_s = u_s$ for all $s > t$. We have thus shown that there exists no finite permutation of ${}_1\mathbf{u}$, and consequently, no alternative path Pareto-dominating a finite permutation of ${}_1\mathbf{u}$.

Hence, ${}_1\mathbf{u}$ is $R^S$-maximal given $x_1$. □

**Acknowledgments.** Presented at the International Conference on Economics of Sustainable Forest Management, Toronto, May 20 – 22, 2004. We thank Shashi Kant and other conference participants for helpful comments. Furthermore, Asheim gratefully acknowledges the hospitality of the Stanford University research initiative on *the Environment, the Economy and Sustainable Welfare*, where parts of this work was done, and financial support from the Hewlett Foundation through this research initiative. Finally, both authors are grateful for financial support from the Research Council of Norway (Ruhrgas grant).

## REFERENCES

Asheim, G.B., Buchholz, W., & Tungodden, B. (2001). Justifying sustainability. *Journal of Environmental Economics and Management,* 41, 252–268.

Chichilnisky, G. (1996). An axiomatic approach to sustainable development. *Social Choice and Welfare* 13, 231–257.

Heal, G.M. (1998). *Valuing the future: Economic theory and sustainability.* New York: Columbia University Press.

Heal, G.M. (2001). Optimality or sustainability. Columbia University (paper presented at the EAERE 2001 Conference).

Krautkraemer, J.A. (1985). Optimal growth, resource amenities and the preservation of natural environments. *Review of Economic Studies*, 52, 153–170.

Mitra, T. (2005). Intergenerational equity and the forest management problem. Chapter 7 of this volume.

Pezzey, J.C.V., & Toman, M.A. (2002). Progress and problems in the economics of sustainability. In T. Tietenberg, H.Folmer, (Eds.), *The international yearbook of environmental and resource economics 2002/2003*. Cheltenham: Edward Elgar.

Sen, A.K. (1970). *Collective choice and social welfare*. Edinburgh: Oliver and Boyd.

Suppes, P. (1966). Some formal models of grading principles. *Synthese*, 6, 284–306.

# CHAPTER 9

# COMPLEXITIES OF DYNAMIC FOREST MANAGEMENT POLICIES


J. BARKLEY ROSSER, JR.

*Department of Economics, James Madison University, Harrisonburg, VA 22807 USA.*
*Email: rosserjb@jmu.edu*



**Abstract.** The complications for forest management policies are considered in light of nonlinearities in the ecological-economic dynamics of forests. Such nonlinearities imply the existence of multiple solutions in forestry management problems specified fully for all dynamic patterns of amenities. Such nonlinearities imply the possibilities of discontinuities and critical thresholds in such systems. Specific policy problems considered include fire management, pest management, and size of cuts.


> "A 'Public Domain,' once a velvet carpet of rich buffalo-grass and grama, now an illimitable waste of rattlesnake-bush and tumbleweed, too impoverished to be accepted as a gift by the states within which it lies. Why? Because the ecology of the Southwest happened to be set on a hair trigger."
>
> - Aldo Leopold, "The Conservation Ethic," *Journal of Forestry*, 1933, 33, 636-637.

## 1. INTRODUCTION

Forest management is a problem of increasing controversy and difficulty in many parts of the world. Whereas once the emphasis was simply on cutting trees and replanting them to maximize the present value of their lumber content, now forest managers face multiple demands from the public around the world. Forests are seen as the homes of diverse species that may be hunted or fished, or should be preserved because of the rarity or uniqueness. They provide positive externalities at the local level due to preventing flooding and soil erosion and at the global level because of their role in carbon sequestration. However, carbon sequestration may conflict with biodiversity preservation, and within the latter, preserving one species may imply not preserving another one. In less developed countries a variety of social issues arise in the management of forests from dealing with aboriginal inhabitants to




*Kant and Berry (Eds.), Economics, Sustainability, and Natural Resources: Economics of Sustainable Forest Management,* 191-206.

providing land for poor farmers who wish to homestead. Many of these issues involve conflicts between groups and goals that are not easily resolved.

Any effort to resolve these conflicts over the proper management of forests ultimately involves the dynamic ecology of forests. Elements impacting this include the role of fire, the role of pest management, and the methods and techniques of cutting trees when there is harvesting of timber, especially regarding the patch size of the cuts. Considerable experience and literature indicate that a variety of complex dynamics are involved in these latter elements. Multiple equilibria exist with the implied possibility of sudden and discontinuous changes in the nature of a forest, much as described in the opening quotation above from famous ecologist, Aldo Leopold. Deep tradeoffs exist between the local stability of forest ecosystems and their global resilience, tradeoffs that manifest themselves in such contradictions as efforts to prevent forest fires can make forest fires worse, and efforts to eradicate pests can make their attacks worse and more destructive. This idea that in ecosystems there might exist such a tradeoff has become very widespread and influential, but it was initially due to C.S. Holling (1973) who derived it from the study of forest ecology.

This paper will review these questions within the general framework of what the complex dynamics of the interaction between forest ecology and forest economics imply for forest management policies. The paper will first review the basic model of optimal forestry rotation, initially from the standpoint of the Faustmann (1849) model of simple timber use of a forest. This will then be modified to account for other uses of the forest as modeled by Hartman (1976), with the possibility of multiple solutions arising in this context due to the different patterns over time of the various uses of the forest, thus implying more than one locally optimal rotation time for growing and cutting of a forest. The paper will then consider a series of more specific issues in the dynamics of forest management, especially the problems of pest and fire management and patch size of cutting (especially the problem of clearcutting), while taking into account the complications introduced by considering hunting, fishing, grazing, biodiversity, carbon sequestration, and social issues. These problems exist in many forests around the world, both in the temperate forests of the higher income countries and in tropical rain forests in developing economies with their more difficult social and economic circumstances.

The existence of these nonlinearities and the related thresholds and discontinuities, and possibly even more complex dynamical patterns implies a more serious consideration of policy problems in ecologic-economic systems (Rosser, 2001). Policymakers must be aware of the interaction of different parts of policies that they might not have been otherwise. Furthermore, they must be especially cognizant of the possibilities of dramatic collapses of systems as thresholds are crossed. The outbreaks of massive forest fires in the western areas of the United States in recent years and the sudden disappearance of tree species due to pest invasions are perhaps the clearest examples of the dangers and difficulties that policies must deal with in a world of complex nonlinear dynamical interactions in forest ecologic-economic systems.

## 2. THE BASIC MODEL

In the English language tradition, Irving Fisher (1907) proposed that the optimal time to cut a tree for timber use is when its growth rate equals the real rate of interest. However this is true only for the generally unrealistic case where there will be no replanting of the forest after the harvest, and, indeed, there will be no use of the land on which the trees were growing whatsoever after they are cut. The correct solution had been solved within the German language tradition much earlier by Faustmann (1849), who found that if one has an infinite time horizon and plans to replant the forest after harvesting, then the optimal time will be sooner than the Fisher solution implies.[1] This is because the rate of growth of trees decelerates with age, so that cutting sooner means that one can get the more rapidly growing younger trees in place sooner again.

To look at these arguments more closely let us identify the following variables: f(t) will be the growth function of timber with t being time, and T being the optimal rotation time for a forest; p will be the price of timber, assumed to be constant;[2] r will be the real discount rate, assumed for now to equal a real market rate of interest, and c will be the marginal timber cost (the cost of cutting down the trees). We note that for the timber growth function, f(t), we can expect $f'(t) > 0$ for a considerable period of time, although eventually a forest will cease to grow and f'(t) will eventually become negative. In many forests there will be a shorter period after planting when f''(t) will also be positive, but it will tend to turn negative much sooner than will f'(t).

The Fisher solution for the optimal rotation time, T, that maximizes the present value for the single planting of a forest without reforestation or any valuable alternative use of the land after harvesting in the deterministic case is given by

$$pf'(T) = rpf(T). \tag{1}$$

Removing the price term from both sides gives the simpler and clearer

$$f'(T) = rf(T), \tag{2}$$

which has the famously intuitive interpretation that the forest (or tree) should be cut when its growth rate equals the real market rate of interest, the argument being that after this the forest owner will deposit the proceeds in a safe financial asset and earn interest.

Allowing for replanting, or an alternative valuable use of the land, shortens the optimal rotation period of time as the owner wishes to replant sooner the more rapidly growing younger trees or to take advantage of the valuable alternative use sooner. Faustmann correctly analyzed this for the infinite time horizon case with a sum of discounted future earnings from the harvesting of the future plantings. This optimal solution for this infinite sum reduces to

$$pf'(T) = rpf(T) + r[(pf(T) - c)/(erT - 1)]. \tag{3}$$

Given that the additional term on the right hand side can be expected to be positive, this means that the growth rate of the forest will be greater than the real rate of interest at time T, which means the forest should be cut sooner than in the Fisher case given the tendency for forest growth rates to decline after awhile.[3]

The next step in the analysis was due to Hartman (1976) who noted that non-timber amenity values should be taken into account as well, with some of these possibly not being marketed, although some might be. If g(t) describes the time pattern of the flow of such amenities, then the Hartman solution for T is given by

$$pf'(T) = rpf(T) + r[(pf(T) - c)/(e^{rT} - 1)] - g(T). \tag{4}$$

To the extent that g(t) remains positive, this implies an offsetting of the modification provided by Faustmann of the Fisher solution; T will now be greater than for forests whose only value is for timber harvest. Indeed, if g(t) is sufficiently large and remains so as t increases, it may be optimal not to harvest the forest at all and to leave it as a permanently old-growth forest. Clearly what elements are entering into the determination of g(t) is very important, and we turn now to a discussion of this question.

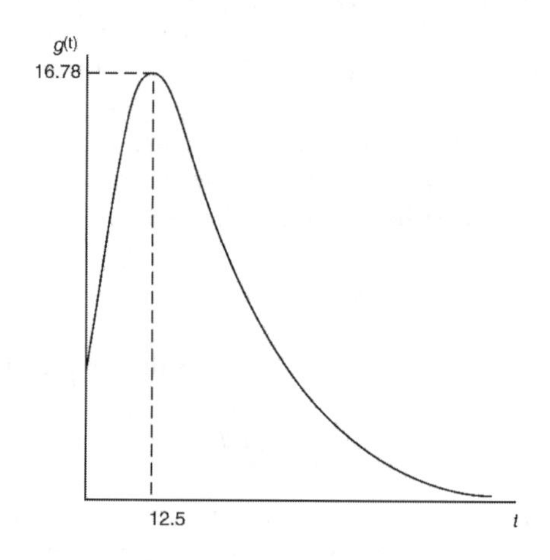


***Figure 9.1.*** *Grazing Benefit Function*

## 3. MODELS OF NON-TIMBER AMENITY VALUES

A simple example that avoids the problems of non-marketed amenities can show how taking account of them can introduce nonlinearities that allow for multiple equilibria and various complexities. Swallow, Parks, and Wear (1990) have examined the case of the National Forest in Western Montana, U.S.A. This forest offers cattle grazing opportunities during the early stages of forest growth before the forest canopy covers the grazing areas. This grazing amenity has been estimated to reach a maximum of \$16.78 per hectare at 12.5 years, with the function taking the following form

$$g(t) = \beta_0 t exp(-\beta_1 t), \tag{5}$$

with estimates of the parameter values being $\beta_0 = 1.45$ and $\beta_1 = 0.08$. The peak grazing value is found at $T = 1/\beta_1$. For this case the grazing benefits function can be depicted as Figure 9.1. When this g(t) is inserted into the Hartman formulation, one gets solutions such as that depicted in Figure 9.2, with MOC representing the marginal opportunity cost and MBD representing the marginal benefit of delaying harvest. The global maximum turns out to be at 73 years for this case, which compares with 76 years for the Faustmann solution due to the earlier benefits from grazing.[4]

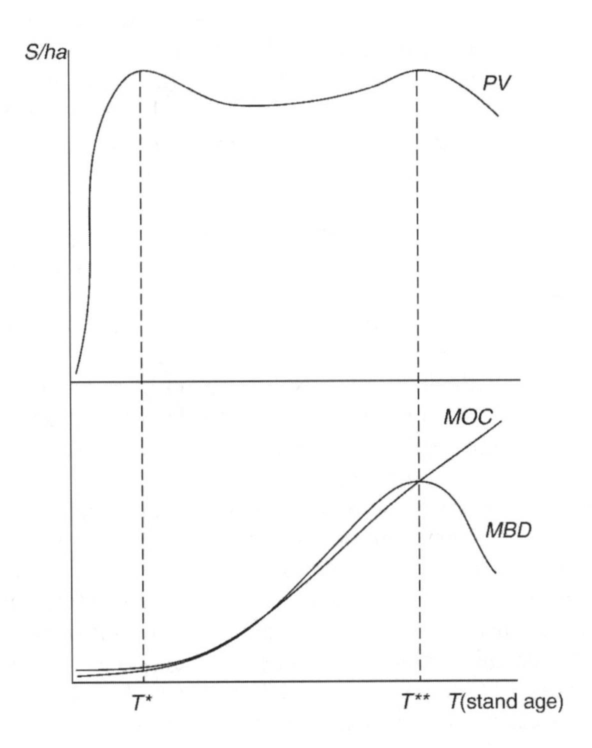


***Figure 9.2.*** *Optimal Hartman Rotation*

Grazing is not the only such non-timber amenity that forests generate, although it is one that has a definite private market value. Other such amenities may involve less clearly marketed phenomena, although in some cases the amenities may be brought to have a market value through appropriate public innovation in institutions managing the forests. Besides grazing, hunting is an activity that may be able to generate income for the private owner of a forest, or possibly a public owner as well, although controlling access to forests by hunters is not always easy, as the long history from the feudal period of aristocrats attempting to control poaching by peasants on their feudal estates demonstrates.[5] In the context of a modern national forest manager such as in the United States, although some of these amenity values may be captured by the state through issuing hunting licenses, most are captured by the hunters themselves. However, the amount of such amenity values have been estimated for many national forests in the U.S. as part of the FORPLAN planning process to determine land use allocation in the U.S. national forests (Johnson, Jones, & Kent, 1980; Bowes & Krutilla, 1985). Similar to hunting, fishing is also a non-timber amenity from forests and faces similar questions regarding measuring the scale of its value.

Somewhat more difficult to measure but very important in public policy discussions and debates is broader biodiversity that involves species that do not provide a direct use for human beings as do those that are harvested somehow, either as timber or as food as with grazing or as the objects of recreational pursuit and killing as with hunting and fishing.[6] A wide variety of indirect amenities are associated with biodiversity in forests (Perrings et. al., 1995), some more potentially marketable, such as potential sources of medicines, than others. An especially controversial case has involved the preservation of spotted owl in the forests of the northwestern U.S. In some less developed countries ecotourism has risen as a way of satisfying the conflict between preserving endangered species or habitats and providing for the economic welfare of indigenous communities that use the habitat. More generally other kinds of recreation besides hunting and fishing in forests are sources of amenity values, including purely aesthetic ones such as people viewing beautiful leaves during the fall season. The measurement of some of these amenity values may involve estimation of "existence values" through a variety of methods, many of these highly controversial. These methods of valuation become even more complicated and controversial when they involve traditional populations in rain forests in poorer nations (Gram, 2001).

Yet another source of non-timber amenities that has increasingly attracted attention is that of carbon sequestration and oxygen generation, with the former viewed as more critical given the problem of global warming, although again the exact value of this amenity is difficult to measure and very controversial. Although there are some distinct complications in the time patterns involved, it appears that in most forests the amenity value of carbon sequestration tends to rise with the length of the rotation period (Alig, Adams, & McCarl, 1998). A crucial aspect of this is that when a forest is cut for timber there tends to be a substantial release of carbon back into the atmosphere. Indeed, the carbon sequestration amenity value can continue to increase even after the forest has not only stopped growing but actually begun to decline. Thus, in contrast to the grazing example considered above, considering carbon sequestration tends to lengthen the time of an optimal rotation within the Hartman framework, although it is likely for specific forests that the amenity values for sequestration are lower than for many other amenities.

A somewhat complicated question arises regarding the relationship between carbon sequestration and biodiversity. It turns out that this very much depends on the forest and also the nature of the reforestation (or afforestation) policies after cutting down the forest for timber harvest purposes occurs. Many observers argue that biodiversity tends to increase with age of forests and thus goes hand in hand with carbon sequestration and that in some areas forests of tree species that support more biodiversity are also better at carbon sequestration as with the longleaf pine native to the U.S. Southeast now largely replaced by the more rapidly growing loblolly pine (Alavalapati, Stainback, & Carter, 2002). Furthermore longer lasting forests with more carbon sequestration may also provide external benefits in terms of reduced soil erosion, flooding, and other environmental benefits (Plantinga & Wu, 2003). However, if reforestation policies involve replacing multi-species and biodiverse forests with mono-species and less biodiverse ones, then a policy oriented towards carbon sequestration may conflict with one oriented toward preserving biodiversity (Caparrós & Jacquemont, 2003).

However, more detailed analysis of some forests as carried out by the FORPLAN process in the United States has revealed that in some areas some of the generalizations listed above do not hold. What is clear is that for a given forest type, the patterns of these non-timber amenities may vary in a much more complicated manner over time as the succession process within a given forest type proceeds. Thus in the deciduous forests found in the George Washington National Forest in Virginia and West Virginia, the patterns of biodiversity and the patterns of hunting and other amenities follow quite a complicated pattern.[7]

The initial pattern after a clearcut in terms of huntable wild animal populations is for deer to reach a maximum population in the neighborhood of five to ten years afterwards. This essentially resembles the pattern for grazing in Western Montana, and what is involved is the deer doing especially well on the edges of areas that have been clearcut at this time afterwards. There is plenty of food for the deer but not so much cover that they cannot do well. In the George Washington National Forest, deer hunting is by far the most popular and thus weighs quite heavily in this amenity value.

A second peak of non-timber amenity value arrives at around 25 years after a clearcut. At this point aggregate biodiversity in terms of the sheer number of different species living in the forest is maximized. This contradicts the widespread view that biodiversity is maximized in old growth forests.[8] At this point the initial growth of oak trees has reached a substantial level, but new species of trees, such as maples, have begun to grow also, and there is a great deal of general undergrowth, the latter especially important. In the George Washington National Forest the hunted species that are especially prevalent at this time tend to wild turkeys and grouse.

Finally there is a later maximum that emerges after about 60 years for bear populations. Bears do well in forests with large fallen-down trees. These older forests have generally lost much of their undergrowth and also have less variety of tree species. Hence they are lower in overall biodiversity, but of course there are people who desire to hunt bears and others who view bears as especially valuable for more general reasons, some of these possibly irrational.[9] Figure 9.3 depicts the pattern just described for the George Washington National Forest for its non-timber amenities, which suggests a more complicated set of possible multiple equilibria than observed for the Western Montana case. Needless to say, it is unsurprising that the Supervisor of the George Washington National Forest would complain that his most difficult problem was adjudicating between those who wished to hunt deer (and thus supported road building and clearcutting) and those who wished to hunt bears (and thus supported preserving old growth forests), with both groups being heavily armed and vigorous in their presentations.

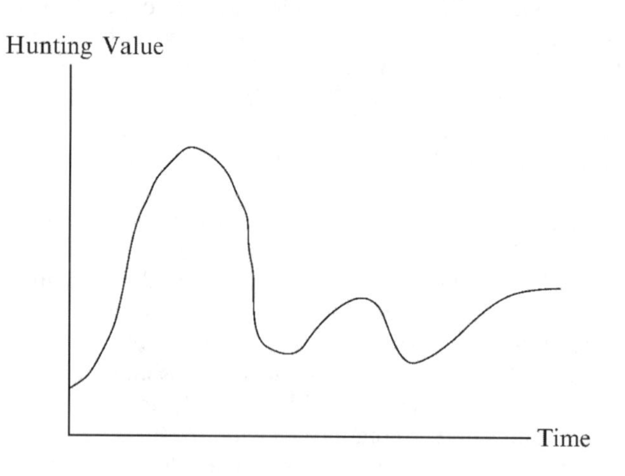


***Figure 9.3.*** *Virginia Deciduous Forest Hunting Amenity*

## 4. MANAGEMENT PROBLEMS BEYOND ROTATION OF TIMBER HARVEST

### *4.1 Fire Management*

Whereas previously we have considered the problem of optimal rotation of timber harvesting while accounting for non-timber amenity values, now we consider an alternative rotation scheme. This involves the use of fire to maximize the persistence of endangered species whose populations peak in a mid-successional stage of a forest ecosystem, much as the wild turkey and grouse in the previous discussion. One such example is eastern bristlebirds in the U.S. (Pyke, Saillard, & Smith, 1995) with Johnson (1992) and Whelan (1995) providing more extended discussions and cases. Stochastic dynamic programming has been used to study optimal fire rotation systems by Possingham and Tuck (1997) and by Clark and Mangel (2000).

Clark and Mangel (pp. 176-181) consider an endangered population wherein habitat quality is given by q(t) since the time of the last fire, r is the litter size, $s_a$ is the probability that an adult survives in the absence of fire, $s_j$ is the probability that a juvenile survives in the absence of fire, and N(t) is the adult population in time t since a fire. The population equation after a fire then becomes

$$N(t+1) = [s_a + s_j r q(t)]N(t). \qquad (6)$$

Letting $r = 2$, $s_a = 0.7$, $s_j = 0.2$, with q(t) reaching a maximum between five and ten years, Clark and Mangel (2000, p. 178) find the trajectory of average population to follow that shown in Figure 9.4.

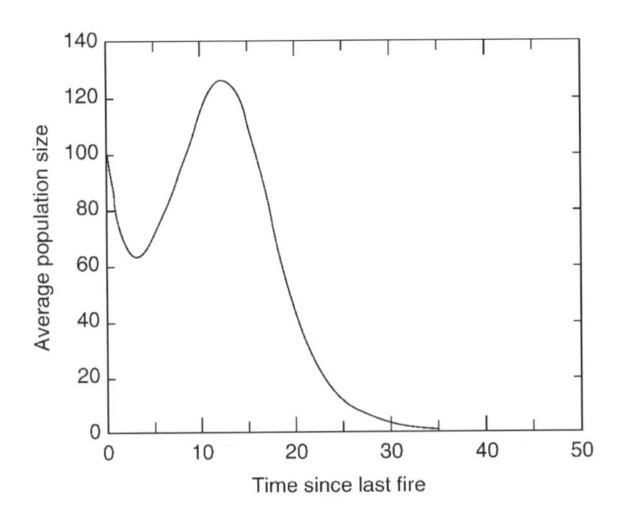


***Figure 9.4.*** *Average Population Path*

Letting a fitness parameter, f, be the percent of the population (both adult and juvenile) that survives a burn, and letting f = 0.8, Clark and Mangel study a case over time of 20 years and population that reaches a maximum of 50 but which must kept at least as great as 3. They compute at each time t the maximum probability of the survival of the species based on starting a fire or not starting a fire for a given population size. Their solution is depicted in Figure 9.5, which divides the time-population space into zones of starting a fire or not starting a fire (Clark & Mangel, 2000, p. 181).

The problem of fire rotation and management has become highly controversial among forest managers within the United States, with the issues going well beyond those of preserving endangered species and involving the costs of the fires themselves and their degree of general destructiveness. Muradian (2001) argues that the relationship between fire frequency and vegetative density is one of multiple states, allowing for the possible of catastrophic dynamics. This follows rather closely the argument of Holling (1973) regarding the tradeoff between resilience and stability. Thus, traditionally policy in the U.S. was to attempt to fight all fires that appeared on national forest or park lands. However, the truly catastrophic fires that have broken out in several national parks, most famously in Yellowstone Park during the 1990s, have made policymakers aware that not allowing any fires at all leads to a dangerous accumulation of underbrush and dead branches and trunks that can lead to a much greater fire when one finally breaks out. The short term stability of fighting all fires leads to the longer term decline in resilience of the forest to catastrophic fire. So a new policy of actively starting fires to maintain resiliency in the forests has been adopted, although this has also become controversial since one of these got out of control in Arizona and ended up destroying property.

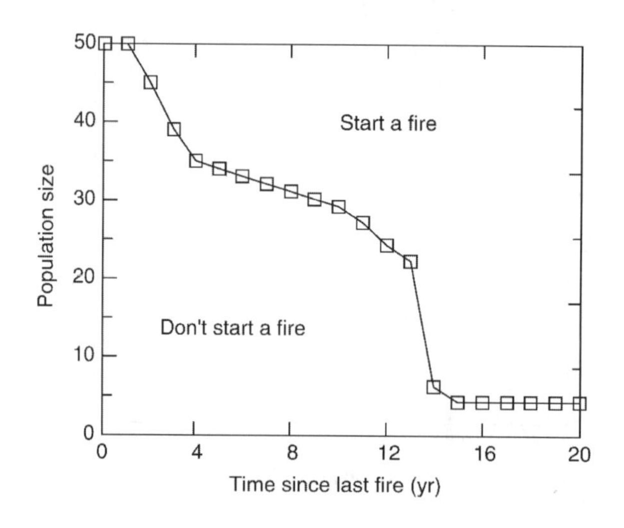


***Figure 9.5.*** *Optimal Fire Management*

### *4.2 Pest Management*

Holling (1965, 1973) initially posed his hypothesis mentioned above after contemplating the dynamics of spruce-budworm outbreaks in western Canadian coniferous forests. Such outbreaks occur in a fairly regular pattern approximately every 40 years or so. Among those looking at this have included May (1977), Ludwig, Jones, and Holling (1978), with Casti (1989) and Rosser (1991) putting the argument into an explicit context of catastrophe theory. What is involved is essentially a three-level predator-prey model, with the budworms feeding on the tree leaves, which grow larger as the trees grow larger, and migratory birds limiting the budworm population by preying on them. The trigger mechanism for the periodic outbreak is that there is an upper limit to the ability of the birds to concentrate in the trees, while the budworms can keep increasing with the leaf size. So, as the bird population becomes limited at a crucial level, they cease being able to limit the budworms which then break out into a rapid increase that in turn triggers a crash in the tree population and thus also a subsequent crash in the budworm population.

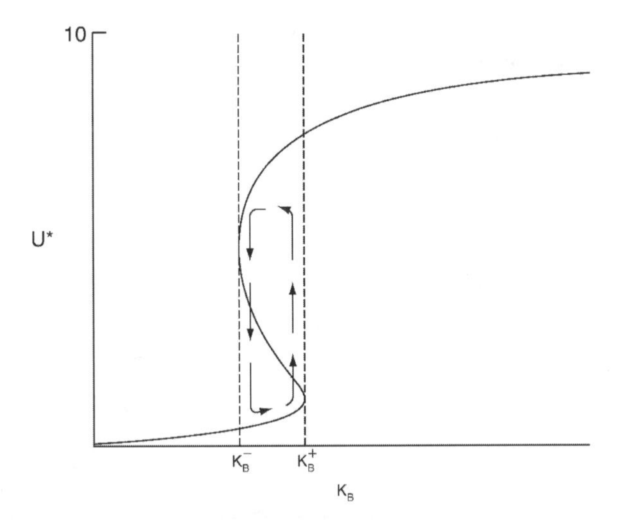


***Figure 9.6.*** *Spruce-Budworm Dynamics*

Following Ludwig, Jones, and Holling, let B equal the budworm population, $r_B$ equal to the natural growth rate of the budworms, $K_B$ equal the budworm carrying capacity (determined by the amount of leaves), $\alpha$ = the predator saturation parameter (a proportion of the budworm carrying capacity), $\beta$ equal the maximum rate of predation by the birds upon the budworms, and $u^*$ being the equilibrium leaf volume. The budworm dynamics in their early stages are given by

$$dB/dt = r_B B(1 - B/K_B) - \beta B^2/(\alpha^2 + B^2). \tag{7}$$

Nonzero equilibria are solutions of

$$(r_B K_B/\beta) = u^*/[(\alpha/K_B)^2 + u^{*2})(1 - u^*). \tag{8}$$

This set of solutions is depicted in Figure 9.6, with the zone of multiple equilibria and associated catastrophic hysteresis loops being that of an infected forest.

Holling, in particular, (1986) has discussed at length policy responses to this problem. Whereas many policymakers are inclined to spray the budworms, this tends to happen to late in the cycle. When they have become visible they are already in the epidemic phase and spraying simply holds the system in that very unstable state. Holling focuses on the larger system, especially the idea of encouraging greater bird population to constrain the budworms. This led him to consideration of how events at great distances might affect the system, e.g. how a failure to maintain wetlands in the U.S. that are used by the birds when they migrate might trigger an outbreak of the budworms by reducing the bird population below a critical size. For

Holling this was example that represents the very essence of "local surprise and global change" in ecosystems.[10]

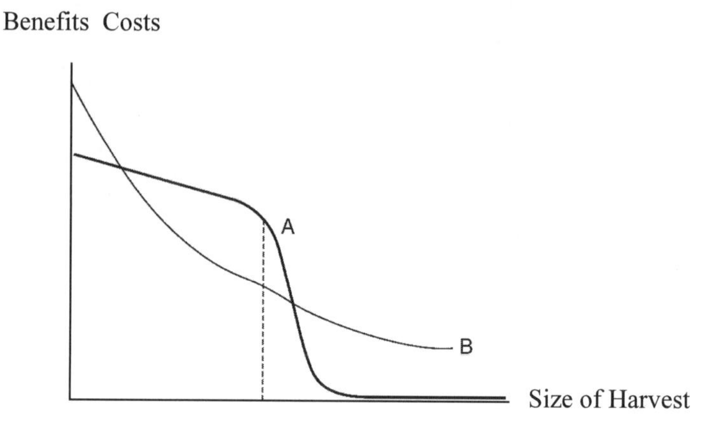


***Figure 9.7.*** *Harvest Cut and Habitat Damage*

*4.3 Patch Size Management*

Finally let us consider the problem of the size of cut areas in forests, which is closely related to the sizes of patches within forests. It has long been known that timber harvesters prefer to harvest by cutting large clearcuts that are then replanted with a single species of all the same age. This is the least expensive method of harvesting for a pure lumber producer unconcerned about any other amenity values of the forest. However, it is known that large clearcuts can reduce the maximum size of patches of trees in a forest and that the survival of species may depend on there being sufficiently large patches of trees to sustain the more fragile species. More particularly it has been argued that there is a nonlinear relationship between habitat destruction (or fragmentation) and largest patch size (Tilman, May, Lehman, & Nowak, 1994; Bascompte & Solé, 1996; Metzger & Décamps, 1997; Muradian, 2001). Below a certain threshold of habitat patch size there is a relative sudden collapse of population for the fragile species.

Figure 9.7 depicts the basic situation involved here. It combines figures found in the above references with the private cost of timber harvesting. The horizontal axis represents the size of the cuts and the vertical axis represents quantities of value units, assuming these can be estimated for the species affected by the timber harvesting. Line A represents the benefits per forest of cutting at a certain scale, the value of the species preserved at that scale of size of cuts, which shows the nonlinear and catastrophically dropping off aspect. It should be kept in mind that the patch size of habitat preserved is inversely related to the size of the timber cut. Line B shows the private costs of timber harvesting per forest, which steadily decline with the size of the cut. Clearly there is an intermediate zone of sizes of cuts where it would be socially superior to cut although these would entail higher costs for the private timber harvester. Unsurprisingly this issue is one that continues to be very controversial in forest management.

## 5. CONCLUSIONS

We have reviewed a variety of complex ecological-economic problems associated with dynamically managing forests around the world. Needless to say we have barely scratched the surface of these issues and it must be noted that we have mostly dealt with fairly stylized cases or specific examples. Many cases and situations may appear to be very different from the ideas presented here. We must also note that we have largely avoided any detailed discussion of issues arising from conflicts or ambiguities about property rights or access, although these are very serious matters in the traditional communities living in the tropical rain forests, and are also issues even in the temperate forests in the more high income countries such as Canada and the United States.

We began by reviewing the basic literature on optimal rotation of a forest being harvested solely for timber use, resulting in the Faustmann solution. This was modified by considering the analysis of Hartman, which allows for accounting for non-timber amenities of the forest, including grazing, hunting, fishing, recreation, social values of traditional communities, carbon sequestration, and more general existence and aesthetic values. A central issue involves the fact that these other amenities may exhibit complicated time patterns over the life of a forest that do not correspond directly with the growth rate of the trees, the factor that underlies decision making about timber harvesting by a private owner for whom timber is the sole value of interest. This leads to the possibility of multiple equilibria and considerable dynamic complications. These complications have led to great difficulties for actual decision-makers responsible for the management of forests that have multiple uses, as in the national forests of the United States.

We also considered a further set of management issues, notably fire management, pest management, and patch size management in regard to the size of cuts made during timber harvesting. In all of these cases the presence of nonlinearities and discontinuities impose heightened difficulties for forest managers. The presence of critical thresholds in all of these cases is a pervasive phenomenon, and one that poses deep problems not easily solved. Although we did not examine cases involving chaotic or other more erratic dynamics in forest management, such phenomena are possible. However they would tend to operate over relatively long time scales, unlike for much shorter-lived biological populations. For forests the more dramatic and compelling problems arise from the discontinuities that are studied more by catastrophe theory, collapses of species populations, sudden collapses of entire ecosystems due to fire or outbreaks of pests, or the more insidious damage of badly managed or timed timber harvesting (Rosser, 2001). Dealing with these problems will challenge forest managers for the foreseeable future.

**Acknowledgements:** The author wishes to thank Tim Allen, Albert Berry, Christophe Deissenberg, Shashi Kant, and Akio Matsumoto for assistance or useful comments. None are responsible for any errors remaining in this paper.

## NOTES

---

[1] Samuelson (1976) provides an account of the historical development of these two approaches. Many other well-known economists advocated the Fisher approach prior to the translation into English of Faustmann's paper in 1968, including Harold Hotelling (1925) and Kenneth Boulding (1935), and in the German language tradition, Faustmann's predecessor, Johann Heinrich von Thünen (1826). However, there were some in English who understood that Fisher was not correct prior to 1968 (Alchian, 1952; Gaffney, 1957). Even without explicitly accounting for replanting the Fisher solution has problems when land rent is positive, indicating the present value of potential alternative uses.

[2] Constancy of the price is a non-trivial assumption. A considerable literature exists that assumes that the price and other forest values follow stochastic processes, usually some variant of Brownian motion. Option theory using Ito's lemma then is used to provide solutions for optimal control stopping problem for various such models under different assumptions (Reed & Clarke, 1990; Zinkhan, 1991; Reed, 1993; Conrad, 1997; Willassen, 1998; Saphores, 2003). Arrow and Fisher (1974) first suggested the use of option theory to deal with the possibility of irreversible loss of uncertain future forest values.

[3] It is possible for there to be multiple solutions to (3) given that forest growth rates initially increase. However any solution when growth rates rise will violate second-order conditions for optimality.

[4] That such multiple solutions can arise due to non-monotonicity of the time pattern of net benefits for forestry cases was first observed by Porter (1982), who was specifically concerned with the wilderness benefits of forests. Prince and Rosser (1985) have noted the link between such patterns in natural resource use and the reswitching question in capital theory. It should be noted that although we mostly focus on amenities that vary over time some may have constant values over time.

[5] It is somewhat ironic that many national forests in Europe were formerly owned by aristocratic families.

[6] Of course in less developed countries hunting and fishing in forests by aboriginal or more generally poor people may well be a major source of their food and not a matter of recreation at all. Kant (2000) provides a discussion of how to integrate the broader socio-economic problems facing traditional communities into an intertemporal optimization framework for forestry management. Especially for aboriginal populations in tropical rain forests, questions of cultural survival enter in as well as more standard ones of economic development or environmental sustainability. For discussions of the co-evolution of culture and ecology in tropical rain forests see Norgaard (1981, 1994).

[7] The source of this information is from the author's own unpublished work on the FORPLAN model developed for use in managing the George Washington National Forest in 1980.

[8] What is true is that in many forests certain endangered species fare well in old growth forests, the spotted owl example in the northwestern forests of the U.S. being a famous example, even if in some of these cases there is less aggregate biodiversity. To the extent that old growth forests become rare, this fact alone can increase the amenity value of the few species that thrive only in such forests.

[9] It is well known that surveys of the public show much greater willingness to pay to preserve large mammals than other species, especially ones perceived to be "cuddly," which is certainly the case for bears (Weitzman, 1992). It is no accident that the World Wild Fund uses Panda bears as a symbol in their fundraising activities, although Pandas are not technically bears but more closely related to raccoons.

[10] A further aspect of this involves studying the relations between the different hierarchical levels of the system, with the question of effects passing from lower levels to higher levels becoming very important in this phenomenon of destabilization due to seemingly minor causes. Among those studying this aspect of such dynamics include Allen and Starr (1982), Holling (1992), and Rosser et al. (1994).

## REFERENCES

Alavalapati, J.R.R., Stainback, G.A., & Carter, D.R. (2002). Restoration of the longleaf pine ecosystem on private lands in the US South: An ecological economic analysis. *Ecological Economics,* 40, 411-419.

Alchian, A.A. (1952). *Economic replacement policy*. Santa Monica: RAND Corporation.

Alig, R.J., Adams, D.M., & McCarl, B.A. (1998). Ecological and economic impacts of forest policies: Interactions across forestry and agriculture. *Ecological Economics,* 27, 63-78.

Allen, T.F.H., & Starr, T.B. (1982). *Hierarchy: Perspectives for ecological complexity*. Chicago: University of Chicago Press.

Arrow, K.J., & Fisher, A.C. (1974). Preservation, uncertainty and irreversibility. *Quarterly Journal of Economics*, 87, 312-319.

Bascompte, J., & Solé, R. (1996). Habitat fragmentation and extinction thresholds in spatially explicit models. *Journal of American Ecology,* 65, 465-473.

Boulding, K.E. (1935). The theory of a single investment. *Quarterly Journal of Economics,* 49, 475-494.

Bowes, M.D., & Krutilla, J.V. (1985). Multiple use management of public forestlands. In Allen V. K. and Sweeny, J.L. (Eds.), *Handbook of natural resources and energy economics (pp.531-569).* Amsterdam: Elsevier.

Caparrós, A., & Jacquemont, F. (2003). Conflicts between biodiversity and carbon sequestration programs: Economic and legal implications. *Ecological Economics,* 46, 143-157.

Casti, J.L. (1989). *Alternate realities: Mathematical models of nature and man.* New York: Wiley-Interscience.

Clark, C.W., & Mangel, M. (2000). *Dynamic state variable models in ecology: Methods and applications.* New York: Oxford University Press.

Conrad, J.M. (1997). On the option value of old-growth forests. *Ecological Economics,* 22, 97-102.

Faustmann, M. (1849). On the determination of the value which forest land and immature stands possess for forestry, English edition, M. Gane (Ed.) (1968), *Martin Faustmann and the evolution of discounted cash flow.* Oxford University Paper 42.

Fisher, I. (1907). *The rate of interest*. New York: Macmillan.

Gaffney, M. (1957). Concepts of financial maturity of timber and other assets. *Agricultural Economics Information Series* 62, Raleigh: North Carolina State College, September.

Gram, S. (2001). Economic valuation of special forest products: An assessment of methodological shortcomings. *Ecological Economics,* 26, 109-117.

Hartman, R. (1976). The harvesting decision when a standing forest has value. *Economic Inquiry,* 14, 52-58.

Holling, C.S. (1965). The functional response of predators to prey density and its role in mimicry and population regulation. *Memorials of the Entomological Society of Canada,* 45, 1-60.

Holling, C.S. (1973). Resilience and stability of ecological systems. *Annual Review of Ecology and Systematics,* 4, 1-24.

Holling, C.S. (1986). Resilience of ecosystems, local surprise and global change. In W.C. Clark & R.E. Munn (Eds.), *Sustainable development of the biosphere (pp.292-317).* Cambridge: Cambridge University Press.

Holling, C.S. (1992). Cross-scale morphology, geometry, and dynamics of ecosystems. *Ecological Monographs,* 62, 447-502.

Hotelling, H. (1925). A general mathematical theory of depreciation. *Journal of the American Statistical Association,* 20, 340-353.

Johnson, E.A. (1992). *Fire and vegetation dynamics: Studies from the North American boreal forest.* New York: Cambridge University Press.

Johnson, K.N., Jones, D.B., & Kent, B.M. (1980). *A user's guide to the forest planning model* (FORPLAN). Fort Collins: USDA Forest Service, Land Management Planning.

Kant, S. (2000). A dynamic approach to forest regimes in developing economies. *Ecological Economics,* 32, 287-300.

Ludwig, D., Jones, D.D., & Holling, C.S. (1978). Qualitative analyses of insect outbreak systems: The spruce budworm forest. *Journal of Animal Ecology,* 47, 315-332.

May, R. M. (1977). Thresholds and breakpoints in ecosystems with a multiplicity of stable states. *Nature,* 269, 471-477.
Metzger, J.P., & Décamps, H. (1997). The structural connectivity threshold: An hypothesis in conservation biology at the landscape scale. *Acta Oecologica*, 18, 1-12.
Muradian, R. (2001). Ecological thresholds: A survey. *Ecological Economics,* 38, 7-24.
Norgaard, R.B. (1981). Sociosystem and ecosystem coevolution in the Amazon. *Journal of Environmental Economics and Management,* 8, 238-254.
Norgaard, R.B. (1994). *Development betrayed: The end of progress and a coevolutionary revisioning of the future.* London: Routledge.
Perrings, C., Mäler, K., Folke, C., Holling, C.S., & Jansson, B. (Eds.) (1995). *Biodiversity loss: Economic and ecological issues*. Cambridge: Cambridge University Press.
Plantinga, A.J., & Wu, J. (2003). Co-benefits from carbon sequestration in forests: Evaluating reductions in agricultural externalities from an afforestation policy in Wisconsin. *Land Economics,* 79, 74-85.
Porter, R.C. (1982). The new approach to wilderness through benefit cost analysis. *Journal of Environmental Economics and Management,* 9, 59-80.
Possingham, H., & Tuck, G. (1997). Application of stochastic dynamic programming to optimal fire management of a spatially structured threatened species. In A.D. McDonald & M. McAleer (Eds.), *Proceedings international congress on modelling and simulation* (Vol. 2). Canberra: Modeling and Simulation Society of Australia.
Prince, R., & Rosser, Jr. J.B. (1985). Some implications of delayed environmental costs for benefit cost analysis: A study of reswitching in the western coal lands. *Growth and Change,* 16, 18-25.
Pyke, G.H., Saillard, R., & Smith, J. (1995). Abundance of eastern bristlebirds in relation to habitat and fire history. *Emu,* 95, 106-110.
Reed, W.J. (1993). The decision to conserve or harvest old-growth forest. *Ecological Economics,* 8, 45-69.
Reed, W.J., & Clarke, H.R. (1990). Harvest decision and asset valuation for biological resources exhibiting size-dependent stochastic growth. *International Economic Review,* 31, 147-169.
Rosser, J.B. Jr. (1991). *From catastrophe to chaos: A general theory of economic discontinuities*. Boston: Kluwer Academic.
Rosser, J.B. Jr. (2001). Complex ecologic-economic dynamics and environmental policy. *Ecological Economics,* 37, 23-37.
Rosser, J.B. Jr., Folke, C., Günther, F., Isomäki, H., Perrings, C., & Puu, T. (1994). Discontinuous change in multi-level hierarchical systems. *Systems Research,* 11, 77-94.
Samuelson, P.A. (1976). Economics of forestry in an evolving society. *Economic Inquiry,* 14, 466-491.
Saphores, J. (2003). Harvesting a renewable resource under uncertainty. *Journal of Economic Dynamics and Control,* 28, 509-529.
Swallow, S.K., Parks, P.J., & Wear, D.N. (1990). Policy-relevant nonconvexities in the production of multiple forest benefits. *Journal of Environmental Economics and Management,* 19, 264-280.
Tilman, D., May, R., Lehman, C., & Nowak, M. (1994). Habitat destruction and the extinction of debt. *Nature,* 371, 65-66.
von Thünen, J.H. (1826). *Der Isolierte Staat in Biehiezung auf Landwirtschaft und Nationaleckonomie.* Hamburg: Perthes.
Weitzman, M.L. (1992). On diversity. *Quarterly Journal of Economics,* 107, 363-406.
Whelan, R.J. (1995). *On the ecology of fire*. New York: Cambridge University Press.
Willassen, Y. (1998). The stochastic rotation problem: A generalization of Faustmann's formula to stochastic forest growth. *Journal of Economic Dynamics and Control,* 22, 573-596.
Zinkhan, F.C. (1991). Option pricing and timberland's land-use conversion option. *Land Economics,* 67, 317-325.

# CHAPTER 10

# NONLINEARITIES, BIODIVERSITY CONSERVATION, AND SUSTAINABLE FOREST MANAGEMENT

JEFFREY R. VINCENT AND MATTHEW D. POTTS

*University of California at San Diego*
*La Jolla, CA 92093-0519, USA*
*Email: jvincent@ucsd.edu; potts@ucsd.edu*

**Abstract.** There are two broad approaches for jointly producing timber and conserving biodiversity in forests: segregated management, in which timber production is emphasized in some parts of the forest and biodiversity conservation in others, and integrated management, in which conservation measures are incorporated into logging regimes. Nonlinearities in forestry production sets affect the relative economic superiority of these two approaches. Such nonlinearities can result from economic, institutional, and ecological factors. They can cause segregated management to be superior to integrated management even in forests comprised of identical stands. The policy relevance of this and other effects of nonlinearities on spatial aspects of forestry management depends, however, on the relative values of biodiversity and timber.

## 1. INTRODUCTION

"Sustainable forest management" (SFM) as commonly defined requires more than the successful regeneration of timber trees after harvest. One of its most important dimensions is the conservation of biological diversity. All certification systems for SFM contain requirements related to biodiversity. The most prominent international certification organization, the Forest Stewardship Council, has declared ten principles of SFM, and two of them pertain to biodiversity conservation (Forest Stewardship Council, 2002). Principle 6 states that "Forest management shall conserve biological diversity and its associated values … and, by so doing, maintain the ecological functions and the integrity of the forest." Principle 9 states that "Management activities in high conservation value forests shall maintain or enhance the attributes which define such forests." Similarly, the International Tropical Timber Organization (1998) has established seven criteria for SFM, with Criterion 5 focusing on biodiversity.



*Kant and Berry (Eds.), Economics, Sustainability, and Natural Resources: Economics of Sustainable Forest Management,* 207-222.

Certification systems allow flexibility in the management approaches used to conserve biodiversity in timber production forests. They explicitly recognize that biodiversity can be conserved either by prohibiting logging in some parts of a forest or by modifying harvesting methods in the areas where logging is allowed. That is, possible approaches include both the spatial segregation of biodiversity conservation from timber production and the integration of conservation measures into logging regimes.[1] For example, Indicator 5.7 of the International Tropical Timber Organization (1998) calls for the "Existence and implementation of management guidelines to … keep undisturbed a part of each production forest" (p. 14), while its Criterion 5 states that "Biological diversity can also be conserved in forests managed for other purposes, such as for production, through the application of appropriate management practices" (p. 13). Common examples of the latter practices include reducing the number of trees felled and adopting directional felling methods or cable logging to reduce damage to residual trees and the forest floor.

In this paper we demonstrate that nonlinearities in forestry production sets affect the relative superiority of spatially segregated and spatially integrated approaches for conserving biodiversity in timber production forests. By "forestry production sets" we refer to the technically feasible combinations of biodiversity and timber that can be produced from a forest for a given level of management inputs. By "nonlinearities" we refer to violations of the standard assumptions about production sets, such as that their boundaries are continuous, smooth, and outward bowed (i.e., convex). Our focus is on the implications of nonlinearities for spatial aspects of forest management. In this regard this Chapter complements Chapter 9 in this volume by J. Barkley Rosser, which focuses on dynamic aspects.

A nonlinear relationship between the amount of biodiversity conserved in a forest and the amount of timber harvested can result from a variety of factors. We consider three: economic, institutional, and ecological. We show that some nonlinearities favor segregated approaches while others favor integrated approaches. We also show that nonlinearities can lead to the counterintuitive result that segregated management can be superior to integrated management even in forest estates comprised of identical stands.[2] That is, the justification for segregated management does not hinge on some forest stands being richer in biodiversity than others. As will be discussed, this particular result illustrates the possibility that nonlinearities can generate multiple forest management equilibria.

This paper is intended to be illustrative, not comprehensive. It does not provide a taxonomic survey of forest management nonlinearities, nor does it provide a complete review of the literature on this topic. Instead, it employs a series of three simple models to illustrate the effects of different types of nonlinearities. These models are drawn from our previous work, with some modifications to make the exposition within this paper as consistent as possible. Although the models' formulations differ, they share the feature that their main results can be depicted in a straightforward way using production possibility frontiers.[3] More complex nonlinear models can of course be formulated. Our intention is to strip away the complexity and to expose the intuition as to how nonlinearities affect the choice between segregated and integrated forest management approaches.

We begin with a general theoretical model that demonstrates how a nonlinearity at the level of an individual forest stand can favour segregated management at the level of the forest estate. This model was first presented in a paper by Vincent and Binkley (1993). The particular nonlinearity in this first model is a nonconvexity. We then consider a model that illustrates how a nonconvexity can result from economic and institutional factors. This second model is from a paper by Boscolo and Vincent (2003), who in addition to presenting it in a theoretical context applied it to tropical rainforest data from Malaysia. We summarize their simulation results, which underline the important point that a nonconvex production set does not necessarily imply that segregated management is superior to integrated management. The relative values of biodiversity and timber also matter.

The third and final model considers nonlinearities resulting from two ecological factors: species' populations being clumped instead of randomly distributed across the forest, and species having minimum viable populations within reserves where logging is prohibited. The first factor favours more integrated management, in the sense of having a large number of small reserves spread across the forest (in the extreme, a refugium within each annual cutting block), while the second favours more segregated management, in the sense of having a small number of large reserves (in the extreme, just a single reserve in one location in the forest). This model is from a paper by Potts and Vincent (2004). In contrast to the previous two models, the production set in this model is convex. Like Boscolo and Vincent (2003), Potts and Vincent applied their model to rainforest data from Malaysia. We show a subset of their results, which, similar to the results in Boscolo and Vincent, illustrate the impact of relative economic values on the relative superiority of more integrated and more segregated management approaches.

We conclude the paper by summarizing the findings in the paper and discussing the implications for SFM, especially in the context of tropical rainforests.

## 2. A GENERAL MODEL OF NONLINEARITIES IN FORESTRY PRODUCTION SETS

Figure 10.1 pertains to a model in which production possibilities at the stand level are affected by management effort.[4] The vertical axis shows the physical output of biodiversity (e.g., the number of species preserved), and the horizontal axis shows the physical output of timber (e.g., cubic meters harvested). The model can be interpreted as being either static or dynamic; in the latter case, the axes are expressed in terms of discounted sums of current and future outputs. The production possibilities frontier shifts from $SM^B$ to IM to $SM^T$ as more management effort is applied to the stand. Intermediate frontiers also exist; we show just these three because they are sufficient to illustrate the difference between segregated and integrated management.

Suppose that the forest contains two stands, which is the minimum number necessary to illustrate this difference. Moreover, suppose that the stands are identical in all respects, with production possibilities as shown in Figure 10.1, and that a given level of management effort is to be allocated between them. If effort is split evenly between the stands, then the production possibilities frontier is identical for each stand and equals IM. If instead effort is split unevenly, then the stand receiving more effort has a frontier of $SM^T$, while the stand receiving less effort has a frontier of $SM^B$.

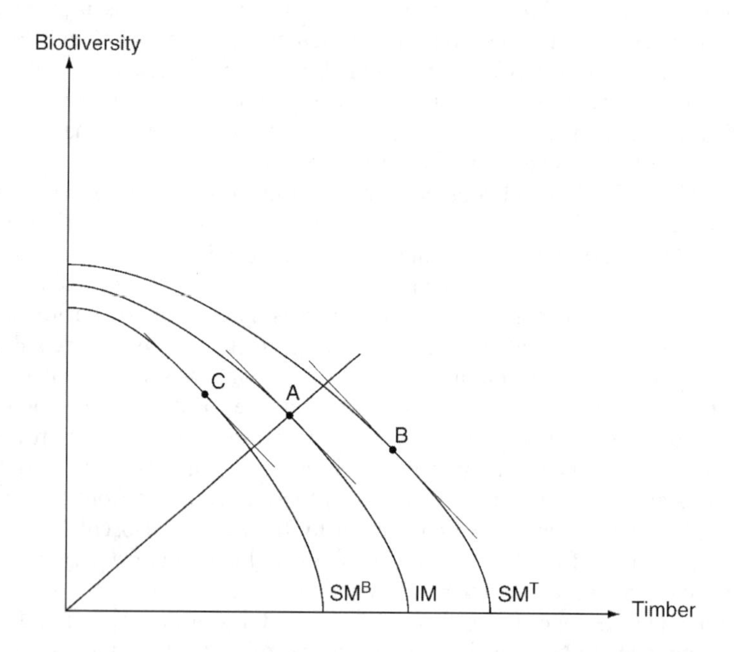


***Figure 10.1.*** *Superiority of Segregated Management in a Two-Stand Model with a Nonconvex Production Set*

Let us make the additional assumption that $SM^T$ and $SM^B$ are equidistant from IM along any ray from the origin, with the distance rising as the biodiversity-timber output ratio falls (i.e., moving clockwise along IM). Equidistance is obviously a special assumption. It is not necessary for the result we are about to prove, but it makes the proof easier. We will discuss the implications of relaxing it.

Suppose that the values of timber and biodiversity are such that the ratio of these values, the relative price line, is tangent to IM at point A. If effort is split evenly between the two stands, then both stands should be managed at this point. This is the integrated management equilibrium (hence the abbreviation IM for the frontier): both stands are managed identically, with each one producing a mix of biodiversity and timber.

This is not the only equilibrium, however. The geometry of the frontiers implies that the price line is tangent to $SM^T$ at a point B that lies to the right of the ray that passes through point A, and is tangent to $SM^B$ at a point C that lies to the left of the ray. These two points represent the segregated management equilibrium (hence the abbreviation SM). The stands are no longer managed in an identical way. Production at point B is relatively more specialized in timber, while production at point C is relatively more specialized in biodiversity. The degree of specialization depicted in Figure 10.1 is not complete, but it would tend more toward complete specialization—only biodiversity conserved in one stand, and only timber produced in the other—if we drew the figure so that the distance of $SM^T$ and $SM^B$ from IM rose more rapidly as one moves clockwise along IM (i.e., if the intercepts of $SM^T$ and $SM^B$ on the timber axis were farther apart).

Which equilibrium is superior? Note that the tangents through points B and C intersect the ray at points beyond where $SM^T$ and $SM^B$, respectively, intersect it. Given that $SM^T$ and $SM^B$ are equidistant from IM, these divergences imply that the aggregate value of production is greater when one stand is managed at point B and the other is managed at point C than when both are managed at point A. Segregated management is superior to integrated management, even though the two stands are identical.

If there are diminishing returns to management effort, so that the outward shift of $SM^T$ away from IM is less than the inward shift of $SM^B$, then this result still holds as long as the outward shift of the former is sufficiently large compared to the inward shift of the latter. That is, it holds as long as the returns to management effort across the two products do not diminish too rapidly. This observation points toward the underlying condition that drives the superiority of segregated management in this model: management effort has a nonconvex impact on production (Helfand & Whitney, 1994). The allocation of effort at the integrated-management equilibrium satisfies the first-order conditions for optimality—the price line is tangent to IM at point A—but not the second-order conditions, which pertain to how rapidly the returns to management effort are diminishing at that point. In contrast, the allocation at the segregated-management equilibrium satisfies both sets of conditions. This nonconvexity creates a diseconomy of scope between biodiversity and timber: starting with one stand managed at B and the other at C, shifting effort from the former to the latter to allocate management more evenly leads to an appreciable decrease in the aggregate production of timber without a sufficiently offsetting increase in the aggregate production of biodiversity. Spatially segregating production of the two products mitigates this diseconomy.[5]

We close by noting that the equilibria depicted in Figure 10.1 are interior solutions. Corner solutions are also possible. If the price line is sufficiently flat, then the equilibria are on the biodiversity axis; if it is sufficiently steep, they are on the timber axis. Given that $SM^T$ and $SM^B$ are equidistant from IM along any ray, including the axes, segregated management offers no advantage over integrated management in such cases. It generates the same aggregate value of biodiversity and timber as does integrated management, not a greater value. The superiority of segregated management in the model therefore depends on not only the nonconvexity in the production set but also on the relative values of biodiversity and timber. As we will see, relative economic values also affect the relative superiority of the two management approaches in the models in the next two sections.

## 3. NONLINEARITIES DUE TO ECONOMIC AND INSTITUTIONAL FACTORS

Perhaps the most surprising feature of the analysis in the previous section is the result that segregated management can be superior to integrated management even when forest stands are identical. How empirically relevant is this result? In a landmark study of the management of U.S. national forests for timber and nontimber values, Bowes and Krutilla (1989, p. 342) reported empirical optimization results in which

> … a higher relative recreational value did not necessarily lead to a longer rotation, even when older stands were generally preferred for their amenity values. Instead, we might see an increasing amount of the area set aside as protected old growth, while a shorter timber rotation cycle was instituted on the remaining sites in the management unit. We found such solutions even when the management area was perfectly homogeneous. Such harvest solutions could not be found if the stands were treated independently.

This is a perfect illustration of the effects of a production nonconvexity as predicted by the model in Figure 10.1. Indeed, in the theory section of their book Bowes and Krutilla (pp. 51-87) attribute such solutions to nonconvexities, although they depict them using isocost curves instead of production possibilities frontiers. They conclude that "It seems likely that specialization of land use, such as we found, is often apt to result in more effective production of such services as wildlife and increased water flow than would be possible from uniform management of a land areas" (pp. 342-343). Subsequent studies by Swallow and Wear (1993) and Swallow, Talukdar, and Wear (1997) offered additional empirical evidence of nonconvexities in forest production sets. In the cases they examined, the nonconvexities resulted from ecological interactions among stands.

Helfand and Whitney (1994) mentioned a common characteristic of production systems that can create a nonconvexity: fixed costs. Boscolo and Vincent (2003) examined the effects of this economic factor—specifically, fixed logging costs—on the joint production of biodiversity and timber.[6] They also examined the effects of institutional factors that can have a similar impact on the production set. Figure 10.2 illustrates how fixed logging costs can make production at the stand level nonconvex. The vertical axis shows the discounted sum of current and future physical outputs of biodiversity,[7] while the horizontal axis shows the net present value of current and future timber harvests (i.e., net income to the forest owner). The production set for the stand consists of two parts. One is the usual set of points bounded by the production possibilities frontier. The other is point A, which is the no-logging point. Fixed logging costs are responsible for the discontinuity between these two parts. They cause the net present value of timber harvests to be negative when harvests are small. These production points lie to the left of the vertical axis between the intercept of the frontier and point A. Although they generate positive outputs of biodiversity, they are dominated by the no-logging point, which generates higher outputs of both biodiversity (A is a larger positive amount) and timber (zero instead of negative). The production set is nonconvex because the tangent from point A to point C lies above points on the frontier between point C and the intercept of the frontier on the biodiversity axis.

Institutional factors can create a similar discontinuity even if fixed costs are negligible. In principle, detailed logging regulations can be designed that minimize the impacts of timber harvesting on biodiversity. For example, specific trees might be selected for felling at specific times, with adjustments made for unique site characteristics and current environmental conditions. Application of such regulations would generate a smooth frontier connecting point A to the point of maximum timber production (i.e., the intercept on the timber axis). But in practice, such regulations are beyond the administrative capacity of many countries, especially developing countries, to monitor and enforce. Such countries instead typically employ much simpler regulations, such as minimum diameter cutting limits and fixed cutting cycles. The result is that for any given timber harvest, the accompanying level of biodiversity conservation is lower than it would have been under the "ideal"—but administratively infeasible—regulations. The frontier thus shifts down, leaving a gap between the no-logging point and points on the frontier that involve some production of timber. The result is again the nonconvex production set depicted in Figure 10.2.

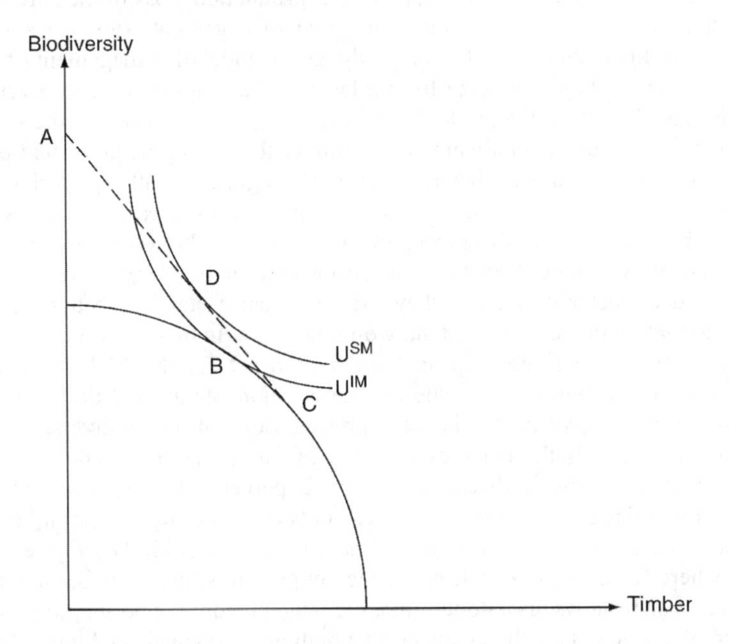


***Figure 10.2.*** *Effect of Nonconvexities due to Fixed Logging Costs and Administrative Constraints in an n-stand Model*

Suppose that the forest contains *n* identical stands, each with the production set shown in Figure 10.2. The aggregate economic welfare generated by the forest is given by U(**B**,**T**), where **B** and **T** are the aggregate outputs of biodiversity and timber across the stands. The price of timber is fixed (e.g., the region or country is a price-taker in timber markets), but the marginal welfare values of biodiversity ($U_B$) and timber income ($U_T$) are not. The welfare function is homothetic, so that U[**B**,**T**] = *n*U[**B**/*n*,**T**/*n*], where U[**B**/*n*,**T**/*n*] is the average welfare value of a stand in the forest. Figure 10.2 shows two indifference curves for the average welfare function. The curve labeled $U^{IM}$ shows the average welfare attained if all stands are managed identically, at point B. This is the integrated-management equilibrium. The segregated-management equilibrium involves managing $\alpha n$ stands at point A, which is completely specialized in the production of biodiversity, and $(1-\alpha)n$ stands at point C, which is more specialized in the production of timber than point B. The average production per stand in this equilibrium is indicated by point D, and average welfare is given by $U^{SM}$, which is higher than $U^{IM}$. Since average welfare per stand is higher under segregated management, so is total welfare aggregated across the *n* stands. As in Figure 10.1, segregated management is superior to integrated management even though all stands are identical.

The nonconvexity in this model is policy-relevant only if point B, the integrated-management equilibrium, lies to the left of point C, as in Figure 10.2. If point B lies to the right, on the convex portion of the production possibilities frontier, then integrated management dominates segregated management. Boscolo and Vincent investigated this issue by constructing a dynamic model of management of a tropical rainforest stand. They drew data from a large-scale, long-term forest inventory plot in Malaysia. They used the model to estimate the production set for the stand. They predicted the outputs of biodiversity and timber that would be generated over a 60-year time period under different harvest regimes (cutting cycles, logging technologies, minimum diameter cutting limits). They measured biodiversity by using an index related to differences in forest structure between the managed stand and an old-growth stand. Structure was defined by the basal area in different species groups and diameter classes.[8] They assumed that a stand with a structure more similar to that of an old-growth stand would be richer in biodiversity.

Figure 10.3 shows the empirical production set for the Malaysian rainforest. Each point in the figure shows the outputs of biodiversity and timber for a given harvest regime. The vertical axis is the present value of the biodiversity index, and the horizontal axis is the net present value of timber harvests. The present values were calculated using a discount rate of 2 percent. The present value of the biodiversity index has a maximum of 7.62 units at the no-logging point,[9] and the net present value of timber harvests has a maximum of $4,744. The figure shows the points where fixed logging costs cause the net present value of timber harvests to be negative. It also shows, as a dotted line, the tangent from the no-logging point to the frontier of the set (i.e., the point corresponding to point C in Figure 10.2). The (inverse) slope of the tangent is US$1,274 per index unit of biodiversity. This is nearly identical to a crude estimate of the current global welfare tradeoff between biodiversity and timber income, US$1,100 per index unit, implied by forest valuation studies reviewed by Lampietti and Dixon (1995).[10] Although the estimates of the slope of the frontier and the welfare tradeoff are obviously approximate, they imply that points B and C nearly coincide for the Malaysian rainforest. Segregated management would appear to offer little advantage compared to integrated management for this particular forest at the current point in time.

On the other hand, a stylized fact in environmental economics is that nonmarket values of the natural environment, such as passive-use values associated with biodiversity, are rising more rapidly than commodity values, such as timber (Krutilla 1967, Fisher, Krutilla, & Cicchetti, 1972). If this is the case, then the biodiversity-timber income indifference curves are becoming flatter over time. The future welfare tradeoff will be higher than the current US$1100 per biodiversity index unit. The integrated-management equilibrium will move to the left of point C, the welfare associated with it will fall below the welfare associated with segregated management, and point D will slide to the left along the tangent as logging is prohibited in more stands.

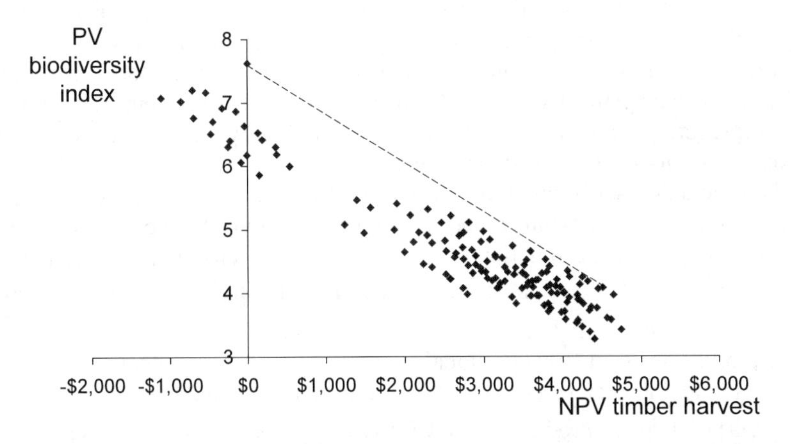


***Figure 10.3.*** *Production Set for a Malaysian Rainforest Stand Harvested under Different Cutting Cycles, Minimum Diameter Cutting Limits, and Logging Technologies (Conventional vs. Reduced Impact)*

Even with these changes, however, the potential gains from segregated management appear to be modest, at least for the forest in Figure 10.3. The nonconvexity is not very large: the intercept of the frontier implied by the points in the figure is only 12 percent below the no-logging point. Moreover, nearly all of this gap is due to administrative constraints on logging regulations as opposed to fixed logging costs. The gap will be negligible in countries with strong institutions, and thus so will be the economic advantages of segregated management.

## 4. NONLINEARITIES DUE TO ECOLOGICAL FACTORS

Potts and Vincent (2004) also use simulation to evaluate the relative superiority of segregated and integrated management of tropical rainforests for biodiversity and timber. Their model differs from the one in Boscolo and Vincent (2003) in several ways. The most important is that they highlight nonlinearities that are due to ecological, not economic or institutional, characteristics of the forest. In contrast to the two previous models, these nonlinearities do not make the forest production set non-convex, but they still affect the choice between the two management approaches.

Potts and Vincent consider a narrower management question than Boscolo and Vincent. Within a forest of $A$ hectares, they ask how large an aggregate area, $a$, should be set off limits from logging, and how many equal-sized reserves, $m$, this protected area should be divided into. The remaining forest area, $A - a$, has negligible biodiversity value but produces timber. The model is static, but it can be given a quasi-dynamic interpretation if one thinks of the forest as being divided into $n$ equal-sized annual harvest areas, or coupes, with the number $n$ being an exogenous parameter and only $1/n^{\text{th}}$ of $A - a$ harvested each year. In this context, integrated management refers to a solution in which $m = n$: a reserve is associated with each coupe. In contrast, under segregated management $m < n$: biodiversity conservation is spatially concentrated into a smaller number of reserves. The model thus allows a gradation of management regimes, from more integrated ($m$ being closer to $n$) to more segregated ($m$ being closer to 1), not a binary choice between the two approaches as in the previous two models.

The production set in this model consists of the set of all possible combinations of the number of species preserved and the production of timber. The number of species preserved, $S$, is a function of both the total protected area and the number of reserves: $S[a,m]$. This is a more direct measure of biodiversity than in the model in the previous section, which used forest structure as a proxy measure. The amount of timber produced is simply proportional to the area of timber production forest, $A - a$. The economic value of reserves is a function of the number of species preserved, $V[S[a,m]]$. Timber production is valued at $p_T$ per hectare. The static welfare function is thus given by $W[a,m] = V[S[a,m]] + p_T(A-a)$.[11] $S[a,m]$ is assumed to be concave,[12] and so $W[a,m]$ is concave too.

Spatial distributions of species' populations and minimum viable populations create nonlinearities that affect the optimal choice of $a$ and $m$. Let us consider these two nonlinearities in turn. Recent ecological studies have found that the majority of tropical tree species have populations that are spatially aggregated, or clumped, instead of randomly distributed throughout the forest.[13] Figure 10.4 illustrates this difference. Clumping is apparently more pronounced in tropical forests than in temperate and boreal forests (Condit et al., 2002). It arises due to two main reasons, habitat heterogeneity and dispersal limitations. For the latter reason, it can occur even in forests with relatively uniform physical characteristics (topography, soils, etc.).

When they are clumped, individual trees of a given species (conspecific individuals) are not equally likely to occur everywhere in the forest. The probability of occurrence is a nonlinear function of the occurrence of other trees of the same species instead of being a fixed number that does not change across the forest. The result is that the expected number of different species occurring within a reserve of a given size is smaller compared to the case of random placement, while the variance in the number of species across reserves is larger. For a given total protected area, these statistical properties favor a reserve system with a larger number of reserves over a system with a smaller number of reserves. That is, they favor more integrated management. These effects of clumping enter the model through the first-order condition for $m$, $S_m = 0$.

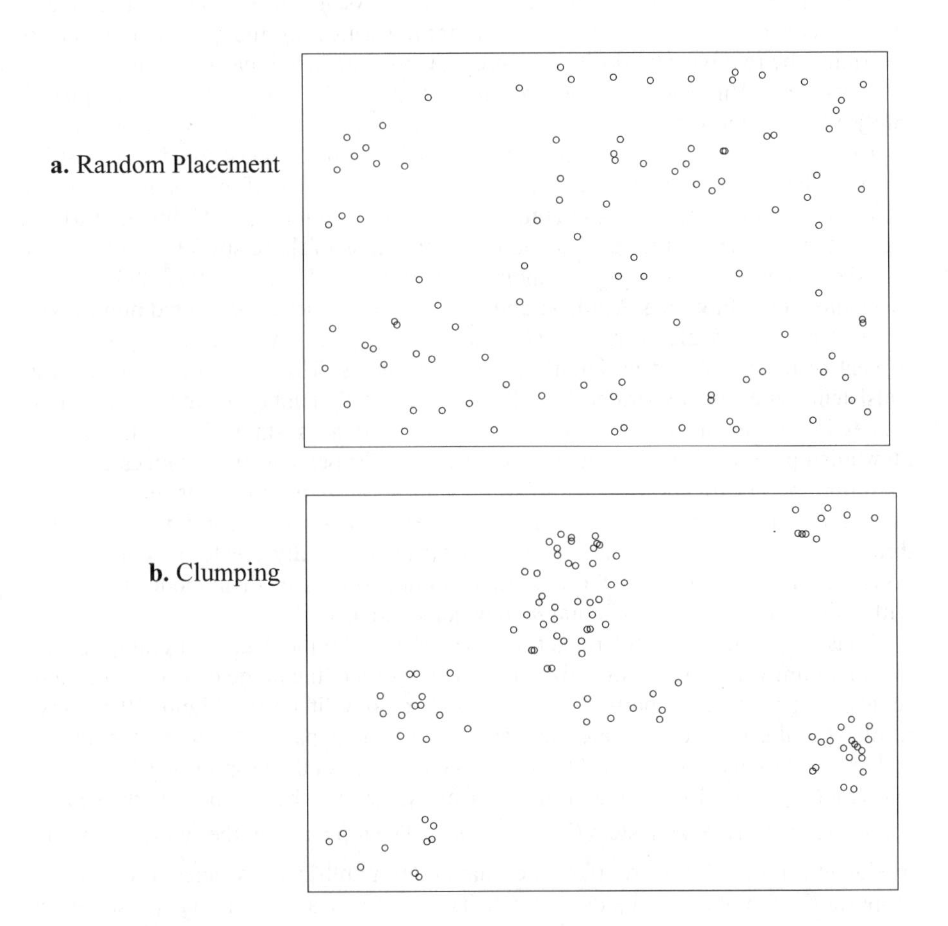


***Figure 10.4.*** *Random Placement vs. Clumping of Trees of a given Species*

The tendency of clumping to favor a larger number of reserves is weaker when the total protected area is smaller, however, because then each reserve might be too small to contain minimum viable populations of very many species. The minimum viable population is the threshold number of species below which a species will not survive. It makes the relationship between survival of a species and population size nonlinear. For a given number of reserves, it favors a reserve system with a larger area over a system with a smaller area. That is, it favors more segregated management—the opposite of clumping. It has this effect through the first-order condition for $a$, $S_a = p_T/V'$. Note that this expression includes the marginal values of biodiversity and timber, $V'$ and $p_T$. The optimal total protected area is larger when biodiversity is more valuable relative to timber. As in the two previous models, relative economic values interact with nonlinearities in the production set to determine the relative superiority of segregated vs. integrated management.

Potts and Vincent construct a simulation model to study the empirical implications of these factors. They analyze a forest of 10,000 hectares, which is the approximate minimum size of a sustained-yield timber concession in Peninsular Malaysia harvested on a 30-year cycle (i.e., $n = 30$). A cycle of this length is typical for tropical rainforests. They use Hubbell's unified neutral model (Hubbell, 2001) to predict the number of tree species and the populations of those species in a forest of this size. They use the negative binomial distribution (He & Gaston, 2000, He & Legendre, 2002, Plotkin & Muller-Landau, 2002) to predict the expected numbers of species and individuals of those species in reserve systems with $a$ ranging up to 25 percent of the forest and $m$ ranging up to 30 reserves. This range of values of $a$ is consistent with the experience of Perak Integrated Timber Complex, the first concession in Peninsular Malaysia to be certified as sustainable by the Forest Stewardship Council, which agreed to reserve 10-20 percent of its concession as a condition of certification.[14] Simulating values of $m$ up to 30 is necessary to determine whether integrated management, $m = n = 30$, is optimal. Although thresholds like minimum viable populations can potentially create nonconvexities, Potts and Vincent find that the empirical production set in their model is convex within the range of values of $a$ and $m$ they considered.

Potts and Vincent consider a set of cases defined by the degree of clumping and the minimum viable population size. Their cases reflect the range of values for these ecological parameters reported in the tropical ecology literature. Figure 10.5 shows results for the case of a scale-independent clumping parameter and a minimum viable population of 100 trees. The horizontal axis shows the proportion of the forest that is logged (i.e., $1 - a/A$), and the vertical axis shows the number of tree species contained in the reserve system (i.e., $S[a,m]$). The top curve in the figure shows the production possibilities frontier: the maximum number of species for a given proportion of timber production forest. The numbers along this curve show the numbers of reserves that yield these maximum values. As expected, the species-maximizing number of reserves increases as the total protected area increases (i.e., as the percentage of timber production forest decreases): the minimum viable population is then less likely to be violated. Yet, the largest number of reserves is only 10, which is far below the perfectly integrated-management solution, 30. The bottom curve shows the number of species protected by 30 reserves.[15] Although the number of species preserved for this reserve number becomes greater as total protected area rises, it remains far below the numbers on the frontier.

In sum, the third model includes two ecological nonlinearities, the clumping of conspecific trees and minimum viable populations. The former favors spatial dispersion of biodiversity reserves, which makes management more uniform (integrated) across the forest, while the latter favors a spatially more concentrated reserve system, which makes management more specialized (segregated). The net result for the case shown in Figure 10.5 is that more segregated management is superior. This is especially true when less of the forest is protected, which occurs when the relative value of biodiversity is lower. Other cases that Potts and Vincent examine indicate, not surprisingly, that segregated management is even more superior when species' population distributions are more random and when minimum viable populations are larger.

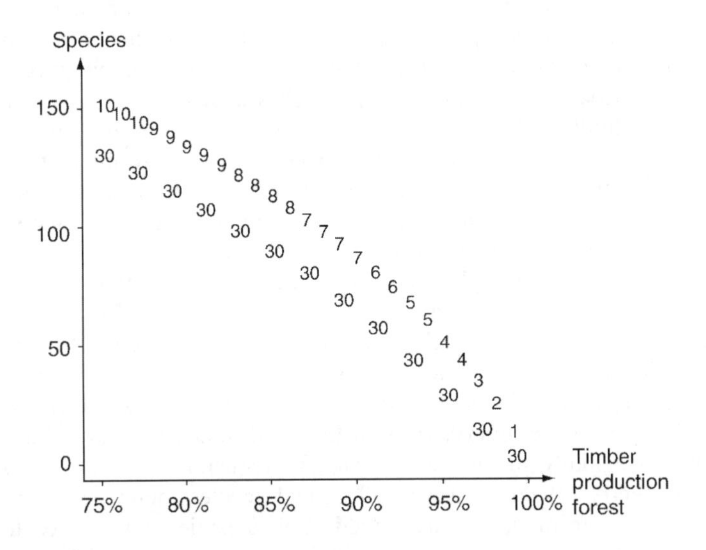


***Figure 10.5.*** *Production Set for a Malaysian Rainforest when the Clumping Parameter is Fixed and the Minimum Viable Population is 100 Trees*

## 5. CONCLUDING REMARKS

We have reviewed three models that illustrate the effects of economic, institutional, and ecological nonlinearities on the relative superiority of integrated and segregated approaches to forest management. We have demonstrated that nonlinearities at smaller spatial scales—for example, nonconvexities in stand-level production sets and clumping of the populations of individual trees—can have important management implications at larger scales. The general tendency appears to be for nonlinearities to favour segregated forest management—emphasizing timber production in some areas and biodiversity conservation in others, even if the stands are identical—although the strength of this tendency depends on forest characteristics and the relative values of biodiversity and timber. The economic and institutional nonlinearities in the empirical example shown in Figure 10.3 are relatively small, and hence so are the potential gains from segregated management in that example. A higher relative value of biodiversity strengthens the tendency toward segregation when nonlinearities result from fixed logging costs or administrative constraints on logging regulations, as in Figure 10.2, but weakens it when nonlinearities result from clumped population distributions, as in Figure 10.5.

A tendency toward segregated management does not necessarily mean complete specialization. In the segregated management equilibrium represented by points B and C in Figure 10.1, there is some output of both timber and biodiversity at both points; it is the proportions of the outputs that differ between the points. Although the segregated management equilibrium in Figure 10.2 includes a proportion $\alpha$ of the stands being managed at the no-logging point A, it also includes $1-\alpha$ of them being managed at point C, where there is some output of both timber and biodiversity. Finally, although in Figure 10.3 we have assumed that biodiversity conservation is negligible in harvested forests, even under this assumption the figure still shows that there is a gradation of approaches depending on how much of the forest is put off limits to logging, with management being more segregated (just a single reserve) when little of the forest is protected and more integrated (multiple reserves, albeit not one in each annual coupe) when more of it is protected. The flexibility that SFM certification systems allow in the choice of segregated and integrated management approaches is therefore appropriate.

Biodiversity is defined explicitly only in the second and third models, and in both cases the definitions are based on numbers of tree species. Despite this focus on just one component of floral diversity, the results from these models have bearing on the conservation of other taxonomic groups found in forests, including animals. Many animals, especially invertebrates, are associated with particular tree species through pollination or herbivory. If as in the third model trees are clumped, then such animals are likely to be, too. In fact, a frequently cited study by Taylor, Woiwod, and Perry (1978) of nearly 90 animal species found that only one had populations that were randomly distributed at all densities. Of course, differences between plants and animals in terms of mobility and minimum viable populations imply that a reserve system that is designed to be optimal for floral conservation is unlikely to be optimal as well for faunal conservation, but in both cases the nonlinearities that underlie Figure 10.5 should have similar qualitative impacts on the relative advantages of more segregated and more integrated approaches.

We close by drawing attention to two nonlinearities that are especially relevant to SFM in tropical forests but have opposite impacts on the relative advantages of the two management approaches. Both were mentioned earlier. Administrative constraints necessitate the use of simpler forest management regulations, which as we have seen tend to favor segregated management. On the other hand, the clumping of tree populations tends to favor integrated management. Given that the topic of this volume is the economics of SFM, it is perhaps useful to close with these reminders that SFM must take into account institutional and ecological factors in addition to purely economic ones.

## NOTES

[1] Other terms that are often used to distinguish between segregated and integrated forest management include specialized vs. uniform management and dominant-use vs. multiple-use management.
[2] We define a stand as a spatially distinct management unit containing a relatively homogeneous forest type.
[3] Other applications of production possibilities frontiers to the analysis of the joint production of biodiversity and timber include Rohweder, McKetta, and Riggs (2000) and Lichtenstein and Montgomery (2003).
[4] For example, the budget that is available to spend on silvicultural operations.
[5] For a discussion of diseconomies of scope between environmental quality and industrial output in the context of industrial pollution, see Baumol and Oates (1988, ch. 8).
[6] They also considered carbon sequestration.
[7] It might seem odd to discount a physical quantity like biodiversity, but in fact economists discount physical quantities all the time. For example, in calculating the net present value of a future timber harvest, economists implicitly discount the quantity of timber harvested, since the value of the harvest, which is what is directly discounted, is the product of the quantity of timber harvested and the price of timber. Biodiversity in Figure 2 and other figures in this chapter must be discounted so that the relative values of biodiversity and timber, which appear in the indifference curves and price lines, are defined consistently.
[8] The index was a "proximity to climax index" that took the structre of a virgin (unlogged) stand as the reference point. It was calculated as 1 minus the root mean squared error between the actual structure of the stand and the structure before logging (i.e., the structure of the virgin stand). The index has a value of 1 for a given period if the forest has the same structure and composition as the virgin forest and 0 if it is bare land.
[9] The biodiversity index is calculated at 5-year intervals. With a 60-year time horizon, the undiscounted value would thus be 13.
[10] See Boscolo and Vincent (2003) for a description of how this tradeoff value was calculated. Essentially, the present value of benefits associated with biodiversity in developing country forests, reported in Lampietti and Dixon (1995), was divided by Boscolo and Vincent's estimate of the present value of the physical biodiversity index for a virgin stand in Malaysia.
[11] With $n$ fixed, the first-order conditions for the quasi-dynamic model discussed in the previous paragraph are equivalent to those for this static model as long as $p_T$ is scaled to be consistent with the area harvested in a given period.
[12] Potts and Vincent demonstrate that this function is concave in their simulation model.
[13] See, for example, He, Legendre, and LaFrankie (1997).
[14] Moreover, capping $a$ at 0.25 in the simulation model is desirable from a statistical standpoint because the negative binomial model provides inaccurate predictions when total protected area accounts for a large portion of the forest.
[15] The integrated-management solution in this model is not an equilibrium. It satisfies the first-order condition for $a$ but not the first-order for $m$, as evidenced by the fact that the curve for $m = 30$ lies within the production possibilities frontier.

## REFERENCES


Baumol, W.D., & Oates, W.E. (1988). *The theory of environmental policy* (2nd edition). Cambridge, U.K.: Cambridge University Press.

Boscolo, M., & Vincent, J.R. (2003). Nonconvexities in the production of timber, biodiversity, and carbon sequestration. *Journal of Environmental Economics and Management,* 46, 251-268.

Bowes, M.D., & Krutilla, J.V. (1989). *The economics of public forestlands*. Washington, D.C.: Resources for the Future.

Condit, R., Pitman, N., Leigh, E.G., Chave, J., Terborgh, J., Foster, R.B., Nunez, P., Aguilar, S., Valencia, R., Villa, G., Muller-Landau, H.C., Losos, E., & Hubbell, S.P. (2002). Beta diversity in tropical forest trees. *Science,* 295, 666-669.

Fisher, A.C., Krutilla, J.V., & Cicchetti, C.J. (1972). The economics of environmental preservation: A theoretical and empirical analysis. *American Economic Review*, 62(4), 605-619.

Forest Stewardship Council. (2002). *FSC principles & criteria of forest stewardship*. Retrieved August 1, 2004, from www.fsc.org/fsc/how_fsc_works/policy_standards/princ_criteria.

He, F.L., & Gaston, K.J. (2000). Estimating species abundance from occurrence. *American Naturalist,* 156, 553-559.

He, F.L., & Legendre, P. (2002). Species diversity patterns derived from species-area models. *Ecology,* 83, 1185-1198.

He, F.L., Legendre, P., & LaFrankie, J.V. (1997). Distribution patterns of tree species in a Malaysian tropical rain forest. *Journal of Vegetation Science,* 8, 105-114.

Helfand, G.E., & Whitney, M.D. (1994). Efficient multiple-use forestry may require land-use specialization: Comment. *Land Economics,* 70, 391-395.

Hubbell, S.P. (2001). *The unified neutral theory of biodiversity and biogeography*. Princeton, N.J.: Princeton University Press.

International Tropical Timber Organization. (1998). *Criteria and indicators for sustainable management of natural tropical forests*. ITTO Policy Development Series No.7. Yokohama, Japan.

Krutilla, J.V. (1967). Conservation reconsidered. *American Economic Review,* 57(4), 777-786.

Lampietti, J.A., & Dixon, J.A. (1995) To see the forest for the trees: a guide to non-timber forest benefits. *Environment Department Working Paper No. 013*. Washington, D.C.: The World Bank.

Lichtenstein, M.E., & Montgomery, C.A. (2003). Biodiversity and timber in the Coast Range of Oregon: Inside the production possibility frontier. *Land Economics,* 79, 56-73.

Plotkin, J.B., & Muller-Landau, H.C. (2002). Sampling the species composition of a landscape. *Ecology,* 83, 3344-3356.

Potts, M.D., & Vincent, J.R. (2004) (in review). Spatial distribution of species populations, relative economic values, and the optimal size and number of reserves. Submitted to *Journal of Environmental Economics and Management.*

Rohweder, M.R., McKetta, C.W., & Riggs, R.A. (2000). Economic and biological compatibility of timber and wildlife production: An illustrative use of production possibilities frontier. *Wildlife Society Bulletin,* 28, 435-447.

Swallow, S.K., Talukdar, P., & Wear, D.N. (1997). Spatial and temporal specialization in forest ecosystem management under sole ownership. *American Journal of Agricultural Economics,* 79, 311-326.

Swallow, S.K., & Wear, D.N. (1993). Spatial interactions in multiple-use forestry and substitution and wealth effects for the single stand. *Journal of Environmental Economics and Management,* 25, 103-120.

Taylor, L.R., Woiwod, I.P., & Perry, J.N. (1978). The density-dependence of spatial behavior and the rarity of randomness. *Journal of Animal Ecology,* 47(2), 383-406.

Vincent, J.R., & Binkley, C.S. (1993). Efficient multiple-use forestry may require land-use specialization. *Land Economics,* 69, 370-376.

# CHAPTER 11

# JOINT FOREST MANAGEMENT: EXPERIENCE AND MODELING

MILINDO CHAKRABARTI

*Department of Economics, St. Joseph's College*
*Darjeeling, West Bengal, India 734 104*
*Email: milindos@hotmail.com*

SAMAR K. DATTA

*Centre for Management in Agriculture*
*Indian Institute of Management, Ahmedabad, India*
*Email: sdutta@iimahd.ernet.in*

E. LANCE HOWE

*Institute of Social and Economic Research, University of Alaska*
*Anchorage, Alaska 99508, USA.*
*Email: elhowe@ uaa.alaska.edu*

JEFFREY B. NUGENT

*Department of Economics, University of Southern California,*
*Los Angeles, CA 90089-0253, USA.*
*Email: nugent@usc.edu*

**Abstract.** The experience with Joint Forest Management (JFM) in different countries has varied considerably, succeeding in limiting deterioration of the forest in some cases but not in others. Inequality within the forest community has also had a tendency to increase. The purposes of this chapter are (1) to review relevant literature on JFM, (2) to develop a multi-purpose model that could be used to identify conditions that can influence the likelihood of success of JFM in improving the welfare of those living and working in forest communities as well as making forest use more sustainable, and (3) to highlight the role of forest externalities and institutional conditions in analyzing the effects of JFM, and (4) to suggest applications and extensions that could provide valid policy implications tailored to specific



*Kant and Berry (Eds.), Economics, Sustainability, and Natural Resources: Economics of Sustainable Forest Management,* 223-252.

circumstances. Although highly simplified, the model is designed so as to be flexible enough to deal with a wide variety of settings in rural areas of developing countries and yet at the same time specific enough to provide some policy conclusions. Even the present highly simplified model demonstrates general conclusions about the efficacy of JFM cannot be drawn without very specific empirical knowledge concerning the behavioral and technological parameters in the model.

## 1. INTRODUCTION

Data from around the world is revealing that substantial portions of the world's forests are quite rapidly disappearing and deteriorating. Frequently, such resources are owned by state or national governments, but can be considered *de facto* common property resources. Many of these common property regimes, however, have deteriorated so as to become rather indistinguishable from open access. Naturally, this has resulted in declining welfare of those in the forest community dependent on these resources.[1] As knowledge of these circumstances has spread, often at the behest of NGOs, governments are increasingly including local user groups in the management process of these forest resources. This is what is called co-management or Joint Forest Management (JFM).

While the details of JFM vary considerably from place to place, a common characteristic is for local communities to receive somewhat greater property rights and influence over local natural resources than under the preceding regimes. Some evaluators have gone so far as to see JFM as a creative and potentially optimal arrangement combining the separate strengths inherent in property regimes of private ownership, direct state control, and communal property so as to help sustain this important natural resource base (Baland & Platteau, 1996).

Current programs range from large game wildlife management in Africa (Bulte & Horan, 2003), fisheries in Japan (Baland & Platteau, 1996; Kenneth, 1989), community woodlots in Ethiopia (Gebremedhin, Pender, & Tesfay, 2003), and forests in Mexico (Klooster, 2000; Munoz-Pina, de Janvry, & Sadoulet, 2002), India (Kumar, 2002; Richards, 2000), China (Hyde, Belcher & Xu, 2003) and Nepal (Agrawal & Ostrom, 2001; Edmonds, 2002), to the management of all village resources in Burkina Faso (Baland & Platteau, 1996). The U.S. government has also experimented with co-management among Arctic Alaskan communities with respect to select marine mammals and large game. In Canada also, there are some 15 different examples of co-management in which the role of the local user group varies widely (Rusnack, 1997).[2]

As a result, JFM is viewed by some as a mechanism that can be counted on to promote the quality of life for the rural poor and at the same time to reduce forest degradation. Nevertheless, the jury is still out on its overall success since the experience seems to have varied from place to place, allegedly depending on institutional and other characteristics (Baland & Platteau, 2001; Bardhan, 2002; Jaramillo & Kelly, 2000; Kumar, 2002; Platteau, 2001). While JFM may lead to efficiency gains relative to pure State management in certain contexts, it may not do so in all. At the same time, moreover, many studies have been less sanguine about its role in reducing poverty and inequality, indeed suggesting that elite groups within the forest communities may capture the bulk of the benefits, quite possibly immiserizing the poor (Klooster, 2000; Kumar, 2002).

While the literature has begun to provide interesting stylized facts based on individual case studies or surveys, the modeling of these circumstances and the ability to evaluate the potential benefits of different features of JFM is still in a relatively primitive stage. The objectives of this chapter are (1) to review the literature on JFM relevant to modeling and assessment, (2) to provide a simple general equilibrium model that captures the stylized facts derived from the existing literature, (3) to highlight the role of forest externalities and institutional conditions in analyzing the effects of JFM, and (4) to suggest applications and extensions of the model that could yield policy implications tailored to the very specific circumstances of individual JFM cases.

Although the model necessarily makes many simplifications, it is designed to capture four important environmental and institutional features highlighted in the literature on JFM, namely, (1) the heterogeneous character of, and inequality within, forest user groups, (2) the influence of such heterogeneity on the degree of dependence on forest resources, the sustainability of forest production and the degree of inequality between the user groups, (3) the effect of JFM on each of these relationships and considerations, and (4) the importance of the quality of the forest and the externalities thereof, and the possible effect of JFM on the effectiveness of regulatory control and property rights over forest land. Section 2 reviews the relevant literature. Section 3 outlines the model. Section 4 derives insights from the model, suggests applications and further extensions.

## 2. FINDINGS FROM THE LITERATURE

### *2.1 Specific Examples of JFM*

In 1989 the Indian Central Government mandated that the individual state governments formally adopt JFM as the primary mechanism through which the State would manage state-owned forest resources. The policy was reportedly motivated by a desire to both reduce environmental degradation (which, according to Kumar (2002), the Central Government attributed largely to local communities using the forests as *de facto* open access property) and to reduce rural poverty. The states, however, were left with a great deal of flexibility with respect to the particular approach they would adopt. In the 26 of 28 states that have formally adopted JFM, the incentives offered by various State Forest Departments to local village forest communities have ranged from wage payments for protective labor services, to in-kind and revenue shares of the non-timber forest products collected, to revenue shares of timber sales, and to combinations of each (Kumar, 2002).

A similar form of JFM was recently adopted in Nepal, though with somewhat less direct government involvement. Due to increasing rates of forest clearance and growing environmental degradation, the Nepalese government began a process of transferring ownership and control of all forests to local communities or "Forest User Groups." The central government provides the user groups with both the framework and resources necessary to reduce resource extraction (Edmonds, 2002).

User-groups in Nepal receive a greater share of the return from successful management in land held as common village property than those in India.

JFM has also occurred in Mexico. As a result of land reform that followed the 1910 peasant-led revolution in Mexico, roughly 80% of Mexico's forests are currently held as *de jure* common property (Klooster, 2000). Yet, only after the legislative changes of the 1980's, did local communities begin to have some autonomy in collectively managing timber resources. Prior to that, the communities were forced to contract with approved logging companies that autonomously made the important production and other decisions. More recently, however, communities were allowed to form cooperatives to harvest and manage logging operations under specified criteria, a context akin to the Nepalese case given that communities both own and manage resources with considerable State oversight. As a result of these changes, several successful examples of JFM have emerged in Mexico.[3]

According to Liu and Edmunds (2003), since 1978 China, too, has undergone a variety of JFM-like reforms. Indeed, the form and pace of these reforms have varied widely over space as well as time. In general, they have involved the devolution of management and control from the central government to the regional and local level and with different degrees of property rights conferred to individuals and groups in forest areas. A special problem that has arisen in the Chinese case has been the credibility of announcements of reform policies inasmuch as the government has from time to time seen fit to reverse some of these partial property rights devolutions on the basis of insufficient new investment by the forest populations in afforestation.

### *2.2 Outcomes and Institutional Features of Successful JFM*

There is at least some empirical evidence supporting the hypothesis that forest resources are managed more efficiently and in a more sustainable way under JFM than under central management. In an excellent empirical study of such programs, Edmonds (2002) tests the robustness of relatively lower mean levels of resource extraction in Nepalese forests managed by "Forest User Groups" relative to areas managed purely by the central government. Using several different estimation techniques, he finds that the difference is indeed robust, supporting the view that Nepalese JFM is more efficient in managing and preserving forest resources than the central government. Consistent with Edmond's (2002) findings, Kumar (2002) finds similar evidence in India. Yet, Kumar argues that the distribution of benefits under JFM has at the same time been highly unequal (a rural elite capturing most of the economic benefits) and that much of the gain in lower resource extraction has come at the expense of the poorest.

In Mexico, Klooster (2000) reports that in seven of the eight cases, community managers have been successful in increasing forest area but also that, in contrast to the Indian case, the distribution of benefits among community members has been relatively equal. Notably, the "successful" communities in Klooster (2000) were primarily the indigenous, ethnically homogenous communities.

Consistent with these findings, Kant and Berry (1998), Kant (2000), and Kumar (2002) argue that with group homogeneity JFM may result in a more efficient outcome both in terms of the sustainability of natural resources and income distribution. The explanation offered is that shared institutions at the community level reduce the degree of moral hazard and adverse selection therein serving as an important element in the stability of JFM. Homogenous groups are more likely to share common goals and values with respect to subsistence harvest amounts, enforcement mechanisms, and the distribution of benefits. Heterogeneity, however, can undermine these mechanisms and shared norms (Baland & Platteau, 1997). But at the same time, as shown by Varughese and Ostrom (2001) with respect to the Nepalese case, heterogeneities, while making collective action more difficult, may not necessarily eliminate effective local collective action when user groups can create rules which account for such heterogeneities (see also Hackett, Schlager, & Walker (1994) for experimental evidence).

Clearly, heterogeneity can take different forms. The two dimensions most frequently identified as affecting JFM outcomes have been intra-community differences in social class/power and income. Kumar's (2002) study points to caste inequality and an unequal distribution of benefits. In India, the group with dominant power essentially ran the village forest committee so that the preferences of that group were reflected in the programs adopted, helping that group to extract a majority of the benefits. Similarly, Platteau (2001) uses an analytic model and descriptive observations to characterize the oft-observed problem of "elite capture," that is, the ability of the dominant group to capture the benefits from a common property arrangement. He argues that this is a significant problem that must be accounted for in setting up appropriate incentive and enforcement mechanisms. Groups may also be homogenous with respect to goals but heterogeneous in terms of income. Cardenas (2003) presents evidence (based on field experiments conducted in rural Colombia) of reduced cooperation when the heterogeneity is based on the unequal distribution of wealth. The impact of wealth inequality is also demonstrated in several chapters of Baland, Bardhan, and Bowles (2001).

A second element identified as important to the success of JFM is the user group's degree of dependence on the resource base (Cardenas, 2003; Kant, 2000; Kant & Berry, 1998). Groups highly dependent on non-timber forest products, for example, are likely to have strong incentives to cooperate with the government or some other entity in managing the forest to achieve and maintain an "optimal" harvest level.

Consistent with these aspects of a successful regime, the particular incentive mechanisms selected by the State can also be of critical importance to the success of JFM. That is, given that a particular forest area is held by the State, the central government must decide on the degree of new local ownership or management, and a particular means for rewarding time spent by community members in cooperation and the enforcement of JFM rules of protection. In the case of Nepal, select communities obtained ownership to village land and the government heavily subsidized enforcement costs (Bromley & Chapagain, 1984; Chakraborty 2001), while communities in village India do not, as a rule, receive common property rights over local forests. Instead, India's state governments have the flexibility to develop incentive schemes that would promote forest agreements between the state and local communities. Such arrangements include providing members of the forest community with a share of profits from harvested forest, direct wage payments for enforcement effort, and /or a share of forest biomass.

As argued by Richards (2000) and modeled by Kant (2000) and Kant and Berry (2001, 1998), optimal resource allocation strategies may differ significantly on a continuum from pure private ownership through State control to open access. Even within a given state or province, community incentives for cooperation may vary significantly according to the type of land tenure, institutions, income inequality and natural resource dependence. In the language of Kant and Berry (1998), a user group in region A may be more heterogeneous (in terms of income or class) than a corresponding group in B but because of greater reliance on the natural resource may have greater incentive to use the resource in a self-sustaining way. Since there are tradeoffs in these respects, the relative success of one group vis-à-vis another may hinge on the details of the incentive system chosen by the JFM. It has also been argued that group homogeneity and greater dependence on the local resource base contribute to greater use of cooperative JFM whereas heterogeneity and independence encourage private property arrangements (Kant & Berry, 1998).

It would certainly appear that the impetus for initiating JFM and early experience with it may have considerable influence on its long-term effectiveness. Given that past state ownership and management has often resulted in very considerable and non-sustainable encroachment and misuse of the forests, to be successful it is obviously important for any new regime like JFM to make clear that past violations of sustainable use will no longer be tolerated. Any strengthening of community norms sanctioning violations and of cooperation in the enforcement of these norms would seem rather certain to raise the probability of success.

Yet, in fact, on these and many other potentially important aspects, the literature is either silent or unclear about the likely effects of other conditions on JFM outcomes. For example, as Ostrom (1999) has noted, virtually no attention has been given as to what to do when the local institutional conditions are quite inimical to rule compliance. One reason for this is the absence of historical/political perspectives in these studies. A partial exception is Agrawal and Ostrom (2001) which noted that the village council-managed forest areas in the Kumaon region of India that developed endogenously in the 1920s and 1930s from local resistance to arbitrary management of these forests by the colonial government have been much more successful in establishing a transparent system of rules and decision making for forest use and sustainability than in the more state-initiated JFM experiments in India or Nepal.

Another partial exception is the work by experimentalists on rule compliance and cooperation in common resource management settings. Several common property experimental game studies have shown that communication tends to improve cooperation and efficiency relative to what can be accomplished by rule sanctioning alone (Ostrom, Walker, & Gardner, 1992; Ostrom, 1999; Cardenas, 2003). Moreover, when through communication local resource users can design and choose their own rules for efficient use and enforcement, Cardenas (2003) and Hackett, Schlager, and Walker (1994) have shown that they may be able to overcome the obstacles to cooperation created by heterogeneity within the group. When communication is not possible as in large groups and forest areas, voting institutions may be an alternative means for accomplishing efficient decision-making and management (Walker, Gardner, & Ostrom 2000) and greater interaction over time can similarly improve efficiency (Hoffman, McCabe, Shachat, & Smith, 1994; Palfrey & Rosenthal, 1994). Also, evenly enforced sanctioning institutions which reward individual appropriators for monitoring have been found to result in more efficient appropriation levels (Casari & Plott, 2003). These, in turn, may be more efficient than state-of-the-art schemes designed by international experts (Ostmann, 1998; Cardenas, Stranlund, & Willis, 2000).

As discussed, we make an effort to explicitly model, albeit in an incomplete manner, the institutional features highlighted above. Heterogeneity is illustrated by differences between the two different forest groups in terms of both income and access on the one hand and the degree of dependence on the resource base on the other. Different levels of enforcement are also assumed in interactions between the Forest Department and the Forest Community.[4] Forest communities may also differ in the extent to which JFM arrangements allow members to extract more non-timber products from the forest and shares in the present value of increased forest biomass.

### *2.3. Literature on Cooperation not Specific to JFM*

Aside from the literature focusing on JFM experiments, there is a very extensive literature of very considerable relevance to JFM issues and modeling both on the relations between deforestation and land tenure and on inter-group cooperation in maintaining common property in the face of the tragedy-of-the-commons threat. If the returns are higher on other uses of the land than for timber and non-timber products, community members will have little incentive to prevent deforestation and conversion of the land to other uses. This is more likely to be the case the more depleted the forest has become and the lower is its ability to generate non-timber products that can benefit the members of the community. Many studies of cooperation have also confirmed the relevance of group characteristics in distinguishing between successful and unsuccessful attempts at defending the integrity of a jointly owned and managed resource[5]. Yet, for the very heterogeneous Terai region of Nepal, Chakraborty (2001) has shown that, despite very unfavorable group characteristics, cooperation in maintaining common property rights without large-scale deforestation can still be possible. This occurred because, thanks to both a rather stable elite-group-based traditional system of authority and a sufficient reserve on remaining government lands for satisfying subsistence needs, the traditional elite was able to exercise sufficient leadership to establish rules and, with the help of the forest department, credibly commit to their enforcement.

## 3. MODEL AND SIMULATION FRAMEWORK

Next we proceed to explain and outline the model whose primary purpose is to identify the key behavioral linkages between two user subgroups (elite and non-elite) within the local Forest Community, and between them and the Forest Department, a Residual Non-forest Sector and the government, with and without

JFM. So as not to reach conclusions about differences between the pre-JFM and JFM cases merely by assumption, we have tried to keep the two cases as similar as possible. The most important difference is that in the pre-JFM case we assume that there is little or no dialogue between the forest department and members of the forest community over the use of protective labor and other decisions whereas under JFM there would be. We also assume that the forest community would share in some of the revenues of the forest department from timber sales in the JFM. While the forest department might well have rather different objectives under the two different arrangements, for simplicity we assume that the forest department is trying to maximize forest biomass subject to its budget constraint.

In the present version of the model, certain simplifying assumptions are made such as that some agents do not play a very substantial role. For example, government is assumed to be rather passive in the model with its resource allocations largely exogenous. Similarly, we have not introduced land as a factor of production in either the forest or non-forest sectors. Finally, the non-forest sector has been assumed to be rural-, rather than urban-, based, hence not requiring rural-urban migration and transport costs to access employment there. Yet, as explained below, these are all assumptions that can be subsequently relaxed so as to come to grips with issues beyond those considered in this chapter.

### *3.1 The Five Sectors*

Although there are five sectors, including two different user groups and the government, Figure 11.1 illustrates only the linkages between three of these, namely, the Forest Department (FD), the Forest Community (FC) as a whole (instead of separately as two different groups), and the Residual Sector (R). As detailed in the analytical model that follows, the figure presents the basic flows of goods, services, incomes, and expenditures among these key groups.

Briefly, the FC provides protective labor, $N_p^s$, and labor to process the forest good, $F^s/c$, to the FD, and labor for production of the market good, $N_r^s$, to the R Sector in exchange for payments $w_p, w_h$ and $w_r$, respectively. Labor income, in addition to a share of FD revenues flow out of the FC back to the R Sector in the form of payments for good $X$, $P_x X_c$, and to the FD in terms of penalties for excess gathering, and payments for the Forest good, $P_f F_c^d$. Income also flows into (out of) the FD (R Sector) through sales (purchases) of timber, $P_f F$. The government also provides transfers to the three sectors through taxes on R Sector profits and FD timber revenues. We assume that these government transfers to both the FC and R Sectors are non-cash, in-kind benefits that contribute directly to utility, whereas transfers from the government to the FD are in cash and so enter the FD's budget constraint. Both gathering by the FC, $N_{df}$, and sales by the FD, $F^s$, reduce the size of the forest and its positive environmental externalities enjoyed by all sectors. In contrast, biomass production under FD management increases the size and/or quality of the forest and hence potential satisfaction of demand for the forest good.

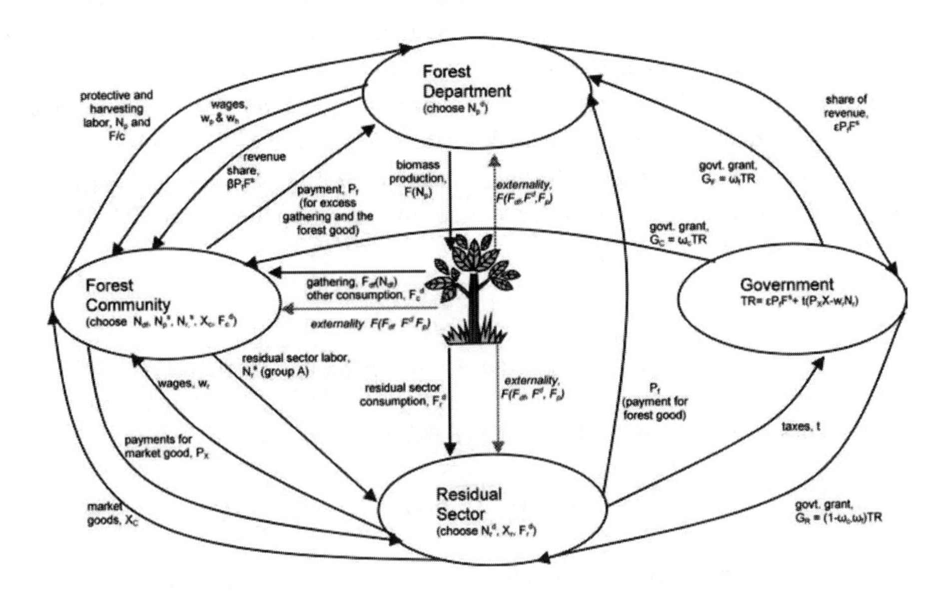


*Figure 11.1* *A Circular Flow Diagram of Joint Forest Management*

### 3.2.1 *The Forest Community (FC)*

Utility for both FC groups is a function of goods collected from the forest, $F_{df}$ (which can be thought of as fuel-wood, non-timber forest products, (NTFP's), or timber), consumption of the processed forest good, $F_c^d$, government grants, $G_c$ (which are in-kind),[6] consumption of the market good produced in the residual sector, $X_c$, and an the positive externality effect, $F_E$. $F_E$, is a function of $F_{df}$, $F^s$, which includes the quantities of the Forest Good purchased by the FC and R sectors, and $F_p$, the increase in the forest due to protective labor supplied by the FC, where,

$$F_E = F_E(F_{df}, F_c^d + F_r^d, F_p),$$

$$F_{E1} < 0, F_{E11} < 0, F_{E2} < 0, F_{E22} < 0, F_{E3} > 0, F_{E32} < 0.$$

The Forest Community's utility can therefore be written as,

$$U_c = U_c(F_{df} + F_c, G_c, X_c, F_E(F_{df}, F_c^d + F_r^d, F_p)) \tag{1}$$

where $X_c, F_c^d, F_p$, and $F_{df}$ are choice variables and $F_r^d$ is demand for the processed forest good in the R sector, and $U_c$ is increasing in all arguments.[7] $F_{df}$ is simply a function of time allocated to removal of timber resources by the FC (which may be legal or illegal), $N_{df}$,

$$F_{df} = F_{df}(N_{df}), \quad F_{df}' > 0, F_{df}'' < 0, \tag{2}$$

Similarly, $F_p$ is a function of time allocated to protective labor, $N_p$,

$$F_p = F_p(N_p), \quad F_p' > 0, F_p'' < 0, \tag{3}$$

#### 3.2.1.a *Forest Community Group A Income Constraints*

In order to measure the effects of group heterogeneity on co-management, we assume two different group types, an "elite" and "non-elite" group. The two groups are distinct in that, as detailed below, the elite group has greater income earning potential than the non-elites.

For the elite group, we assume that consumption spending on market goods, $P_x X_{c,A}$ and the processed forest good, $P_f F_{c,A}^d$, is constrained by wage earnings from the R Sector, $w_r N_r^s$, plus income earned from the FD, specifically, the sum of labor payments for protective services $w_p N_{p,A}^d$, payments for timber harvesting $w_h \alpha F^s / c$

(where alpha, ($\alpha \le 1$) is an exogenously determined share of FD harvest employment going to elites) and a share of revenues from FD sales based in part on their contribution to protective labor services, $(N^s_{p,A}/(N^s_{p,A}+N^s_{p,B}))\beta P_f F^s$. The exogenously given parameter beta (where $\beta < 1$), is the share of FD revenues going to the Forest Community. The FC elite are also forced to pay fraction gamma (where $\gamma \le 1$) for the amount removed that is in excess of $F_L$, i.e., what is stipulated (by cooperative agreement or unilateral limit set by the government), and which can be thought of as a percentage of the biomass produced. The budget constraint is therefore,

$$\begin{aligned} & w_r N^s_r + w_p N^d_{p,A} + w_h \alpha \frac{F^s}{c} + \frac{N^s_{p,A}}{N^s_{p,A}+N^s_{p,B}} \beta P_f (F^d_{c,A} + F^d_{c,B} + F^d_r) \\ & = P_x X_{c,A} + P_f F^d_{c,A} + \gamma \eta(.) H(.) \end{aligned} \tag{4}$$

In the pre-JFM case, the FC does not receive a share of FD sales in this budget constraint.

The $\eta(.)$ term in equation (4) is a proxy for the effectiveness of enforcement of the agreement on maximum biomass removal. We assume that the probability of being caught for taking too much out of the forest increases with the magnitude of that removed above the allowable ceiling. Hence, $\eta(.)$ is a function of $(F_{df,A}+F_{df,B}-F_L)$ where $\frac{d\eta}{dF_{df}} > 0, \frac{d^2\eta}{dF_{df}^2} > 0$. We also assume that the penalty for excessive clearing, $H(F_{df,A}+F_{df,B}-F_L)$ is a function of the amount of this excess where $\frac{dH}{dF_{df}} > 0, \frac{d^2 H}{dF_{df}^2} > 0$. If the rules against excess clearing are perfectly enforced, $\eta(.) = 1$, and the FC would pay for all harvesting beyond the agreed upon amount. On the other hand, if $\eta(.) = 0$, the FC would face no penalty for removal of timber. Similarly, a lower (higher) value of $H$ is associated with a higher (lower) order of forest clearance than that specified by $F_L$. It is likely that enforcement would be greater under JFM than under pure state control because of greater incentives for local collective action under JFM.[8]

### *3.2.1.b Forest Community Group B Income Constraint*

The "non-elite's" problem is identical to the elite's except that the non-elite: i.) do not supply labor to the residual sector and do not receive wages from the R sector; ii.) receive $\frac{N^s_{p,B}}{N^s_{p,A}+N^s_{p,B}} \beta P_f F^s$ as their share of sales of F; iii.) pay the share $(1-\gamma)$

as penalty; and iv.) provide the share $(1-\alpha)$ of harvest labor. As for the elite group, in the pre-JFM case, the non-elite group does not receive a share of FD sales.

### *3.2.1.c Labor Market Constraints of Both Sectors*

Labor time supplied by both FC groups is constrained by total time. Time is allocated to protective service $N^s_{p,A}, N^s_{p,B}$, timber harvesting activities, $\frac{F^s}{c}$, collection of fuel-wood or non-timber forest products, $N_{df}$, and group A supplies labor to the R Sector, $N^s_r$. The labor market constraint is therefore,

$$N_{TOT} = N_{TOT,A} + N_{TOT,B} \tag{5}$$

where,

$$N_{TOT,A} = N^s_{p,A} + \alpha \frac{F^s}{c} + N^s_r + N_{df,A}$$

$$N_{TOT,B} = N^s_{p,B} + (1-\alpha)\frac{F^s}{c} + N_{df,B}$$

### *3.2.2 The Residual Sector (R)*

The owners of capital and managers that constitute the households in the R Sector seek to maximize utility, which is a function of the consumption of the processed forest good, $F^d_r$, consumption of the market good, $X_r$, the government grant, $G_r$, and the "environmental purity externality." Utility in the R Sector can therefore be written as:

$$U_r(F^d_r, G_r, X_r{}^d, F_E(F_{df}, F^d_r + F^d_c, F_p)) \tag{6}$$

where the choice variables are $F^d_r$ and $X_r{}^d$; and, $U_r$ is increasing in all arguments. Spending on the consumption good in the R Sector, $P_x X_r$, and the forest good, $P_f F^d_r$, is constrained by after-tax profits earned in the production of X, $(P_x X - w_r N^d_r)(1-t)$. The budget constraint for the R Sector is therefore:

$$(1-t)(P_x X - w_r N^d_r) = P_f F^d_r + P_x X_r{}^d \tag{7}$$

where the production of X is simply a function of labor,

$$X = X(N_r^d), \quad \frac{dX}{dN_r^d} > 0, \frac{d^2X}{(dN_r^d)^2} < 0 \tag{8}$$

*3.2.3 The Forest Department (FD)*

The FD's costs of protective labor and timber removal services, $w_p N_p^d + w_h \frac{F^s}{c}$, are the same in both the pre-JFM and JFM cases. Similarly, its income includes transfers from the government, $G_f$ and timber sales in both cases. Under JFM, however, total timber sales, $P_f F^s$, are shared with both the government (which gets the share ε) and the FC (which gets β), the share $(1-\varepsilon)(1-\beta)P_f F^s$ remaining with the FD. In the pre-JFM case the FD makes no transfers to the FC, so its share remains $(1-\varepsilon)P_f F^s$. In the JFM case, the FD budget constraint is:

$$G_f + (1-\varepsilon)(1-\beta)P_f F^s + \eta(.)H(.) = w_p N_p^d + w_h \frac{F^s}{c} + C \tag{9}$$

where $N_p^d$ is the choice variable and C is a given fixed cost.

In JFM, we assume that the FD seeks simply to maximize the end-period biomass defined as:

$$F_{TOT} = \overline{F} + F_p(N_p) - F^s - F_{df} \tag{10}$$

where F is the initial stock of forest, $F_p$ is the production of new forest and $F_S$ and $F_{df}$ are the sales of the forest good by the FD and collection for own use of the good by the FC.

*3.2.4 Government Grants*

The Government grant received by each stakeholder is a fraction of its tax receipts from the R Sector, $t(P_x X - wN_r)$, plus its share of timber sales, $\varepsilon P_f F$. Thus, the total tax revenue collected by the Government is

$$t(P_x X - wN_r) + \varepsilon P_f F \tag{11}$$

The FC receives a share, $\omega_c$, the FD $\omega_f$, and the R Sector $(1-\omega_c-\omega_f)$.

### 3.3 Constrained Optimum by Sector

#### 3.3.1.a The Forest Community: Group A

The elite group in the FC chooses $N_{df,A}, N_r^s, N_{p,A}^s, F_{c,A}^d$ to maximize utility as described in equation (1). After forming the Lagrangian, where $\lambda_1$ is the multiplier on the elite's labor time constraint and $\lambda_2$ is the multiplier on its budget constraint, the first order conditions are given in Table 11.1. In the pre-JFM period first order conditions are identical except that $\beta = 0$, implying that there is no income benefit from sales of the forest good and changing equations (19) and (20), as shown in Table 11.1.

As shown in the table, in choosing the quantities of *defacto* labor, protective labor, and demand for the forest good to maximize utility, the FC sets marginal benefits equal to marginal costs. In the case of *defacto* labor, $N_{df,A}$, the direct utility benefit is set equal to the negative externality effect on utility, the shadow price of time and money, and the penalty for over-extraction (see equation (16)). In contrast, marginal benefits for protective labor include a positive externality benefit, direct wage benefits, and a share of revenue from the forest department which are set equal to the shadow price of time (equation (19)). Because consumption of the forest good, $F_{c,A}^d$, is also modeled as reducing biomass, the first order condition is similar to the case of *defacto* labor. Marginal benefits realized in consuming more $F_{c,A}^d$ include a direct utility benefit and a greater share of forest department revenue which are set equal to the indirect utility loss and the price of $F$ (equation (20)).

#### 3.3.1.b The Forest Community: Group B

Similar to the elite, the non-elite (group B) choose $N_{df,B}, N_{p,B}^s, F_{c,B}^d, X_{c,B}$ (but not $N_r^s$) to maximize utility as described in equation (1). In this case, $\lambda_3$ is the multiplier on the time constraint and $\lambda_4$ is the multiplier on the income constraint. First order conditions in both cases are identical to group A's when accounting for differences in multipliers and differences with respect to the inequality parameters. Note that group B must pay the share $(1-\gamma)$ of the penalty incurred due to timber withdrawal not only by itself but also by the elite group A. In addition, group B's share of forest department revenue is determined by $\dfrac{N_{p,B}^s}{N_{p,A}^s + N_{p,B}^s}$ which changes appropriate elements of equations (19) and (20). As with group A, in the pre-JFM case $\beta = 0$, as the FD does not share revenue with the forest community.

### *3.3.2 The Residual Sector*

The R Sector seeks to maximize utility as described in equation (6) subject to a budget constraint. Forming the Lagrangian, where $\lambda_5$ is the multiplier for the budget constraint, first order conditions are given in Table 11.1. None of the conditions for the R Sector differ between the pre- and post-JFM cases.

As shown in Table 11.1, in choosing the optimal level of $F_r^d$ the FC sets the direct utility benefit equal to the price of consumption in addition to the indirect negative utility loss. And in choosing the optimal amount of labour, $N_r^d$, the marginal revenue product of labour is equal to the wage rate.

### *3.3.3 The Forest Department*

The FD chooses $N_p^d$ to maximize biomass (equation (10)) subject to their income constraint (equation (9)). Forming the Lagrangian and taking first order conditions, we present results in Table 11.1, where $\lambda_6$ is the multiplier for the budget constraint. Equation (24) indicates that the forest department maximizes biomass by acquiring protective labor to the point that the marginal product of protective labor is equal to the wage times the multiplier of the income constraint, an outcome that is also identical to that in the pre-JFM case.

### *3.4 Equilibrium Conditions*

The model is closed with the following equilibrium conditions

$$N_r^d = N_r^s \text{ to determine } w_r \tag{12}$$

$$N_p^d = N_{p,A}^s + N_{p,B}^s \text{ to determine } w_p \tag{13}$$

$$X(N_r^{\ d}) = X^d_{\ c,A} + X^d_{\ c,B} + X^d_{\ r} \text{ to determine } P_x \tag{14}$$

$$F^s = F_r^d + F_{c,A}^d + F_{c,B}^d \text{ to determine } P_f \tag{15}$$

### *3.5 Pareto Optimality Conditions*

Pareto optimality conditions are given in Table 11.2. We maintain the notation for the constraints described and in addition we use $\lambda_7$, $\lambda_8$, and $\lambda_9$ to reference the multipliers on the utility functions for forest community groups A and B, and the residual, respectively. For ease of reference a comprehensive list of variables used in the model along with definitions is given in Appendix 11.1.

***Table 11.1.*** *First Order Conditions in the JFM Environment*[a]

**A. Forest Community (Group A)**

$$\frac{dL}{dN_{df,A}}: U^1_{c,A} F_{df,A}{}'(N_{df,A}) + U^4_{c,A} F^1_E F_{df,A}{}'(N_{df,A}) - \lambda_1 - \lambda_2 F_{df}{}'(N_{df,A}) \gamma [\eta'(.) H(.) + H'(.) \eta(.)] = 0 \qquad (16)$$

$$\frac{dL}{dN^s_r}: -\lambda_1 + \lambda_2 w_r = 0 \qquad (17)$$

$$\frac{dL}{dX_{c,A}}: U^3_{c,A} - \lambda_2 P_x = 0 \qquad (18)$$

$$\frac{dL}{dN^s_{p,A}} \ \underline{JFM\ case}: \ U^4_{c,A} F^3_E F_{p,A}{}'(N^s_{p,A}) - \lambda_1 + \lambda_2 (w_p + \beta P_f (F^d_{c,A} + F^d_{c,B} + F^d_r)(\frac{N^s_{p,B}}{(N^s_{p,A} + N^s_{p,B})^2})) = 0 \qquad (19)$$

$$\underline{pre\text{-}JFM\ case}: \ U^4_{c,A} F^3_E F_{p,A}{}'(N^s_{p,A}) - \lambda_1 + \lambda_2 w_p = 0 \qquad (19)'$$

$$\frac{dL}{dF^d_{c,A}} \ \underline{JFM\ case}: \ U^1_{c,A} + U^4_{c,A} F^2_E - \lambda_2 P_f (1 - \beta \frac{N^s_{p,A}}{N^s_{p,A} + N^s_{p,B}}) = 0 \qquad (20)$$

*Table 11.1 (cont.)*

*Table 11.1 (cont.)*

<u>*pre-JFM case*</u> : $U^1_{c,A} + U^4_{c,A} F^2_E - \lambda_2 P_f = 0$ (20)'

**B. Residual Sector**

$$\frac{dL}{dF^d_r} : U^1_r + U^4_r F^2_e - \lambda_5 P_f = 0 \tag{21}$$

$$\frac{dL}{dN^d_r} : \lambda_5 (1-t)(P_x X'(N^d_r) - w_r) = 0 \tag{22}$$

$$\frac{dL}{dX_r} : U^3_r - \lambda_5 P_x = 0 \tag{23}$$

**C. Forest Department**

$$\frac{dL}{dN^d_p} : F_p'(N^d_p) - \lambda_6 w_p = 0 \tag{24}$$

***Table 11.2.*** *Pareto Optimality Conditions in the JFM Environment*

**A. Forest Community (Group A)**

$$\frac{dZ}{dN_{df,A}}: -F_{df,A}{}'(N_{df,A}) - \lambda_1 - F_{df,A}{}'(N_{df,A})(\eta(.)\frac{dH}{dF_{df,A}} + H(.)\frac{d\eta}{dF_{df,A}})(\lambda_2\gamma + \lambda_4(1-\gamma))$$

$$+\lambda_6 F_{df,A}{}'(N_{df,A})(\eta(.)\frac{dH}{dF_{df,A}} + H(.)\frac{d\eta}{dF_{df,A}}) + \lambda_7 F_{df,A}{}'(N_{df,A})U_A^1 + F_E^1 F_{df,A}{}'(N_{df,A})(\lambda_7 U_A^4 + \lambda_8 U_B^4 + \lambda_9 U_r^4) = 0 \quad (25)$$

$$\frac{dZ}{dN_{r,A}^s}: -\lambda_1 + \lambda_2 w_r = 0 \quad (26)$$

$$\frac{dZ}{dX_{c,A}}: \lambda_7 U_A^3 - \lambda_2 P_x = 0 \quad (27)$$

$$\frac{dZ}{dN_{p,A}^s}: -\lambda_1 + \lambda_2 w_p - \beta P_f (F_{c,A}^d + F_{c,B}^d + F_r^d)\frac{N_{p,B}^s}{(N_{p,A}^s + N_{p,B}^s)^2}(\lambda_4 - \lambda_2)$$

$$+F_E^3 F_p{}'(N_{P,A}^S)(\lambda_7 U_A^4 + \lambda_8 U_B^4 + \lambda_9 U_r^4) = 0 \quad (28)$$

*Table 11.2 (cont.)*

*Table 11.2 (cont.)*

$$\frac{dZ}{dF^d_{c,A}}: -\lambda_2 P_f + \frac{P_f \beta}{N^s_{p,A} + N^s_{p,B}}(\lambda_2 N^s_{p,A} + \lambda_4 N^s_{p,B}) + \lambda_7 U^1_A + F^2_E(\lambda_7 U^4_A + \lambda_8 U^4_B + \lambda_9 U^4_r) = 0 \quad (29)$$

**B. Forest Community (Group B)**

$$\frac{dZ}{dN_{df,B}}: -F_{df,B}{}'(N_{df,B}) - \lambda_3 - F_{df,B}{}'(N_{df,B})(\eta(.)\frac{dH}{dF_{df,B}} + H(.)\frac{d\eta}{dF_{df,B}})(\lambda_4(1-\gamma) + \lambda_2\gamma)$$
$$+\lambda_6 F_{df,B}{}'(N_{df,B})(\eta(.)\frac{dH}{dF_{df,B}} + H(.)\frac{d\eta}{dF_{df,B}}) + \lambda_8 F_{df,B}{}'(N_{df,B})U^1_B + F^1_E F_{df,B}{}'(N_{df,A})(\lambda_7 U^4_A + \lambda_8 U^4_B + \lambda_9 U^4_r) = 0 \quad (30)$$

$$\frac{dZ}{dX_{c,B}}: \lambda_8 U^3_B - \lambda_4 P_x = 0 \quad (31)$$

$$\frac{dZ}{dN^s_{p,B}}: -\lambda_3 + \lambda_4 w_p - \beta P_f(F^d_{c,A} + F^d_{c,B} + F^d_r)\frac{N^s_{p,A}}{(N^s_{p,A} + N^s_{p,B})^2}(\lambda_2 - \lambda_4)$$
$$+F^3_E F_p{}'(N^S_{P,B})(\lambda_7 U^4_A + \lambda_8 U^4_B + \lambda_9 U^4_r) = 0 \quad (32)$$

*Table 11.2 (cont.)*

*Table 11.2 (cont.)*

$$\frac{dZ}{dF^d_{c,B}}:-\lambda_4 P_f+\frac{P_f\beta}{N^s_{p,A}+N^s_{p,B}}(\lambda_4 N^s_{p,B}+\lambda_2 N^s_{p,A})+\lambda_8 U^1_B+F^2_E(\lambda_8 U^4_B+\lambda_7 U^4_A+\lambda_9 U^4_r)=0 \quad (33)$$

**C. Residual Sector**

$$\frac{dZ}{dF^d_r}:\frac{\beta P_f}{(N^s_{p,A}+N^s_{p,B})}(\lambda_2 N^s_{p,A}+\lambda_4 N^s_{p,B})-\lambda_5 P_f+\lambda_9 U^1_r+F^2_E(\lambda_7 U^4_A+\lambda_8 U^4_B+\lambda_9 U^4_r)=0 \quad (34)$$

$$\frac{dZ}{dN^d_r}:\lambda_5(1-t)[P_x X'(N^d_r)-w_r]=0 \quad (35)$$

$$\frac{dZ}{dX^d_r}:-\lambda_5 P_x+\lambda_9 U^3_r=0 \quad (36)$$

**D. Forest Department**

$$\frac{dZ}{dN^d_p}:F_p'(N^d_p)-\lambda_6 w_p=0 \quad (37)$$

## 4. ANALYSIS AND SUGGESTIONS FOR SOME POSSIBLE SIMULATION EXPERIMENTS

Since the main purpose of the model was to examine the effects of JFM on various behavioral outcomes, the first question we put to the model is "Is the introduction of JFM sufficient to increase forest biomass and the welfare of the forest community?"

### *4.1 The Impact of JFM on Forest Biomass and FC Welfare: Simulation*

Because of the model's several institutional parameters $\alpha, \beta, \gamma, \varepsilon, t$, and its behavioral and technological functions, each of which would have to be converted from the above general specifications to ones of appropriate functional forms with realistic parameter values, any definitive answer to the question must await the completion of these tasks, each of which is well beyond the objectives of this chapter.

Nevertheless, we can get a hint at the qualitative answer by comparing the relevant first order conditions of the JFM case with those of the pre-JFM case. For example, from equation (19) it can be seen that the sharing of revenues from timber sales with the elite of the FC would give such community members an additional benefit in supplying protective labor to the FD. The same would be true for the non-elite of the FC. In both cases also the incentive for illegal or unauthorized collecting from the forest $(F_{df})$ would fall. As a result, $N_p^s$ would rise and both $N_{df}$ and $F_{df}$ would fall. From these effects, it would seem rather clear that biomass would increase with the introduction of JFM.

But, at the same time, from the first order condition for the demand for the forest good by the elite $(F_{c,A}^d)$, given by equation (20), with $\beta > 0$ it can be seen that the relative cost of purchasing the forest good $F$ would fall. The same would be true for the non-elite group B. Hence from this effect the demand for $F$ by all sectors would rise, thereby having the opposite effect on forest biomass and the environment. Hence, it is clear that the answer to the question is ambiguous and would depend on the relative size of these two opposite effects that, in turn, would depend not only on the sharing parameter $\beta$ but also on the various parameters of the utility and production functions.

One useful simulation exercise that could be conducted after functional forms and suitable parameter values were chosen, therefore, would be to see how the answer would vary with the choice of $\beta$. Similarly, it would be useful to see how the answer would depend on the parameters of the utility functions, such as differences in the strength of the taste for $F$ (relative to $X$) and its sensitivity to both relative price and income changes.

Another set of simulation experiments might well investigate the effects of heterogeneity between the elite and non-elite groups. This could be done by varying the differences between the elite and non-elite groups with respect to (1) their relative preferences for the forest good relative to the residual sector good, (2) the magnitudes of their preference for the positive environmental externality and its sensitivity to $F_{df}$, (3) their shares in the penalty paid for excess clearing/collecting, (4) the distribution of FC revenues from timber sales, (5) the productivity of protective labor, (6) the penalty and probability of being detected for excess gathering, (7) government grant allocations and (8) the degree of access to employment in the R sector.

In this way, once suitable parameter values had been selected, simulation methods could be used to generate specific, testable hypotheses about how the effects of JFM would be expected to vary depending on environmental circumstances including the tastes and preferences of the different groups, institutional conditions concerning the values of $\alpha, \beta, \gamma, \varepsilon,$ and $t$, and the magnitudes of the environmental externalities. Then, with further data collection on the values of these parameters across actual and otherwise comparable JFM experiences, one could formally test the hypotheses. These could include subtler and more nuanced versions of those suggested in the literature survey given in Section 2 above. For example, one could test hypotheses concerning the extent to which greater dependence on the forest good by non-elite members of the FC would offset their disadvantages in terms of wealth and access to employment in the rural sector, with and without JFM.

### *4.2 A Comparison of Pareto Optimal Conditions to the Benchmark Case*

Another objective of the model is to determine to what extent the typical features of JFM contribute to achieving outcomes that are closer to the Pareto-optimal solutions. Some progress toward the fulfillment of this objective can be accomplished by comparing the first-order conditions (FOC) in Table 11.1 with the Pareto optimality conditions (POC) in Table 11.2 for corresponding decision variables in the above formulation. For example, to what extent do the FOCs derived from the JFM case move the solutions of key variables like forest biomass and the total utility achieved by each group within the FC toward the POC solutions. For the forest community, it can readily be seen from equations (17) and (26) that the first order condition with respect to $N_r^s$ in the JFM case is of the same form as that for Pareto optimality. Yet, for most of the other decision variables, this is not the case, even under JFM.

For example, from (16) it can be seen that the FOC under JFM includes neither all of the negative externality effects on utility nor all of the negative income effects of $N_{df}$ relative to the corresponding POC equation (25). In the POC case, each group within the forest community accounts for negative utility impacts on all other sectors (recall $F_E^1 < 0$) in addition to the negative income effects on the other forest community group. Consequently, under JFM, the model's FOC predicts levels of $N_{df}$ that are greater than the levels implied by the POCs.

Since $X_{c,A}$ only affects the forest community group A, the FOC for $X_{c,A}$ in the JFM case, equation (18) is of the same form as corresponding POC equation (27).

Furthermore we see that the predicted levels of protective labor under JFM are sub-optimal from equations (19) and (28). Relative to the POC, in (19) the elite group in the FC does not take into account the positive externality benefits accruing to the other sectors and neither does it account for the income benefits realized by the other forest community group with increased levels of $N_p^s$. Since in the pre-JFM framework, the FC does not receive a share of income from sales of the processed forest good $(\beta = 0)$, the levels of $N_p^s$ which increase forest biomass and eventually income are likely to be even lower.

A final important difference for the FC is in the level of purchases of the processed forest good. From equations (20) and (29) in the JFM case the FC does not account for the negative externality effects on other sectors ($F_E^2 < 0$) as is required for the POC. Neither does it account for the positive income benefits realized by the other FC group (as the other group receives a share of FD revenues). But, since similar comparisons apply to the pre-JFM case, in this respect, JFM does not bring the solution closer to Pareto optimality in itself.

For the residual sector, from equation (21) in the JFM case and equation (34) in the Pareto optimal case we can see that in deciding $F_r^d$ under JFM the residual sector does not take into account the negative externality effect from $F_r^d$. On the other hand $F^s$ is increasing in $F_r^d$ and consequently it has a positive effect on the income of the FC which gets a share of FD sales. But, this positive effect is not taken into account in the JFM case. At this level of generality, therefore, it cannot be ascertained whether $F_r^d$ under JFM will be lower or higher than the Pareto optimal level. The FOCs for $N_r^d$ and $X_r$ are of the same form since $N_r^d$ appears only in the budget constraint for the residual sector and $X_r$ appears only in the utility function.

For the FD, we note from (24) and (37) that the choice of protective labor is at the Pareto optimum. In each case the FD sets the marginal benefits of protective labor equal to its costs and is not accounting for external benefits. In contrast, as already discussed, the external benefits of protective labor are taken into account by the FC in its' choice of protective labor.

### *4.3 Other Findings and Extensions with Respect to Inequality*

In terms of inequality, there are three levels wherein inequality affects the equilibrium values of choice variables. First, suppose that $\gamma < 0.5$, which implies non-elites share a greater burden in paying penalties for illegal $N_{df}$. One could imagine such a scenario, for instance, in a case where elites possess greater authority in determining which community members would pay sanctions for illegal extraction. In such a case, based on (16), the levels of $N_{df,A}$ are likely to be higher but at the same time those of $N_{df,B}$ would be lower.

Second, and more importantly, as shown in (19) the greater the participation of non-elites in protective labor, the lower will be the benefits to elites. Assuming that wage differentials between the R sector and the FD are enough to compensate elites for relatively greater levels of $N_{p,B}$ (and consequently a lower share of FD revenues accruing to elites) elites would be no worse off. If, however, wage differentials from the residual sector to do not compensate elites for this difference, elites could increase their share of protective labor, reducing the share provided by non-elites and therefore indirectly increasing the amount of time non-elites spend on $N_{df}$. It is the assumptions (a) that non-elites cannot work in the residual sector and (b) that the distribution of forest department revenues would be based on the share of labor provided to the forest department that drive this result. In contrast, the effect of higher levels of $N_{p,B}$ in the Pareto optimal case are less negative for the elite group (see equations (28) and (32)) and depend on the relative weights of $\lambda_2$ and $\lambda_4$.

Finally, as already discussed, the unequal share of revenues from the forest department also affects the equilibrium values of the demand for the forest good as shown in equation (20). Assuming greater levels of inequality, for instance with elites providing a greater share of protective labor, demand for the forest good by elites is higher relative to the case where elites provide a lower share of protective labor.

In summary, these results imply that in cases where the wage differential between the R sector and the FD is likely to be low (or even negative), elites will likely provide a greater share of protective labor. This will lead to greater levels of production of the processed forest good and at the same time greater levels of labor allocated to unauthorized collection of the forest good ($N_{df}$) by non-elites. If, however, $\gamma$ is low enough, $N_{df}$ by non-elites will be significantly reduced, but at the expense of their consumption. The model therefore predicts that inequality can potentially constrain efficient levels of $N_{df}$. If elites are able to constrain illegal extraction by non-elites (i.e. a low $\gamma$), then JFM results in greater gains in biomass. Yet, at the same time, if greater gains in biomass are realized under these conditions, such gains come at the expense of biomass consumption by the non-elites.

Useful conclusions for policy should also be derivable. If besides gaining a share in timber sales in JFM, which as noted above is a stimulus for greater supply of $N_p$ and reduced supply of $N_{df}$, the forest community were asked to share in the fixed or other costs of planting new forest, to what extent would this policy offset or strengthen the aforementioned effects of JFM on forest biomass? Would there be some combination of cost sharing and revenue sharing (consistent with budget balancing by all agents) that would yield optimal results as in some of the sharecropping literature?

How much additional benefit in terms of the desired objectives of JFM would be achieved by introducing into the JFM case additional mechanisms to induce communication and collaboration between the two FC groups and between the FD and these groups? If the model could be used to identify conditions under which the non-elite Group B could be immiserized, it should be possible to identify policies such as tax policies and institutional rules that would reverse this and thereby prevent immiseratization.

*4.4 Extensions*

As noted above, the model in its present form is highly simplified. Several extensions of the model deserve high priority in future research. One of these is to introduce land into the model, specifically into the production functions of the F and R sectors and also its rental and purchase into the budget constraints of the relevant parties.

In view of the findings of Rozelle, Huang and Benziger (2003) and Zhang, Uusivuori, Kuuluvainen, and Kant (2003) concerning the strikingly different effects of relative price changes and property rights on natural forest stock and managed forests, it would also be important to disaggregate F into these two types and possibly also into various other tree types because of their differential utility for using sectors such as furniture, paper and home construction. By the same token, the model could then be usefully extended in the direction of having various additional forest-using sectors as well as forest-competing sectors like agriculture and urbanization. Likewise, with forest-competing sectors included, it might be desirable to introduce other inputs and outputs of both forestry and agriculture into the model. If so, it would then be possible to examine the effects of subsidies and taxes on these inputs and outputs. (Note for example, that numerous scholars such as Repetto and Gillis, 1988 and Binswanger, 1991 suggest that implicit subsidies to agriculture are a major contributor to deforestation).

Each one of these extensions would open the model up to new policy uses. With land included, one could examine how the differential assignment of property rights (i.e., modeled as decision-making power over land use) would affect outcomes. What kinds of property rights allocations – allocations to specific agents, or commonly to two or more agents – would be most beneficial? With competition from agriculture and more interdependencies between the different agents and markets, would the standard policy pronouncements concerning agricultural subsidies still hold? If so, would they be strengthened or weakened? Suppose that the optimal property rights allocations are infeasible, what second-best policies should be adopted? Instead of simply treating the environmental benefits of forest biomass as an externality, conceptually the model could be modified so as to make the environmental benefits marketable. If so, in what direction and to what extent would this affect the potential benefits of JFM? Finally, how might the answers to these questions vary by institutional circumstances? These are all questions that an appropriately extended model of this sort could address.

**Acknowledgements** An earlier version of this chapter was presented to the Conference on the Economics of Social Forest Management at the University of Toronto May 20-22, 2004. The authors express their appreciation for comments on the paper at the conference, on a related paper presented at the 78th Annual Meeting of the Western Economic Association International, Denver, Colorado, July 11-14, 2003 and especially the written comments of Albert Berry and Shashi Kant. They also express thanks to Rahul Nilakantan for his excellent research assistance.

## APPENDIX 11.1. SYMBOLS AND DEFINITIONS*

| Symbol | Definition |
|---|---|
| **Utilities** | |
| $U_c$ | Utility of the forest community |
| $U_r$ | Utility of the residual sector |
| **Labor** | |
| $N_p$ | Protective labor |
| $\frac{F^S}{c}$ | Labor to harvest process the forest good |
| $N_r$ | Labor for production of the market good |
| $N_{df}$ | Labor for defacto gathering of the forest good |
| **Wage Rates** | |
| $w_p$ | Wage rate for protective labor |
| $w_h$ | Wage rate for harvesting and processing of forest good |
| $w_r$ | Wage rate for production of market good |
| **Prices** | |
| $P_f$ | Price of the forest good |
| $P_x$ | Price of the market good |
| **Quantities** | |
| $X_c$ | Quantity of market good consumed by forest communities |
| $X_r$ | Quantity of market good consumed by residual sector |
| $F_c$ | Quantity of forest good consumed by the forest communities |
| $F_r$ | Quantity of forest good consumed by the residual sector |
| $F_L$ | Quantity of forest good approved for removal by forest communities and forest department |
| $\overline{F}$ | Quantity of initial forest biomass |
| $F^S$ | Quantity of forest good supplied |
| **Parameters** | |
| $\alpha$ | Share of FC A in forest department (FD) harvest employment |
| $\beta$ | Share of FD revenues going to the forest community as a whole |
| $\gamma$ | Fraction of the fine for excess removal paid by community A |
| $t$ | Tax rate on profits of residual sector |
| $\varepsilon$ | Share of forest department revenues going to the government |

| | |
|---|---|
| $\omega_c$ | Share of forest community in government revenue (grant) |
| $\omega_f$ | Share of forest department in government revenue (grant) |
| **Functions** | |
| $\eta$ | Probability of getting caught for excess removal |
| $H$ | Penalty for excess removal |
| $F_E$ | Environmental externality |
| **Government Grants** | |
| $G_c$ | Grant to the forest community |
| $G_r$ | Grant to the residual sector |
| $G_f$ | Grant to the forest department |
| **Lagrange Multipliers** | |
| $\lambda_1$ | Shadow price of time for community A |
| $\lambda_2$ | Shadow price of income for community A |
| $\lambda_3$ | Shadow price of time for community B |
| $\lambda_4$ | Shadow price of income for community B |
| $\lambda_5$ | Shadow price of money for residual sector |
| $\lambda_6$ | Shadow price of money for forest department |
| $\lambda_7$ | Lagrange multiplier for utility of community A |
| $\lambda_8$ | Lagrange multiplier for utility of community B |
| $\lambda_9$ | Lagrange multiplier for utility of residual sector |

* Superscripts 'd' and 's' indicate demand and supply respectively. Subscripts $U_{c,i}$ indicates utility of i$^{th}$ community (I = A, B), $X_{c,i}$ indicates consumption of market good by the i$^{th}$ community (I = A, B), and $F_{c,i}$ indicates consumption of the forest good by the i$^{th}$ community (I = A, B)

## NOTES

---

[1] Sandalwood forests in the Indonesian province of Nusa Tengara Timur provides an interesting and rather telling example of how, in the absence of democracy, even decentralization of forest ownership and forest policy can result in destruction of the resource. See Marks (2002).

[2] See Kruse et al. (1998) for a comparison of Canadian and US caribou co-management programs.

[3] See Becker and Leon (1998), and Smith (2000) for South American examples.

[4] Under joint management, the Forest Community may be allowed to remove a certain percentage of biomass production without penalty. But, forest products removed beyond that specified amount are subject to fines. Both the fine and the percentage of removal allowed are flexible.

[5] Many of these are of the type identified by Olson (1962) and Ostrom (1990). For example, the ability of a group to have successful collective action in promoting the commons is higher the longer the members of the group have resided together in the same area, the more homogeneous they are in their backgrounds, the more they have different though not necessarily conflicting goals, the less unequal they are in their

income and wealth, the smaller the group, the better they can observe each other's actions, and the more they can trust each other.

[6] Apart from government in-kind transfers to the FC, some Indian state governments mandate certain percentages of profits earned by Village Forest Committees be allocated to community development projects (Kumar, 2002).

[7] The first two elements in the externality function, $F_E$, however, enter that function negatively.

[8] For instance, see Ostrom, Walker and Gardner (1992) and Hackett, Schlager and Walker (1994) who present experimental evidence consistent with this hypothesis for a similar commons situation.

[a] The numerical superscripts 1, 2 …in these expressions represent the first derivatives with respect to the first, or second or other arguments of the relevant function while ʻindicates the first derivative when there is only one argument in the function.

## REFERENCES

Agrawal, A., & Ostrom, E. (2001). Collective action, property rights, and decentralization in resource use in India and Nepal. *Politics and Society*, 29(4), 485-514.

Baland, J., Bardhan, P., & Bowles, S. (Eds.) (2001). *Inequality, collective action and environmental sustainability*. Forthcoming: under review Princeton University Press and Russell Sage Foundation. Available electronically at: http://discuss.santafe.edu/sustainability/papers

Baland, J., & Platteau, J. (1996). *Halting degradation of natural resources: Is there a role for rural communities?* New York: Oxford University Press.

Baland, J., & Platteau, J. (1997). Wealth inequality and efficiency in the commons. Part I: The unregulated case. *Oxford Economic Papers*, 49, 451-482

Baland, J., & Platteau, J. (2001). Collective action on the commons: The role of inequality. In J. Baland, P. Bardhan, S. Bowles (Eds.), *Inequality, collective action and environmental sustainability*. Forthcoming: under review Princeton University Press and Russell Sage Foundation. Available electronically at: http://discuss.santafe.edu/sustainability/papers

Bardhan, P. (2002). Decentralization of governance in development. *Journal of Economic Perspectives*, 16(4), 185-205.

Becker, C. D., & Leon, R. (1998). Indigenous forest management in the Bolivian Amazon: Lessons from the Yuracare people. Working Paper, CIPEC. Available electronically at: http://dlc.dlib.indiana.edu/documents/dir0/00/00/00/17/dlc-00000017-00/becker.pdf.

Binswanger, H.P. (1991). Brazilian policies that encourage deforestation in the Amazon. *World Development*, 19(7), 821-829.

Bromley, D.W., & Chapagain, D.P. (1984). The village against the center: Resource depletion in South Asia. *American Journal of Agricultural Economics,* 66, 868-873.

Bulte, E., & Horan, R.D. (2003). Habitat conservation, wildlife extraction, and agricultural expansion. *Journal of Environmental Economics*, 45, 109-127.

Cardenas, J. (2003). Real wealth and experimental cooperation: Experiments in the field lab. *Journal of Development Economics*, 70, 263-289.

Cardenas, J., Stranlund, J., & Willis, C. (2000). Local environmental control and institutional crowding out. *World Development*, 28(10), 1719-1733.

Casari, M., & Plott, C.R. (2003). Decentralized management of common property resources: Experiments with a centuries-old institution. *Journal of Economic Behavior and Organization*, 51, 217-247.

Chakraborty, R.N. (2001). Stability and outcomes of common property institutions in forestry: Evidence from the Terai region of Nepal. *Ecological Economics*, 36(2), 341-353.

Edmonds, E. (2002). Government-initiated community resource management and local resource extraction from Nepal's forests. *Journal of Development Economics*, 68, 89-115.

Gebremedhin, B., Pender, J., & Tesfay, G. (2003). Community natural resource management: The case of woodlots in Northern Ethiopia. *Environment and Development Economics*, 8, 129-148.

Hackett, S., Schlager, E., & Walker, J. (1994). The role of communication in resolving commons dilemmas: Experimental evidence with heterogeneous expropriators. *Journal of Environmental Economics and Management*, 27, 99-126 and reprinted in J. Shogren (Ed.), *Experiments in Environmental Economics*, 2003. Vermont: Ashgate Publishing Limited.

Hoffman, E., McCabe, K., Shachat, K., & Smith, V. (1994). Preferences, property rights, and anonymity in bargaining games. *Games and Economic Behavior*, 7, 346-380.

Hyde, W.F., Belcher, B., & Xu, J. (Eds.) (2003). *China's forests: Global lessons and market reforms*. Washington, D.C.: Resources for the Future.

Jaramillo, C., & Kelly, T. (2000). Deforestation and property rights over rural land in Latin America. In K. Keipi (Ed.) *Forest Resource Policy in Latin America*, 2000. Washington DC: Inter-American Development Bank. Available electronically at: http://www.iadb.org/sds/doc/1411eng.pdf.

Kant, S. (2000). A dynamic approach to forest regimes in developing economies. *Ecological Economics*, 32, 287-300.

Kant, S., & Berry, A. (1998). Community management: An optimal resource regime for tropical forests. Working Paper UT-ECIPA-BERRY-98-01, Department of Economics, University of Toronto. Available electronically at: http://www.economics.utoronto.ca/ecipa/archive/UT-ECIPA-BERRY-98-01.pdf

Kant, S., & Berry, A (2001). A theoretical model of optimal forest resource regimes in developing economies. *Journal of Institutional and Theoretical Economics,* 157(2), 331-355.

Kenneth, R. (1989). Solving the common-property dillemma: Village fisheries rights in Japanese coastal waters. In F. Berkes (Ed.), *Common property resources: Ecology and community-based sustainable development*. London: Belhaven Press.

Klooster, D. (2000). Institutional choice, community, and struggle: A case study of forest co-management in Mexico. *World Development,* (28)1, 1-20

Kruse, J., Klein, D., Braund, S., Moorehead, L., & Simeone, B. (1998). Co-management of natural resources: A comparison of two management systems. *Human Organization*, 57(4), 447-458.

Kumar, S. (2002). Does þarticipation' in common pool resource management help the poor? A social cost-benefit analysis of Joint Forest Management in Jharkhand, India. *World Development*, 30(5), 763-782.

Liu, D., & Edmunds, D. (2003). Devolution as a means of expanding local forest management in south China: Lessons from the past 20 years. In W.F. Hyde, B. Belcher, J. Xu (Eds.), *China's forests: Global lessons and market reforms* (pp. 27-44). Washington, D.C.: Resources for the Future.

Marks, S.V. (2002). NTT sandalwood: Roots of disaster. *Bulletin of Indonesian Economic Studies* 38, (2), 223-240.

Munoz-Pina, C., de Janvry, A., & Sadoulet, E. (2002). Recrafting rights over common property resources in Mexico: Divide, incorporate, and equalize. Working Paper, University of California, Berkeley. Available electronically at: http://are.berkeley.edu/sadoulet/papers/CarlosEDCC.pdf

Olson, M. (1962). *The logic of collective action*. Cambridge: Harvard University Press.

Ostmann, A. (1998). External control may destroy the commons. *Rationality and Society*, 10(1), 103-122.

Ostrom, E. (1990). *Governing the commons: The evolution of institutions for collective action.* Cambridge: Cambridge University Press.

Ostrom, E. (1999). Coping with tragedies of the commons. *Annual Review of Political Science*, 2, 493-535.

Ostrom, E., Walker, J., & Gardner, R. (1992). Covenants with and without a sword: Self-governance is possible. *American Political Science Review*, 86(2), 404-417.

Palfrey, T.R., & Rosenthal, H. (1994). Repeated play, Cooperation and coordination: An experimental study. *Review of Economic Studies*, 61, 803-836.

Platteau, J. (2001). Community-based development in the context of within group heterogeneity. In J. Baland, P. Bardhan, S. Bowles (Eds.), *Inequality, collective action and environmental sustainability.* Forthcoming: under review Princeton University Press and Russell Sage Foundation. Available electronically at: http://discuss.santafe.edu/sustainability/papers

Repetto, R., & Gillis, M. (1988). *Government policies and the misuse of forest resources*. Cambridge: Cambridge University Press.

Richards, M. (2000). "Can sustainable tropical forestry be made profitable? The potential limitations of innovative incentive mechanisms. *World Development*, 28(6), 1001-1016.

Rozelle, S., Huang, J., & Benziger, V. (2003). Forest exploitation and protection in reform China: Assessing the impacts of policy and economic growth. In W.F. Hyde, B. Belcher, J. Xu (Eds.), *China's forests: Global lessons and market reforms* (pp. 109-134). Washington, D.C.: Resources for the Future.

Rusnack, G. (1997). *Co-management of natural resources in Canada: A review of concepts and case studies.* Working Paper. Managing Natural Resources in Latin America and the Caribbean, IDRC, Ottawa.

Smith R. (2000). Community-based resource control and management in Amazonia: A research initiative to identify conditioning factors for positive outcomes. Available electronically at: http://dlc.dlib.indiana.edu/archive/00000352/00/smithr041000.pdf

Varughese, G., & Ostrom, E. (2001). The contested role of heterogeneity in collective action: Some evidence from community forestry in Nepal. *World Development*, 29(3), 747-765.

Walker, J.M., Gardner, Herr, A., & Ostrom, E. (2000). Collective choice in the commons: Experimental results on proposed allocation rules and votes. *The Economic Journal*, 110, 212-234.

Zhang, Y., Uusivuori, J., Kuuluvainen, J., & Kant, S. (2003). Deforestation and reforestation in Hanan: Roles of markets and institutions. In W.F. Hyde, B. Belcher, J. Xu (Eds.), *China's forests: Global lessons and market reforms* (pp. 135-150). Washington, D.C.: Resources for the Future.

# CHAPTER 12

# POST-NEWTONIAN ECONOMICS AND SUSTAINABLE FOREST MANAGEMENT

SHASHI KANT

*Faculty of Forestry, University of Toronto*
*33Willcocks Street, Toronto, Canada M5S 3B3*
*Email: shashi.kant@utoronto.ca*

**Abstract.** This chapter synthesizes the contents of this volume, and provides an overview of a new paradigm of economics, to which I assign the term Post-Newtonian Economics. To put the synthesis in perspective, first the main cause of the current status of Newtonian or neo-classical economics–increasing returns due to information contagion–is discussed. Second, direct and indirect correspondences between the different concepts discussed in the ten chapters of this volume and Kant's basic principles of the economics of sustainable forest management are established. Finally, the basic differences between Newtonian and Post-Newtonian economics are discussed.

## 1. INTRODUCTION

The mainstream of economics—which has been termed Walrasian or neo-classical economics, but that I would like to call Newtonian economics[1]—is a good example of "positive feedbacks" and "generalized increasing returns" or "increasing returns due to information contagion."[2] The main features of positive feedback systems—path-dependence, "lock-in" due to small historical events, and inefficiencies—are some of the main characteristics of Newtonian economics. The concepts of positive feedbacks and increasing returns may still be unacceptable to many Newtonian economists, as they were in 1980s,[3] but this only proves the path-dependence of Newtonian economics. Classical economists, as Arrow (1994) remarks, were well aware about the concept and importance of increasing returns:

> The opening chapters of Adam Smith's Wealth of Nations put great emphasis on increasing returns to explain both specialization and economic growth. Yet the object of study moves quickly to a competitive system and a cost-of-production theory of value, which cannot be made rigorous except assuming constant returns. The English school (David Ricardo, Juan Stuart Mill) followed the competitive assumptions and quietly dropped Smith's boldly-stated proposition that "the division of labour is limited by the extent of the market", division of labour having been shown to lead to increased productivity. (Arrow, 1994 p. IX)



*Kant and Berry (Eds.), Economics, Sustainability, and Natural Resources: Economics of Sustainable Forest Management,* 253-267.

Arrow's observation provides evidence of the historical events leading to "lock-in" in Newtonian economics. The concept of increasing returns was not unknown to neo-classical economists, and it also finds a page or half in every undergraduate or graduate micro-economics textbook. The concept, however, was treated like a pathological specimen in a labelled jar (Arthur, 1994), which was paraded to economics students as an anomaly; this treatment of increasing returns resulted into professional synergies in the economics profession leading to positive feedback and increasing returns, due to information contagion, to the profession, in terms of increased intellectual output that neglected realities. What a tragedy–the concept, which was neglected by neo-classical economists for analytical convenience, has been the main source for the current state of Newtonian (neo-classical) economics. Arthur (1994) realized this state of neo-classical economics, but did not describe, at least explicitly, as an outcome of increasing returns:

> The assumptions economists need to use vary with the context of the problem and cannot be reduced to a standard set. Yet, at any time in the profession, a standard set seems to dominate. These are often originally adopted for analytical convenience but then become used and accepted by economist mainly because they are used and accepted by other economists. .. I am sure this state of affairs is unhealthy. (Arthur, 1994 p. XIX)

One of the main features of positive feedback systems is inefficiencies, and neo-classical economics is full of inefficiencies.

> It encourages use of the standard assumptions in applications where they are not appropriate. And it leaves us open to the charge that economics is rigorous deductions based upon faulty assumptions. (Arthur, 1994, p. xix)

> Its defining assumptions precluded analysis of many key aspects of economic progress, among them the exercise of power, the influence of experience and economic conditions on people's preferences and beliefs, out of equilibrium dynamics, and the process of institutional persistence and change. (Bowles, 2004, pp. 7-8)

These inefficiencies are due to its "locked-in" position in *Chicago man,* which is convenient, successful, unnecessarily strong, but false (McFadden, 1999), and a single *Equilibrium,* which is conceptually simple, analytically strong, but difficult, if not impossible, to exist. Neo-classical economics, due to its locked-in position, has been caught up in a "rational fool's trap", and all efforts to take it out of the trap have faced almost impenetrable resistance. The experimental observations from human behavior, markets, and institutions reported by behavioral economists, psychologists, and other streams of economics, have been termed as anomalies,[4] and behaviors that violate the stringent canons of formal rationality have been treated as idiosyncratic, unstable, or irrational. However, this outcome is also a result of one of the main characteristics of positive feedback systems: stable equilibria cannot be displaced by small deviations from the equilibrium due to self-reinforcing forces, and only wars, revolutions, climate change, strikes, or any other external shock of the similar magnitude, can move the system from one equilibrium to another. External shocks with large magnitudes, but not enough magnitude to move the system from the existing equilibrium to new equilibrium, will result in punctuated equilibrium. In the last two decades, external shocks from emerging streams of new economics, which I call Post-Newtonian economics, such as complexity theory, economics of increasing returns, behavioral economics, agent-based modelling, experimental economics, evolutionary economics, and evolutionary game theory have intensified, and there are some early signs of new punctuated equilibrium.[5] I am sure that combined and continuous external shocks from all emerging streams of economics will be able to transform the dominant paradigm of economics from Newtonian economics to Post-Newtonian economics.

The main objective of this volume is to provide a foundation for the economics of sustainable forest management, but another objective is to bring together the contributions from different streams of economics—complexity theory, social choice theory, behavioral economics, and post-Keynesian consumer theory—and to provide a holistic perspective of emerging concepts of new economics. Direct and indirect correspondences between the different concepts discussed in the ten chapters of this volume and Kant's basic principles of the economics of sustainable forest management are established in the next section. The basic differences between Newtonian and Post-Newtonian economics are discussed, and the similarities between Kant's basic principles and the main features of Post-Newtonian economics are highlighted in section 3.

## 2. THE BASIC PRINCIPLES OF THE ECONOMICS OF SFM

Kant (2003) argues that the basic idea behind SFM is to manage forests in such a way that the needs of the present are met without compromising the ability of future generations to meet their own needs, and economic models of SFM should be able to capture both orientations—individualistic as well as altruistic and/or commitment—of an individual's behavior. The incorporation of such behavior will be possible in economic models that are based on a "both-and" principle rather than an "either-or" principle. Under the umbrella of the "both-and" principle, Kant (2003) proposes four sub-principles of the economics of SFM: the principles of existence, relativity, uncertainty, and complementarity. Kant (2003) concludes that the two dominant requirements of the economics of SFM are a consumer choice theory different than *Chicago man* and the economics of multiple equilibria. The ten chapters of this volume provide strong evidence of an emerging new consumer choice theory and the need for an economics of multiple equilibria, and many chapters confirm, directly or indirectly, the relevance of Kant's four sub-principles to emerging economic thoughts.

### *2.1 Consumer Choice Theory*

The three chapters (chapters 4, 5, & 6) of the second part of this volume are focused on consumer choice theory, but many elements of a new consumer choice theory are discussed in other chapters also. In chapter 2, David Colander identifies a similarity in the changes that are occurring across the emerging streams of economics in allowable assumptions, from the holy trinity of rationality, greed, and equilibrium to a new holy trinity of purposeful behavior, enlightened self-interest, and sustainability. Colander enumerates the contributions of behavioral economics—reference-dependent preferences, the replacement of expected utility theory with prospect theory, hyperbolic discounting, cognitive heuristics, theories of social preferences, and an adaptive learning model—and discusses the concept of "libertarian paternalism" which takes into an account agent's ill-formed preferences or an individual's choices that are influenced by default rules, thus demonstrating that preferences are endogenous. Colander argues that as a result the story of economics will shift from the story of reasonably bright agents in information-rich environments to the story of reasonably bright individuals in information-poor environments. The concept of libertarian paternalism is similar to the concept of procedural rationality, discussed by Marc Lavoie in Chapter 4. The focus of Ali Khan, in Chapter 3, is on putting Kant's four sub-principles and the general theory of inter-temporal resource allocation into a broader interdisciplinary framework that the subject demands, but in doing so, he touches on many aspects related to consumer choice theory such as conversation between generations, ecosystem capital, and the social rate of discount,[6] and raises an important issue about economic models with a zero time preference:

> Thus, it is not surprising that research on models with a zero time-preference, analytically difficult to begin with, abruptly ceases in the eighties. The current conventional wisdom is to see it as "dispensable and misdirected". The effects of this wisdom are pervasive. The biographies of standard textbooks in the field such as those of ... simply ignore the earlier literature on the extension of Ramsey's undiscounted setting. (Chapter 3, p. 55-56)

Similar to the case of neglect of increasing returns, neglect of models of zero-discounting provides another example of path-dependence, and hence possible inefficiencies, in Newtonian economics. In addition, the different asymptotic properties of optimal paths for undiscounted and positively discounted cases of a strictly concave utility function also raise questions about the dominant version of consumer choice theory of neo-classical economics. In short, Khan's discussion on all these aspects demands a fresh and a different look at the consumer choice theory of neo-classical economics.

In the second part, chapter 4, Marc Lavoie discusses seven principles of post-Keynesian consumer choice theory—procedural rationality; the principle of satiation, the principle of separability, the principle of subordination, the principle of the growth of needs, the principle of non-independence and the heredity principle—and the key consequences of these principles, that the utility index cannot be represented by a scalar and the notions of gross substitution and trade-offs, which are central to neo-classical economics, become a minor phenomenon. Lavoie argues that the preference map of forest values is full of lexicographic preferences, and substitution effects are totally wiped out, the axiom of continuity does not hold, and the Archimedes axiom that every thing has a price, becomes irrelevant in the presence of lexicographic preferences. Jack Knetsch, in chapter 5, provides evidence for choices based on mental accounts, disparity between the valuation of gains and losses, and disparity between discounting future gains and losses. Knetsch argues that valuations of gains and losses call for different measures, and the choice of appropriate measures will depend on the reference state, the directions of change, and the expected state, or norm. Knetsch observes that the observed differences between behavioral findings and standard economic theory reflect real preferences that are not well modelled by the axioms of standard consumer choice theory. Colin Price, in chapter 6, and Knetsch, in chapter 5, observe that people do not use a single constant rate to discount the value of all future outcomes for all periods, and discuss the implications of different discounting protocols, which challenges the consumer choice theory of neo-classical economics.

The two chapters in the third part also challenge the outcomes of the consumer choice theory of the neo-classical economics, and demand a fresh outlook on consumer choice theory. Tapan Mitra, in chapter 7, confirms economic optimality, given the objective of intergenerational equity, of the principle of maximum sustained yield by using a social welfare relation which is weaker than the one induced by the overtaking criterion. Similarly, in chapter 8, Geir Asheim and Wolfgang Buchholz prove that the stock-specific sustainability constraints are also justified from an economic perspective. The outcomes of both the chapters may not be acceptable to neo-classical economists, who believe strongly in discounted utilitarianism.

### *2.2 Economics of Multiple Equilibria*

The convexities of production and utility (or consumption) functions, perfect markets, frictionless functioning of markets, absence of increasing returns and externalities, and no market failure due to uncertainties are essential ingredients of the economics of *Chicago man* and a single (*General*) *Equilibrium*. However, in real life, most of these essential ingredients are not available, and the ten chapters in this volume provide multi-dimensional evidence for the absence of these ingredients and the presence of multiple equilibria. Colander, in chapter 2, identified the new holy trinity of purposeful behavior, enlightened self-interest, and sustainability, and argued that the sustainability literature fits into models with multiple equilibria, with equilibria selection mechanism, and with some equilibria being preferred to others. Similarly, periodic optimal paths—in all cases, and specifically a linear utility function, other than the strictly concave period-by-period utility function—and dissimilar asymptotic properties for undiscounted and positively discounted cases of the strictly concave utility function, discussed by Khan in chapter 3, support the case of multiple equilibria. In the case of post-Keynesian consumer choice theory, all seven principles reject the idea of a single equilibrium, and as Lavoie, in chapter 4, states: the multiplicity of equilibria, or the belief that models must be open-ended, is a characteristic feature of post-Keynesian economics. In thc scenario of the existence of lexicographic preferences, the existence of multiple equilibria is natural. Similarly, various features of behavioral economics, discussed in chapter 5, such as mental accounts, reference-dependent preferences, differences between the valuation of losses and gains, different measures for different values, and different discounting rates for gains and losses will result in multiple equilibria. The replacement of a single, and constant, rate of discount for all situations and for all periods by situation and period specific discount rates, discussed in chapter 6, will result in multiple equilibria.

The economic optimality of maximum sustained yield rotation, discussed in chapter 7, and stock-specific sustainability constrained, discussed in chapter 8, also support multiple equilibria. Chapter 9 and 10 are specifically focused on multiple equilibria due to non-linearities in the production and management systems of forest resources. Barkley Rosser, in chapter 9, identifies various sources of non-linearities in forest production systems, such as production of various non-timber products, and in forest management systems, such as pest and fire management and patch size of cutting. The discussion of Jeffrey Vincent and Matthew Potts, in chapter 10, is focused on the implications of non-linearities for spatial aspects of forest management, and a discussion of economic, institutional, and ecological sources of non-linearities in forest production systems.

With this discussion of the two broad features, I move to a discussion of the four sub-principles of the economics of SFM. As I mentioned in the sub-section 2.1, Khan, in chapter 3, focused on putting these principles in a broader and inter-disciplinary framework; I will start the discussion of each principle with his observations, and then move to the related contents from other chapters. In short, Khan has rightly labelled these principles together as an "ethics of theorizing" and rightly observed "that these four sub-principles draw attention to the broader interdisciplinary framing that the subject demands, and emphasizes, rather than a particular theory, the theoretical principles that go into its theorizing."

### *2.3 The Principle of Existence:*

In Kant (2003), I emphasized the existing situations under this principle, and the word "situations" would require a broad interpretation including practices, models in operation, basins (in Colander's, chapter 2, terminology), and norms. Khan starts the discussion of this principle with the following observation:

> The first principle can be read in two opposing ways: first, to take account of existing conditions so as to change them, and not to avoid facing them simply because they have survived so long into the present; or secondly, to take account of them in a way that is resistant to change and reads their survival as an equilibrium that is not stable but desirable.. (Chapter 3, p. 41)
>
> It is this identification with Burkean conservatism that leads Kant to argue for forest rotation based on the annual allowable cut as opposed to Faustmann's rotation. (Chapter 3, p. 41)

I agree with Khan's conclusion that the principle of existence is gesturing towards an "ethics of theorizing", but the above observations provide an interesting example of outcomes based on incomplete knowledge and the diversity of the frames of thought process. When I proposed this principle, I had a face-to-face communication between a forest manager and a forest economist in mind,[7] and my idea was not to suggest either a conservative or a radical approach, but a call for self re-examination, by economists themselves, of so-called economically efficient models suggested by neo-classical economists[8]. As proven by Tapan Mitra in chapter 7 and in his previous work, forest rotation based on maximum sustained yield, which is also known as a forester's rotation, is economically efficient from the perspective of inter-generational equity. In forestry literature, it is a common observation that for a regulated forest, Faustmann's rotation reduces to the rotation of maximum sustained yield for a zero rate of time preference. I was trying, therefore, to point out that there may be very good economic factors in the existing situations and/or practices which do not fit in economic models based on *Chicago man* and a single *equilibrium*, and neo-classical economists might have ignored those economic factors for the sake of their mathematical convenience and elegance of their models, as has been the case with zero time preference and increasing returns. In a way, I was hinting towards recent developments in emerging streams of economics, as confirmed by Colander, in Chapter 2:

> The resulting system is admired not for its efficiency, nor for any of its static properties; the resulting system is admired for its very existence. Somehow the process of competition gets the pieces of the economy to fit together and prevents the economy from disintegrating into chaos. Observed existence, not deduced efficiency, is the key to the complexity story line. (Chapter 2, p. 25)

In addition, the principle of existence is also evident in Colander's description of sustainability in complexity theory:

> Sustainability means keeping within the existing basis of attraction, and not going to another that is considered less desirable. Within a complex system a "rational choice" is much harder, and indeed impossible, to specify. It is multiple levels of the system, not only the individual, that are optimizing, so the individual is the result of lower-level optimization at the physiological level, is himself optimizing, and is a component of higher level systems which are themselves optimizing, and competing for existence. (Chapter 2, p. 27)

The three principles of Post-Keynesian consumer choice theory—procedural rationality, non-independence, and heredity—and many features of behavioral economics—mental accounting or budgeting, reference-dependent preferences, and evolution of preferences—also, directly or indirectly, provide indications of the principle of existence. Existing situations and/or practices may be outcomes of these principles or features, and those outcomes need not be economically inefficient, as Lavoie, in chapter 4, observes:

> The fact that procedurally rational agents often do use compensatory procedures or do not behave as if they were approximating regression analysis or expected utility theory to arrive at their decision does not mean that these agents are erroneous or suffer from some biases. Rather, as Gigerenzer (2000, ch. 8) has demonstrated, non compensatory procedural rules can arrive at the right decision just as often, when such a decision exists, and much more efficiently than compensatory one. (Chapter 4, p. 72)

### *2.4 The Principle of Relativity*

The principle of relativity, as per Kant (2003), suggests that an optimal solution is not an absolute but rather a relative concept. Khan draws parallels between the principle of relativity and Wittgenstein's binary of absolute and relative, and the appropriation of Wittgenstein's binary by Keynes to distinguish between absolute and relative needs.

Khan rightly observes that "there is an important overlap, a common orientation if one prefers, between Kant's principle of existence and his principle of relativity." These two principles require a simultaneous reading. However, the principle of relativity should not be read only for making a distinction between absolute and relative, the broadest interpretation of "relative" will also be the part of this rule. In this sense, the principle of relativity, as per my reading of emerging streams of economics, seems embedded in Colander's (chapter 2) new holy trinity of purposeful behavior, enlightened self-interest, and sustainability. I also think that Colander is hinting at the principle of relativity in his following observation:

> Efficiency is not an end in itself, it is a means to an end; efficiency only has a meaning when one specifies what the goals are, whose goals they are, how the goals are to be weighted, and what methods we have in resolving conflict among goals. (Chapter 2, p. 26)

Finally, Colander confirms the relative or contextual aspect of economics in the following observations:

> Looking at broader issues in social welfare theory, it is very clear that the work is contextual—it can only be understood within a much broader framework of thinking about institutions, social wellbeing, and social welfare. (Chapter 2, p. 25-26)

Colander's discussion of different basins of attractions in complex systems and sustainability as a means of keeping within the existing basin of attraction, and not going to another basin that is considered less desirable, is also, at least implicitly, an indication of the principle of existence and the principle of relativity.

Similar to the principle of existence, many principles of Post-Keynesian consumer choice theory—procedural rationality, subordination of needs, non-independence and heredity—and various features of behavioral economics—reference-dependent preferences, the dependence of choice on the order in which they are made, differences between the valuation of losses and gains, different measures for different values, different discounting rates for gains and losses, and prospect theory—also emphasize the relevance and importance of the principle of relativity.

### *2.5 The Principle of Uncertainty*

The principle of uncertainty suggests that due to uncertainties in social and natural systems, an individual may never be able to maximize his outcomes, and will always search for positive outcomes. Somehow, this uncertainty aspect of this principle missed critical examination by Khan, probably due to his focus on an ethics of theorizing. However, the complexity story, discussed in chapter 2, supports this principle. I am sure that when Colander describes the complexity story as the story of *reasonably bright individuals in an information poor environment*, he is including uncertainty as one of the sources of a poor information environment. Similarly, his description of the complexity story as a never-ending story in which every answer simply raises new questions, and the hope of control gives way to a realization that the best we can hope for is to muddle through, is indicating the non-availability of maximizing outcomes and a search for positive and better outcomes. The focus of

Colander's discussion is on the complexity of systems, and I believe, based on my reading of his text, that he has assumed uncertainty as an inherent property of complex systems. However, the principle of uncertainty has received explicit recognition in the post-Keynesian consumer choice theory, behavioral economics, evolutionary economics, and ecological economics.

Lavoie, in chapter 4, argues that "true uncertainty" which is also called "Keynesian uncertainty" or "Knightian uncertainty" is one of the three distinguishing features (historical time and the importance of aggregate demand being two others) of post-Keynesian economics, and it has received serious attention in environmental economics and ecological economics. Lavoie writes:

> This is linked in particular to the importance which is given to true or fundamental uncertainty, ...post-Keynesians have long emphasized the need to distinguish between fundamental uncertainty and probabilistic risk. The future is uncertain, not only because we lack the ability to predict it, which is tied to epistemological uncertainty and procedural rationality, but also because of ontological uncertainty–the future itself is in the making and the decisions that we are to take will modify its course (Rosser, 2001). When private agents take decisions that affect them directly, fundamental uncertainty leads them to adopt a course of action that will generate safety.... The precautionary principle in environment is clearly tied to fundamental uncertainty. (Chapter 4, p. 68-69)

Lavoie further places a high importance on uncertainty when describing the design of policies:

> Instead of trying to demonstrate that x percent of subjects fail to behave in accordance with the standard neo-classical axioms of rationality, one should provide evidence describing actual behavior. Also, when designing policies, the behavior of the agents should be modelled as is, rather than as it should be if the world were devoid of information limits and fundamental uncertainty. (Chapter 4, p. 73)

One other dimension of the principle of uncertainty is human behavior, which I did not mention in Kant (2003), and none of the chapters in this volume has considered this. As it is discussed in the principle of complementarity, every individual is selfish as well as altruistic; the same individual may behave selfishly or altruistically in the same circumstances but at different periods. For example, an individual's behavior with respect to his kids and spouse may vary from one end of selfishness to the other end of altruism at different periods of time, holding all other things constant. Hence, the incorporation of uncertainty in human behavior in economic models is another challenge to future economists, and may require the use of some of the tools of quantum physics.

### *2.6 The Principle of Complementarity*

The principle of complementarity, as per Kant (2003), suggests that human behavior may be selfish as well as altruistic, people can have economic values as well as moral values, and people need forests to satisfy their lower level needs as well as higher level needs. Khan locates these binaries—economic/moral, lower/higher, and selfish/altruistic—in the work of Wittgenstein and Keynes, and adds two additional aspects: (i) the character of the resource that is to be allocated, the extent to which the natural is implicated in the social, and (ii) the agency doing the allocation, and the extent to which it is public and thereby divorced from the private. The principle of complementarity is fundamental to many emerging streams of economics: agent-based modelling, complexity theory, post-Keynesian consumer choice theory, and behavioral economics. For example, in agent-based modelling, every agent is not a *Chicago man*, and agents can be selfish as well as altruistic. Many principles of post-Keynesian consumer theory—the separability of needs, the subordination of needs, and the growth of needs—confirm the principle of complementarity. Similarly, the recognition of the existence and the importance of lexicographic preferences requires the acceptance of the principle of complementarity. The existence of increasing and decreasing returns to scale adds another perspective to the principle of complementarity.

I would close this section with Khan's conclusion that the four sub-principles draw attention to the broad interdisciplinary framing that the subject demands. There is no doubt that the subject of economics, and for that matter all the subjects of social sciences, demands an interdisciplinary approach. All social sciences deal with humans, and the compartmentalized approach to social sciences, which tries to divide a living human being into different components that have no connections and interactions with each other, is an approach which may be possible only with a dead body and not with a living being. An interdisciplinary approach is not only necessary, it is essential. However, in the economics profession, the first step may be to take an inter-stream (or intra-subject) approach because, in many situations, economists from one stream do not know or recognise what is going on in other streams of economics. In addition, neo-classical or Newtonian economists do not even want to acknowledge the developments which are going on in the emerging streams of economics.

## 3. POST-NEWTONIAN ECONOMICS

*Chicago man*, as McFadden (1999) observed, has become an endangered species; behavioral economics has severely restricted his maximum range, and he is not safe even in markets for concrete goods which was his prime habitat. McFadden (1999) issued a call to evolve *Chicago man* in the direction of *Kahneman-Tversky (K-T) man* by adopting those features needed to correct the most glaring deficiencies of *Chicago man*, and to modify economic analysis accordingly. Thaler (2000) predicted that *homo economicus* will evolve into *homo sapiens* who will have characteristics of less IQ, slow learning, heterogeneity, human cognition, and more emotions. Colander (2000a) declared the death of the term neo-classical economics and the birth of the new millennium economics.[9] Ormerod (2000) expected the re-birth of economics in the 21st century to give us a much better understanding of the world. Brian Arthur realised the need for a new paradigm of economics long ago, when he started working on the economics of increasing returns, and identified many differences between the standard approach and the complexity approach to economics (Colander, 2000b). In addition to these specific calls, all the emerging streams of economics, such as behavioral economics, complexity theory, evolutionary economics, evolutionary game theory, experimental economics, and the economics of increasing returns, have been contributing to the emergence of a new paradigm of economics which I have termed Post-Newtonian economics. The new paradigm will be fundamentally different from the Newtonian economics, as summarized in Table 12.1. The relevance of Kant's four sub-principles to the main features of Post-Newtonian economics is quite clear, but I leave a specific discussion on this issue for some future paper.

***Table12.1.*** *Main Differences between Newtonian and Post-Newtonian Economics*

| Feature | Newtonian Economics | Post-Newtonian Economics |
|---|---|---|
| **Holy Trinity** | Rationality, greed, and equilibrium | Purposeful behavior, enlightened self interest, and sustainability (non-equilibrium and multiple equilibrium) |
| **Agent** | *Chicago man* and rational fool (homo economicus), and homogeneous agents | *K—T man* and social agent (homo-sapiens), and agents are heterogeneous as well versatile |
| **Rationality and information** | Mathematical or constructivist rationality and full information | Procedural and/or ecological rationality, and incomplete information |
| **Preferences** | Exogenous (as imposed by economists), self-regarding, and fixed preferences | Endogenous, reference-dependent, self as well as other-regarding and/or social preferences |
| **Needs and Wants** | No difference between needs and wants | Difference between needs and wants, satiable needs, hierarchy of needs, and growth of needs |
| **Learning and emotions** | No learning and no emotions | Learning from others, frequency-dependent learning, and emotions may produce a behavioral response. |
| **Actions of others and social interactions** | Market clearing prices and contractual exchanges | Agents interactions through market and non-market mechanisms, non-contractual social obligations |
| **Utility** | Scalar utility, expected utility theory | Vector utility, prospect theory and libertarian paternalism |
| **Uncertainty** | Risk | True or Keynesian uncertainty |
| **Elements** | Quantity and Prices | Patterns and Possibilities |
| **Principle** | Maximizing | Satisfying |

*Table 12.1 (cont.)*

*Table 12.1 (cont.)*

| Feature | Newtonian Economics | Post-Newtonian Economics |
|---|---|---|
| **Modeling** | Modeling of Decision Outcome | Modeling of Decision (cognitive) Process |
| **Returns to Scale** | Constant and decreasing returns to scale | Constant, decreasing, and increasing returns to scale as well generalized increasing returns |
| **Feed-backs** | Negative | Negative as well as positive feedback, lock-in, path dependence, inefficiencies |
| **Institutions** | Either no institutions, or formal institutions, are represented by a budget constraint , no role of informal institutions, institutions do not change | Outcomes are dependent on institutional setting, optimal institutions are not freely available; role of formal as well as informal institutions, and institutions evolve over time. |
| **Time, Age, and Generations** | Positive Discounting, No role of age and generations | Zero, positive, and negative discounting, individuals can age, generational turnover becomes central, age structure of population change, and generations carry their experiences. |
| **Equilibrium** | General equilibrium | Multiple equilibria and non-equilibrium |
| **Society** | Aggregation of homogenous agents | Heterogenous agents, similar populations may have different norms, tastes, and customs, resulting in local homogeneity and global heterogeneity. |
| **Solutions** | Closed form solutions | Simple closed form solutions are not necessary; indeed, any solutions that are susceptible to simple interpretations may not exist. |
| **Subject** | Structurally simple, deterministic, stable | Structurally complex, structures are constantly coalescing, decaying, and evolving. All this is due to externalities leading to jerky motions, increasing |

*Table 12.1 (cont.)*

*Table 12.1 (cont.)*

| Feature | Newtonian Economics | Post-Newtonian Economics |
|---|---|---|
| | | returns, transaction costs, and structural exclusions |
| **Approach** | Tool driven | Problem and issue driven |
| **Foundation** | Non-cooperation | Cooperation |
| **Basis** | Newtonian Physics | Quantum Physics and Evolutionary Biology |
| **Nirvana** | Possible if there are no externalities and all had equal abilities | Not possible, externalities and inequalities are driving forces, systems constantly unfolding |
| **Sustainability** | Sustainability of neo-classical economics and neo-classical economists | Sustainability of society |

## 4. CONCLUSIONS

The sustainability of global systems (social as well as natural) is a prerequisite for the existence of economics as well as economists. Hence, the goal of a new paradigm of economics should be "the sustainability of global systems" as opposed to the goal of "sustainability of Newtonian (neo-classical) economics and economists". The main elements of the emerging paradigm of economics, post-Newtonian economics, seem focused on the sustainability of global systems. However, a higher-level integration of all the emerging streams of economics and collective action by economists associated with these streams are necessary ingredients for the structural specifications, establishment, and growth of Post-Newtonian economics. In this volume, we have tried a partial integration of some emerging streams of economics with respect to sustainable management of forest resources only. A comprehensive development of post-Newtonian economics will require many such efforts at different levels, and the recognition of such efforts. I believe that the establishment of an International Association of Post-Newtonian Economics, supported by regional chapters, may be a step in the right direction.

## NOTES

---

[1] Economists may debate the appropriateness or non-appropriateness of these terms, but I am sure that every economist understands which stream of economics is being addressed here. Similar to Hamilton (1970), I prefer the term Newtonian Economics because I believe that the concept of equilibrium came from Newtonian physics.

[2] Generally, increasing returns are associated with economies of scale in production, but the term refers more broadly to any situation in which the payoff for taking an action is increasing in the number of people taking the same action Bowles (2004, p.12) termed it "generalized increasing returns". Arthur and Lane (1993) also identified "information feedbacks" or what they called "information contagion" as a source of positive feedback and increasing returns.

[3] In March 1987, when Brian Arthur visited his old university, Berkeley, one most respected economist commented "Well, we know that increasing returns don't exist." Other most respected economist observed "Besides, if they do, we could not allow them. Otherwise every two-bit industry in the country would be looking for a handout." (Arthur 1994, p. xi)

[4] Richard H. Thaler, with his colleagues, has a published a series of anomalies in The Journal of Economic Perspectives, starting with the first volume of the journal. The series includes the anomalies related to the January effect, the winner's curse, cooperation, inter-temporal choice, preference reversals, the endowment effect, loss aversion, and status quo bias, the flypaper effect, and the equity premium puzzle.

[5] Some of the references in this direction are Auyang (1998), Bowles (2004), Camerer, Loewenstein, & Rabin (2004), Colander (2000b), Friedman (2004), Kahneman and Tversky (2000), Schmid (2004), and Smith (2000).

[6] Ali Khan, in Chapter 3, discusses all the four sub-principles of Kant and these three issues related to consumer choice theory in detail. Hence, I will not discuss these aspects in any detail here, and readers may like to refer back to Chapter 3. However, I will discuss some of his key observations related to the four sub-principles in the respective sub-sections.

[7] In one international conference, after a well-established forest economist finished his presentation about optimal forest rotation, one practising forester asked him about the optimal rotation age at which he should harvest a particular type of forest. The answer of the forest economist was that it will depend upon so and so. The forester again asked, tell me the age at which I should harvest, and the forest economist did not have any answer. This incidence forced me to think about the economic optimality of Faustmann's rotation.

[8] In this regard, Smith's (1985) observation is very useful: "The early polling of economist on Allais, Ellsberg, Second price, and other such "paradoxes" makes it clear that economists will get it "wrong" about as often as the sophomore subject until he or she has had considerable time to think and analyze. Incidentally, this observation provides an answer for that somewhat mythical business-man who asks, "If you' re so smart why ain't you rich?". My classmate, Otto Eckstein, didn't get rich by equating price to marginal cost."

[9] Colander (2000) also discussed the non-appropriateness of many other terms such as "new Classical", "mathematical economics", and "the era of modeling" to describe the recent development in economics.

## REFERENCES


Arthur, W. B. (1994). *Increasing returns and path dependence in the economy.* Ann Arbor: University of Michigan Press.

Arthur, W. B., & Lane, D. A. (1994). Information contagion. In W. B. Arthur (ed.) *Increasing returns and path dependence in the economy*, (pp. 69-97). Ann Arbor: University of Michigan Press.

Auyang, S. Y. (1998). *Foundations of complex systems theories in economics, evolutionary biology, and statistical physics*. Cambridge: Cambridge University Press.

Bowles, S. (2004). *Microeconomics: Behavior, institutions, and evolution.* New York: Russel Sage Foundation and Princeton: Princeton University Press.

Colander, D. (2000a). The death of neoclassical economics. *Journal of the History of Economic Thought*, 22(2), 127-143.

Colander, D. (2000b). Introduction. In D. Colander (Ed.), *The complexity vision and the teaching of economics* (pp.1-28). Cheltenham: Edward Elgar.

Camerer, C.F., Loewenstein, G., & Rabin, M. (2004). *Advances in behavioral economics.* New York: Russell Sage Foundation , and Princeton: Princeton University Press.

Friedman, D. (2004). *Economics lab: an intensive course in experimental economics.* London: Routledge.

Hamilton, D. (1970). *Evolutionary economics.* Albuquerque: University of New Mexico Press.

Kant, S. (2003). Extending the boundaries of forest economics. *Journal of Forest Policy and Economics*, 5, 39-58.

Kahneman, D., & Tversky, A (Eds.) (2000). *Choices, values, and frames*. Cambridge: Cambridge University Press

McFadden, D. (1999). Rationality for economists? Journal of Risk and Uncertainty, 19(1-3), 73-105.

Ormerod, P. 2000. *Death of economics revisited.* Keynote address given to the Association of Heterodox Economists, June 29, 2000. Retrieved September 2, 2004, from www.volterra.co.uk/ Docs/dofer.pdf

Schmid, A. A. (2004). *Conflict and cooperation: institutional and behavioral economics.* Malden: Blackwell Pub.

Smith, V. L. (1985). Experimental economics: reply. *The American Economic Review,* 75(1), 265-272.

Smith, V. L. (2000). *Bargaining and market behavior: essays in experimental economics.* Cambridge: Cambridge University Press.

Thaler, R. (2000). From homo economicus to homo sapiens. *The Journal of Economic Perspective,* 14(1), 133-141.

# INDEX